Strengthening Resilience in Post-disaster Situations

Strengthening Resilience in Post-disaster Situations

Stories, Experience and Lessons from South Asia

Editors:

Julian Gonsalves
Priyanka Mohan

ACADEMIC FOUNDATION
NEW DELHI

International Development Research Centre
Ottawa • Cairo • Dakar • Montevideo • Nairobi • New Delhi • Singapore

First published in 2011
by

ACADEMIC FOUNDATION
4772-73 / 23 Bharat Ram Road, (23 Ansari Road),
Darya Ganj, New Delhi - 110 002 (India).
Phones : 23245001 / 02 / 03 / 04.
Fax : +91-11-23245005.
E-mail : books@academicfoundation.com
www.academicfoundation.com

and

INTERNATIONAL DEVELOPMENT RESEARCH CENTRE
PO Box 8500, Ottawa, ON, K1G 3H9
Canada
E-mail: info@idrc.ca
www.idrc.ca

© International Development Research Centre 2011

This work was carried out with the aid of a grant from the Wellcome Trust, London, UK, Canada's International Development Research Centre (IDRC), and the UK Government's Department for International Development. The views expressed herein do not necessarily represent those of IDRC, the Wellcome Trust, or the Department for International Development.

ALL RIGHTS RESERVED.
No part of this book, including its cover shall be reproduced, stored in a retrieval system, or transmitted by any means, electronic, mechanical, photocopying, recording, or otherwise, without the prior written permission of the copyright holder(s) and/or the publishers.

Cataloging in Publication Data--DK
   Courtesy: D.K. Agencies (P) Ltd. <docinfo@dkagencies.com>

Strengthening resilience in post-disaster situations : stories, experience and lessons from South Asia / editors, Julian Gonsalves, Priyanka Mohan.
   p. cm.
ISBN 9788171889044
ISBN (e-book) 9781552505359

   1. Disaster victims--Rehabilitation--South Asia. 2. Coastal zone management--South Asia. 3. Emergency management-- South Asia. 4. Hazard mitigation--South Asia. 5. Risk assessment--South Asia. 6. Resilience (Personality trait)-- South Asia.   I. Gonsalves, Julian. II. Priyanka Mohan. III. International Development Research Centre (Canada)

DDC 363.3480954   23

Cover Photo: Coast of Tamil Nadu (Julian Gonsalves)
Typeset by Italics India, New Delhi.
Printed and bound in India.

Contents

Foreword 11
Acknowledgement 13
Introduction 15
Acronyms 21

Part 1
Coastal Threats and Challenges

1. Rehabilitating Livelihoods in Tsunami-Affected Coastal Communities in Asia 31
2. Ecosystem Threats and Impacts in Coastal Areas 47
3. Coastal Erosion and its Management 55
4. Tsunami Impacts and Coastal Land Use Issues 73
5. Coastal Planning and Regulatory Issues: Post-Tsunami 85
6. Trends in Land and Resource Use and Land Tenure 99
7. Management Challenges and Approaches for Sustainable ICM 111
8. Integrated Coastal Zone Planning 117
9. Establishing Resilient Coastal Ecosystems for Sustainable Livelihoods 139

Part 2

Disaster Risk Reduction: Key to Adaptation

10. Reducing Disaster Risk: A Challenge for Development. . . . 155
11. Linking Poverty, Vulnerability, and Disaster Risk 167
12. The Impact of Climate Change on the Vulnerable. 175
13. Understanding Adaptation and Mitigation. 183
14. Demistifying Terminologies and Definitions:
    Same Terms, Different Meanings? 193
15. A Framework on Addressing Climate Change
    Adaptation and Vulnerabilities. 205
16. Framework for Disaster-Resistant
    Sustainable Livelihoods. 213
17. Adaptation: The Context for Change 223
18. Looking at Climate Change Adaptation
    from Various Realms. 233
19. Assessing Resilience and Vulnerability:
    Principles, Strategies and Actions 245
20. Analytical and Assessment Tools and Methods 255
21. Participatory Community Risk Assessment 261
22. Integrating Ecosystems Management with DRR 277
23. Ecosystem-based Disaster Risk Reduction 283
24. Indicators for Use in Ecosystem-based
    Disaster Risk Reduction 291
25. Healthy Ecosystems and their Role
    in Disaster Risk Reduction 299
26. Adaptation to Climate Change 309

Contents

27. Capitalising on Similarities: Bridging Differences between CCA and DRR 319
28. Disaster Risk Reduction and Climate Change Framework: The Cordaid Lens 333
29. Enhancing Coping Mechanisms in Resilience Building 339
30. Supporting Local Capacities: From Jargon to Impact 345

Part 3
Building Capacities: The Path to Community Resilience

31. Emergence of Resilience 353
32. Understanding Resilience in Coastal Areas 365
33. Towards Creating a Resilient Community 379
34. Learning to Respond to Disasters: Community Empowerment 387
35. Community Resilience: A Social Justice Perspective 393
36. Boosting Community Resilience 405
37. Community Resilience and Community-based Management 413
38. Information and Communications Technology for Disaster Risk Reduction 423
39. Management of Coastal Resources through Village Level Planning 433
40. Development Planning using Spatial Data 443
41. Towards a Better Assessment Framework: The Case of Wanduruppa 453
42. Building Resilient Communities: Lessons from Cordaid 469

43. Community Disaster Resilience Fund:
    Early Insights and Recommendations 477
44. Capacity Building Interventions:
    Drawing Lessons from the Field 487
45. A Role for Customised Financial Mechanisms 497
46. Village Information Centres:
    Harnessing the Potential of Technology 507
47. Considerations for a Village Resource Centre 515
48. Transforming the Value Chains through
    Business Development Approach 521
49. Planning Coastal Revegetation Programmes 529
50. Role of Coastal Bioshield in
    Strengthening the Coastal Resilience 537
51. Non-Mangrove Bioshields in Coastal Areas 543

Part 4
Risk Reduction Experiences from the Field

52. An Introduction to the SRTAC Project 553
53. Networking and Partnerships in Rural Sri Lanka 563
54. Village Development Planning Processes in Sri Lanka . . . 575
55. Local Participation in Mangrove Management 583
56. Unleashing Women Power in Tamil Nadu 595
57. Exposure Visits: A Tool for Gender Mainstreaming 607
58. Community Disaster Resilience Fund:
    Learning from a Pilot Initiative 615
59. Building Bonds, Breaking Bondages
    in the Gulf of Mannar 627

60. Insurance: Investment towards Security and
    Resilience in Palakayatippa 637

61. Community-Managed Microfinance:
    The Case of Danavaipeta, Andhra Pradesh. 645

62. Community Participation: The Case of
    the Andaragasyaya Canal. 653

63. Revival of Agriculture in
    Sorlagondi Village, Andhra Pradesh 659

64. Transfering Ownership to Communities:
    Flood Mitigation in Ambalantota 665

65. The Importance of Providing Solutions that Really Work. . . 671

66. Collective Management of Resources
    in Village Development. 677

67. Reviving Traditional Paddy Farming
    in Andaragasayaya 685

68. Collective Action for Eco-Shrimp Farming
    in Sorlagondi 691

69. Facilitation makes a World of Difference
    in Danavaipeta 703

70. Weaving in Technology to the Coir Industry:
    The Thalalla Experience 711

71. Applying Market-Systems Approaches in Wanduruppa. . . . 721

72. Integrated Mangrove and Fishery Farming:
    A Model for Saline Transformed Lands 729

73. Raising Mangrove Nurseries in
    Muthuregunathapuram Village 739

74. Viability of Bioshield Development in Sri Lanka 749

75. Participatory GIS for Village Development. 755

76. Village Resource Centres:
    The Spokes of the Information Village 761

77. Village Resource and Knowledge Centre
    as Hubs for Disaster Preparedness 769

78. Livelihood Avenues: Reducing Economic Vulnerabilities . . . 779

79. Customised Process-Oriented
    Facilitation of Project Monitoring 783

    Information Resources 793

Foreword

The tsunami of December 26, 2004 was a wake-up call. It underlined the importance of planning for coastal areas in an integrated manner with concurrent attention to the landward and seaward sides of the shoreline. Since tsunamis have been rare in South Asia, there was not adequate preparedness to meet the challenges of the post-tsunami period. It was therefore an act of vision on the part of International Development Research Centre (IDRC) and Canadian International Development Agency (CIDA) to have organised a regional programme for rehabilitating the livelihoods in tsunami-affected areas in India and Sri Lanka. In this endeavour, M.S. Swaminathan Research Foundation (MSSRF) participated in a collaborative programme with institutions in Sri Lanka. The studies carried out under this project showed how we can convert calamities into opportunities for strengthening the coping capacity of local communities in relation to tsunamis, coastal storms, cyclones and other natural disasters.

Drs Priyanka Mohan and Julian Gonsalves have taken the trouble of putting together the data and experiences gained both in India and Sri Lanka under the project 'Strengthening Resilience in Tsunami-Affected Communities of India and Sri Lanka'. The book has been appropriately called a source book, since it provides a wide range of information relevant to facing challenges like the tsunami in future. We now experience extreme and unpredictable weather events probably associated with climate change. They emphasise the need for anticipatory action to avoid serious damage and hardship during adverse weather conditions. After a meticulous screening of all the papers received, the editors have chosen 79 papers for this source book. They have been grouped under the following categories:

- Coastal threats and challenges.
- Disaster risk reduction: Key to adaptation.

- Building capacities: The path to community resilience.
- Risk reduction experiences from the field.

Thus, the book covers the conceptual and practical aspects of disaster management. It contains information on methods of checkmating potential adverse events. A large proportion of human populations have their habitation near coastal areas. Coastal ecological security is therefore an essential requirement for the livelihood security of coastal communities. The book provides guidelines towards creating resilient communities capable of adaptation to new situations.

The book also stresses the importance of bioshield development in Sri Lanka and India based on both mangrove and non-mangrove species. In 1930, Mahatma Gandhi launched the Salt Satyagraha at Dandi in Gujarat for emphasising that seawater, which constitutes 97 per cent of the world's water resource, is a social resource and hence should not be taxed. The research undertaken following the tsunami of December 26, 2004 has led to the standardisation of methods of seawater farming for coastal area prosperity. Seawater farming involves agroforestry techniques like the promotion of silvi-aquaculture farms. Priyanka Mohan has brought together the rich experience of a large number of stakeholders who depend on coastal resources for their livelihood. I am glad that the rich experience gained following the tsunami of 2004 has been chronicled carefully. We owe a deep debt of gratitude to Priyanka Mohan, Julian Gonsalves, Phet Sayo and the other members of IDRC for preparing this publication whose value is immense since it is based on real life experience.

— M.S. Swaminathan
Member of Parliament (Rajya Sabha);
Chairman, M.S. Swaminathan Research Foundation,
Chennai

Acknowledgement

This Source book was made possible because of the project 'Strengthening Resilience in Tsunami Affected Communities of India and Sri Lanka', developed at a time when many lost their lives and livelihood from a huge disaster. The beneficiaries to this effort are the most vulnerable communities living in the remote coastal villages. Their willingness to participate and their contribution to activities for reducing their vulnerability to disasters and improving on degrading environments, helped a diverse group of practitioners in documenting the rich experiences and learning.

Institutions such as M.S. Swaminathan Research Foundation (MSSRF), Sarvodaya and Practical Action, continue to be the pillars of strength in adopting scientific based approaches and strategies and building capacities of grassroot organisations (Praja Pragathi Seva Sangam (PPSS), People's Action for Development (PAD), and Society for Participatory Research and Integrated Training (SPRIT) while working with the local communities. The project's success is dedicated to these institutions who led field project implementation for over three years. Their efforts have helped coastal communities to relive their lives with a lot more courage and hope. As important likewise to be recognised and acknowledged are the authors whose previously published work are featured in this source book. Their contributions have been edited and repackaged so as to meet the needs of the primary audience of this book: development practitioners, trainers, and educators.

This is the most appropriate juncture to thank the team responsible for the making of this book. The core team led this effort through a series of online consultations with the authors. The entire book has been meticulously planned, reviewed, and revised. For more than six months, primary and secondary sources were gathered. This team was led by Julian Gonsalves who helped guide the development and production of the source book. Our gratitude extends to the entire team of editors: Claudia Svetlana Cabrera, Josephine Bo, Bernadette Joven, Kashinath Hiremath, Kim Charlene Escobin, Angelina Ibus, Lily Ann Lando, Serlie Jamias; and administrative support: Renell Pacrin, Jenelyn Dogelio and Dulce Dominguez. We thank Celso Amutan for taking up the task of designing and laying out of the final version of this book.

While compiling the first drafts, each article was reviewed by a diverse group of experts, including authors from South Asia. We have received rich and insightful feedback from authors and external experts who reviewed the content critically. Constructive suggestions helped Julian Gonsalves to finalise this source book. We are very grateful and thank John Twigg, Ritesh Kumar, Buddhi Weerasinghe, Nihal Atapattu, Shalini Kala, Chandran Puthiyottil, Sudaramoorthy Olaganathan, Chintha Munasinghe, Manoj Silva, Nancy J. Anabel, John David, Ramona Miranda, Vajira Hettige, Ranasinghe Perera, Bhathiya Kekulandala, J.D. Sophia, Stephen McGurk, K. Jayakumar and Phet Sayo for taking their time out in making this effort more fruitful. The illustrations in the book added great value to the text and we take this opportunity to thank the artists Alexyz Fernandes, Justin Lobo, and Ariel Lucerna.

For all those who prefer to watch films on community change, there is a set of 12 films that are designed to complement this book. This visual documentation was prepared by Jason, a photographer who believes in connecting the lives of poor through his videos. We are happy to have had him in this team and ever grateful for his contribution.

The outcomes of this project—the source book and the complementary videos—were the result of hardwork and patience, involving a heavy reliance on electronic and occasional face to face interaction, for many of those who were associated with its conceptualisation, planning and execution. We acknowledge their contributions and are very grateful to them for their generous contributions to understanding better the practical dimensions of community resilience and disasters.

— Priyanka Mohan
New Delhi
India

Introduction

The recent increase in the number and intensity of disaster events has raised concerns about people's safety, lives, and livelihoods. In response, projects are designed as recovery and rehabilitation measures at the community level. Rich and diverse field experiences and lessons are consequently generated. These experiences or lessons are often not documented well enough for sharing with the wider community of practitioners, researchers, local administrators and even policymakers. Often this is because field workers rarely write. Sometimes what is written focusses on accomplishments and outputs, not on the processes, success factors and particular challenges that had to be overcome. Field workers and practitioners usually require time to reflect and need guided assistance in writing. As a result, valuable field-generated knowledge remains untapped or underutilised.

Researchers and academics, on the other hand, by the nature of their work, have the analytical skills and have devoted considerable time to generate knowledge, understand, and then synthesise the lesson from disaster risk reduction (DRR) and, increasingly related areas of climate change adaptation (CCA). New concepts, frameworks, terminologies, and tools for implementing DRR and CCA are evolving. However, these articles, written for the research and policy administrator community are either too complex or inaccessible to the practitioner community.

So new information is constantly being generated at different levels to help us better understand the general area of resilience building, whether it be in a pre/post-disaster situation, or, in response to increasing threats from climate change. It can be said that there is: (1) no dearth of information or research materials on disaster management, livelihood enhancements, adaptation, though there is a need to encourage documentation of more of evidence-based experiences by practitioners; and (2) a significant amount of information available via the Internet, which unfortunately is not accessible to the practitioner, because it is either too complex, academic, or technical.

Purpose and Audience of the Source Book

This Source book was developed to address some of the issues raised earlier and is aimed at the development-practitioner including extension workers, trainers, educators and local government officials. The source book serves as a collection of articles (based on primary and secondary sources), where every article stands on its own, and, therefore can be read independently of each other, without necessarily having to cross-reference or to read the entire collection. A reader can pick up an article of interest, learn, understand, analyse the content, and apply the information as required. Each article is richly illustrated as well to increase their user friendliness and to serve as training handouts. Materials can be used in planning, in undergraduate education and for local development planning. Sources are included so readers can directly contact the original contributors of the articles.

Source Book Content

This Source book on 'Transition towards Resilience: Coping with Disasters' intends to help to reduce the gaps between the concepts, frameworks and field practice. The book also features key concepts and experiences in resilience and disaster management, of relevance to community driven approaches, drawn from both primary and secondary sources.

Part I mainly focusses on coastal issues and threats, post-tsunami issues, and related risks/vulnerabilities that coastal communities are usually challenged with. Part II and III deal with the definitions, concepts and different frameworks that are used in disaster risk reduction, livelihood strengthening and climate change adaptation. Today, vulnerability is increasingly understood in a wider context that aims to integrate two previously separated areas of disaster mitigation and climate change adaptation with a stronger emphasis on resilience building. Part IV of the sourcebook aims to share the overall concept of resilience-building through disaster mitigation. Here, specific, rich, and diverse cases and synthesis of lessons, mostly from the post-tsunami project experiences, are shared.

This source book is a product of dedication and time invested by project partners involved in the IDRC-CIDA funded project. The source book also features repackaged materials from secondary sources which have permitted their use for educational and non-commercial purposes. Hopefully, this source book will further increase the access to these valuable materials. The editors deeply appreciate the opportunity to repackage and simplify these materials for use by the practitioner community.

The Source Book Development Process

An outline for the source book development was initially prepared by the source book's editors (building on the outputs from an external evaluation). The list was shared with project partners and other potential contributors. Primary articles, written by project partners were sourced. Articles were revised, edited, rewritten, restructured, merged, or dropped based on feedback.

The editors subsequently conducted internet-based research for articles from secondary sources. Over 150 articles were screened for their usefulness, short-listed and then identified for repackaging (shortening and illustrating). After seven months of working with production teams, each of the draft manuscript was ready for sharing with key reviewers. An informal review committee was set up, consisting of authors and a few external experts. The draft manuscript was assessed for the relevance of the content, illustrations, and chapter title. Each of the shortlisted articles were thoroughly reviewed and simultaneously, the artwork was also worked upon. A consultative process was used in the implementation of this project.

The editors coordinated with more than 50 people including, authors, contributors and reviewers. One of the main editors and the production teams were based in the Philippines while the other main editor was located in India. The authors were located in South Asia, Europe and Canada. This network of project contributors was mostly connected via

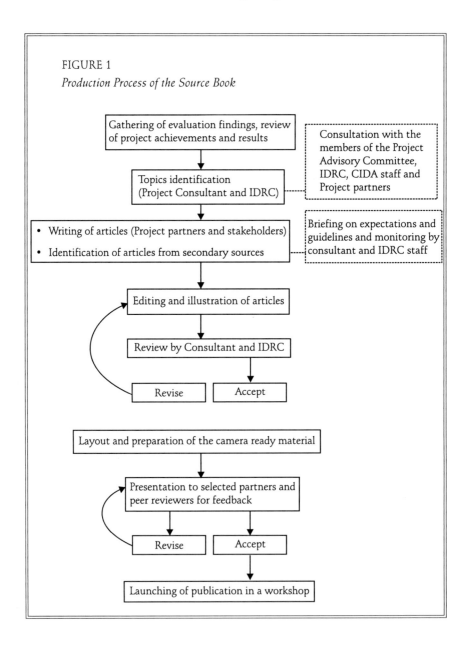

FIGURE 1
Production Process of the Source Book

web-based tools such as emails, skype and phone calls. The editors from the core team met thrice in India for planning, executing and monitoring the progress of the tasks. The process is shown in Figure 1.

The editors are grateful to the contributors of the primary and secondary articles in this collection, for generously making available their knowledge resources to an audience of practitioners. The editors expect that this collection will serve as a motivation for other efforts to document field experiences and to relate field-based efforts to available literature. The idea of linking concepts and practice remains an important challenge in many fields but especially so in the area of DRR, climate change and resilience building.

— Priyanka Mohan (IDRC)

— Julian Gonsalves (Consultant)

Acronyms

AP	Andhra Pradesh
ARC	Agriculture Research Center
ADB	Asian Development Bank
ASEAN	Association of Southeast Asian Nations
ASSEFA	The Association of Sawa Seva Farm
BDO	Block Development Officer
CBO	Community Based Organisation
CUG	Closer User Group
CCD	Coastal Conversation Department
CEA	Central Environmental Authority
CZMP	Coastal Zone Management Plan
CCA	Coastal Conservation Act
CHA	Consortium of Humanitarian Agencies
CIDA	Canadian International Development Agency
CDC	Canal Development Committee
CDRF	Community Disaster Resilience Fund
CCD	Covenant Centre for Development
CBDRR	Community Based Disaster Risk Reduction
CCA	Climate Change Adaptation
CFC	Chlorofluorocarbons
CMDRR	Community Managed Disaster Risk Reduction

CCAC	Coast Conservation Advisory Council
CRMP	Coastal Resource Management Project
CRZ	Coastal Regulatory Zone
CCR	Coastal Community Resilience
CRM	Coastal Resource Management
CIESIN	Center for International Earth Science Information Network
CNPS	Coastal and Native Plant Species
DFC	Department of Forest Conservation
DWLC	Department of Wildlife Conservation
DS	Divisional Secretariat
DMC	Disaster Management Centre
DRR	Disaster Risk Reduction
DSD	District Secretariat Division
DESMIO	District Economic Social Mobilisation Organisation
DRI	Disaster Risk Index
ESCAP	Economic and Social Commission for Asia and the Pacific
EC	Executive Committee
EJF	Environmental Justice Foundation
ETRP	Emergency Tsunami Reconstruction Project
FFEMA	Fisher Friend Mobile Application
FWP	Fixed Wireless Phone
FAO	Food and Agricultural Organization
FD	Forest Department
FRP	Fibre Reinforced Plastic Boats

Acronyms

FDO	Fisheries Development Officer
FISHERR	Financial Initiative for Sustainable Human Economic Resource Regeneration
FCD	Forest Conservation Department
GSM	Global System for Mobile Communications
GPRS	General Pocket Radio Services
GN	Grama Niladhari
GB	General Body
GoSL	Government of Sri Lanka
GND	Grama Niladhari Division
GoI	Government of India
GHG	Greenhouse Gas Emissions
GDP	Gross Domestic Product
GEF	Global Environment Facility
GPA	Global Programme of Action (for the Protection of the Marine Environment from Land-Based Activities)
GESAMP	Joint Group of Experts on the Scientific Aspects of Marine Environmental Protection
GS	Grama Seweka
HH	Household
HFA	Hyogo Framework for Action
HHDC	High Human Development Countries
ICT	Information Communication and Technology
INCOIS	National Centre for Ocean Information Services
ISRO	Indian Space Research Organisation
IFFCO	Indian Farmers Fertiliser Cooperative Limited

IEC	Information Education Communication
IUCN	International Union for Conservation of Nature
IPCC	Intergovernmental Panel on Climate Change
IMFFS	Integrated Mangrove Fishery Farming System
IDRC	International Development Research Centre
IDB	Industrial Development Board
ITDG	The Intermediary Technology Development Group (now known as Practical Action)
IRA	Insurance Regulatory Authority
IDS	Institute for Development Support
ILRI	International Livestock Research Institute
IIED	International Institute for Environment and Development
IFAD	International Fund for Agricultural Development
ICM	Integrated Coastal Management
IMF	International Monetary Fund
INGOs	International Non Governmental Organisations
IOTWS	Indian Ocean Tsunami Warning System
ICZM	Integrated Coastal Zone Management
ICMAM	Integrated Coastal and Marine Area Management
ICAM	Integrated Coastal Area Management
IMCAM	Integrated Marine and Coastal Area Management
IMMA	Integrated Management of Maritime Affairs
IOC/UNESCO	Intergovernmental Oceanographic Commission
ICARM	Integrated Coastal Area and River Basin Management
JMM	Joint Mangrove Management

Acronyms

KW	Knowledge Worker
KSA	Kanchan Seva Asharam
KRRC	Kanyakumari Rehab Resource Centre
LCD	Liquid Crystal Display
LED	Light-Emitting Diode
LG	Livelihood Groups
LHDC	Low Human Development Countries
LODRR	Livelihood Options for Disaster Risk Reduction
MSSRF	M.S. Swaminathan Research Foundation
MOU	Memorandum of Understanding
M&E	Monitoring and Evaluation
MoENR	Ministry of Environment and Natural Resources
MSPs	Minimum Support Prices
MPEDA	Marine Products Export Development Authority
MDGs	Millennium Development Goals
MA	Millennium Ecosystem Assessments
MPA	Marine Protected Areas
JTNVA	Jamsetji Tata National Virtual Academy
NGOs	Non Governmental Organisation
NPPD	National Physical Planning Department, Sri Lanka
NaCSA	National Centre for Sustainable Aquaculture, India
NFFC	National Federation of Fisheries Cooperatives
NARA	National Aquatic Resources Research and Development Agencies
NADRR	National Alliance for Disaster Risk Reduction

NDMA	National Disaster Management Authority
NCC	National Craft Council
NAPAs	National Adaptation Programmes of Action
NRC	National Research Council
NCRC	NGO Coordination Resource Centre
NCCF	National Calamity Contingency Fund
NCCFI	National Cooperative Consumers' Federation of India Limited
OB	Office Bearers
ODA	Official Development Assistance
OECD	Organisation for Economic Co-operation and Development
OCHA	UN Office for the Coordination of Humanitarian Affairs
PAS	Public Address System
PRA	Participatory Rural Appraisal
PRI	Panchayati Raj Institution
PAD	People's Action for Development
PS	Pradeshiya Sabha
PMSD	Participatory Market System Development
PMCA	Participatory Market Chain Analysis
PMM	Participatory Market Mapping
PPSS	Praja Pragathi Seva Sangam
PAG	Project Advisory Group
PTEI	Post-Tsunami Environmental Initiative
REN	Rural Enterprise Network

RF	Revolving Fund
RVC	Rural Volunteers Centre
SMS	Short Message Service
SSS	Sarvodaya Shramadana Society
SHG	Self-Help Group
SPRIT	Society for Participatory Research and Integrated Training
SEEDs	Sarvodaya Economic Enterprise Development Services Limited
SFD	State Fisheries Department
SDC	Sarvodaya District Centre
SRI	System of Rice Intensification
SSS	Sanghamitra Service Society
SAVA	Saurashtra Voluntary Actions
SDC	Swiss Agency for Development and Cooperation
SNAP	Strategic National Action Programme
SMART	Specific, Measurable, Achievable, Relevant and Timely
TN	Tamil Nadu
TAFREN	Task Force to Rebuild the Nation
TRINet	The Resource and Information Network for the Coast
TEEB	The Economics of Ecosystems and Biodiversity
UNEP	United Nations Environmental Programme
UNDP	United Nations Development Programme
UNICEF	United Nations International Children's Fund

UNFCCC	United Nations Framework Convention on Climate Change
UNISDR	United Nations International Strategy for Disaster Reduction
UN	United Nations
UNTRS	United Nations Tsunami Response System
VRC	Village Resource Centre
VKC	Village Knowledge Centre
VHF	Very High Frequency
VDS	Village Development Society
VIC	Village Information Centre
VLI(s)	Village Level Institutions
VDMC	Village Development Management Committee
VDC	Village Development Committee
VDP	Village Development Plan
VFO	Village Farmer Organisations
VCA	Vulnerability and Capacity Assessment
WRMIDMC	Walawe River Mouth Integrated Disaster Management Committee
WRI	World Resources Institute
WCMC	World Conservation Monitoring Center
WB	World Bank
WTO	World Trade Organization
WCFSD	World Commission on Forests and Sustainable Development
WWF	World Wildlife Fund for Nature

Part 1

Coastal Threats and Challenges

The first part of the source book consists of a selection of articles that address the issue of coastal threats and challenges. Coastal erosion, for example, is widespread and influenced by a range of factors, including natural processes, human activities and natural disasters. The protective functions and services of coastal systems are often threatened by livelihood and development activities. Recent disasters such as the tsunami in South Asia have provided opportunities for development workers to learn from reconstruction and rehabilitation efforts. The need for incorporating disaster concerns into coastal management and development is now recognized. It is also now better understood that land-use influences coastal-community livelihoods and the overall environment. Land tenure has affected rehabilitation efforts such as shoreline stabilization and shelter-related investments. Conflicts have arisen because of competing land use issues in coastal areas. Stronger emphasis is now being laid on integrated coastal zone planning, integrated coastal management, the role of enhancing ecosystem resilience, and local planning for disaster preparedness.

1

Rehabilitating Livelihoods in Tsunami-Affected Coastal Communities in Asia

Households in coastal communities undertake a range of activities in order to cope financially and reduce high economic dependency on natural resources.

Source:

R. Pomeroy, B. Ratner and S. Hall *(WorldFish Center, P.O. Box 500 GPO 10670 Penang, Malaysia)*; J. Pimoljinda *(Coastal Habitats and Resources Management Project (CHARM) Thailand)*; V. Vivekanandan *(South Indian Federation of Fishermen Societies. SIFFS Central Office, Karamana Trivandrum 695 002, India)*.

1. Introduction

Tens of millions of people live in coastal communities in India, Indonesia, Sri Lanka and Thailand. The majority of these people are highly dependent for livelihood, income and food security on coastal fisheries, coastal and marine habitats, agriculture, aquaculture and forestry. In Aceh province and on Nias Island, for example, fisheries provide employment to over 94,000 full- and part-time fishers, or nearly 20 per cent of the total coastal population. The fishery sector in Sri Lanka provides direct employment to about 250,000 people, with around one million people in households dependent on the sector. In the state of Tamil Nadu, India, there are 591 fishing villages and 362 landing centres, where around 700,000, mostly small-scale, fishers operate 59,000 vessels.

Most of these coastal communities are poor. Rural coastal communities in these four countries generally have a higher percentage of people living below the poverty line than the national average (Whittingham *et al.*, 2003). The high dependence on natural resources makes these communities particularly vulnerable to changes in resource condition. Silvestre *et al.* (2003) concluded that there was substantive degradation and overfishing of coastal fish stocks, and excessive fishing capacity in the region; resources have been fished down to 5-30 per cent of their unexploited levels. Such declines have increased poverty among coastal fishers. Overfishing has reduced the contribution of coastal fisheries to employment, export revenue, food security and rural social stability in these nations. Furthermore, as a result of human activities that contribute to mangrove removal, siltation and population explosion, coastal habitats are also degraded and become less productive.

The coastal communities face a growing degree of insecurity as a result of poverty and declining fisheries resources, and of factors such as high population growth, limited alternative livelihoods, limited access to land, economic and political marginalisation, unsustainable land use practices and development, competition and conflicts over resources, health burdens, and civil strife. Even prior to the tsunami, few coastal communities could see a way out of the growing insecurity caused by these multiple and complex factors.

The impact of the tsunami on rural coastal communities, particularly poor fishing and aquaculture households, was disproportionate in comparison to other groups of people in the region. The fishing and aquaculture households in the rural coastal communities made up an estimated one quarter of all fatalities. The impact of the tsunami was greatest on the poor, as they have the fewest resources and their ability to recover is the weakest. While some have been able to adapt and shift to other livelihoods post-tsunami for many, their situation just got worse. Not only were lives lost, but so too were household and productive assets (such as boats, ponds, marketing facilities and jetties). These losses reduced the ability of household to earn income and sustain livelihoods. In some areas, whole communities were destroyed. Many coastal households have been additionally affected by a range of secondary impacts, including:

1. Saltwater intrusion onto agricultural land in the coastal zone.
2. Interruption of other economic activities which previously contributed to livelihood strategies.
3. Continuing requirements to repay debts and loans while having no ability to earn income.
4. Disruption of informal sources and mechanisms of credit/savings, and loss of savings kept in households rather than banks.
5. A lack of demand for fish from coastal waters.
6. Psychological impacts on fishing communities.
7. Damage to coastal and marine ecosystems that supported the livelihoods of fishers and fish farmers.

2. Understanding coastal livelihoods

- Rehabilitation of coastal livelihoods after the tsunami needs to look beyond a return to the status quo and address the root causes of vulnerability of coastal people and communities; it needs to build both resilience to cope with future threats and ability to exploit opportunities.
- Adopting this approach requires understanding the diversity of coastal people and communities, especially in relation to their livelihood strategies.
- It also requires understanding the means by which households adapt to reduce their risks, the incentives that drive the decisions of resource users and the sources of their vulnerability to stresses and shocks.

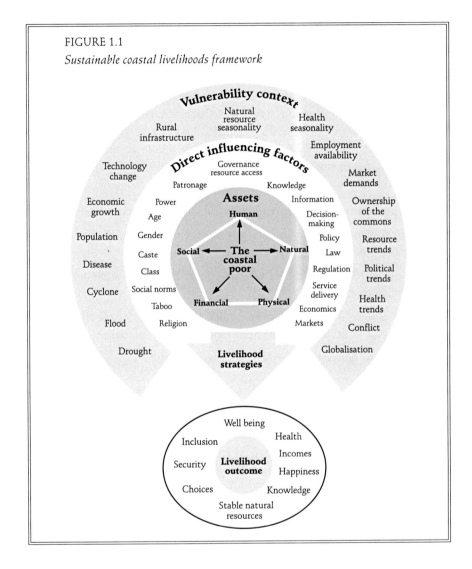

FIGURE 1.1
Sustainable coastal livelihoods framework

2.1. Diversity

- The dominant livelihood in many coastal communities, capture fishing is not the only livelihood.
- When fishing and agriculture are accounted for, all the other livelihoods (ranging from fish-processing to tourism) combined can employ an equal or greater number of people in many coastal communities.

2.2. Adaptation

- Many households in coastal communities undertake a range of activities in order to cope financially and reduce the risks associated with high economic dependency on natural resources.
- Fishing itself is a diverse occupation, with most fisheries in Asia being both multi-species and multi-gear in nature. Existing livelihood strategies may be modified or new strategies adopted to meet changing conditions.
- The household livelihood strategy mix will depend upon season, access to the resource (whether fishing areas or farm land), access to capital, skill base, education and risk preference. Coastal residents

Not everything can be done at once, a process for assessing and deciding on rehabilitation actions needs to take into account issues of both staging and scale:

What actions are feasible now and what steps are needed to address more fundamental problems over the longer term?

What can be done at the local community level and what challenges are more systemic, requiring policy or institutional change at national or even international levels?

Recognising that short-term, uncoordinated action can be detrimental to sustainable long-term rehabilitation, it is also vital that any process be grounded in a longer-term strategic plan. For example, a process for rehabilitating coastal livelihoods may involve the following seven steps:

1. Defining the target area.
2. Community entry and integration.
3. Assessments of resources, needs and opportunities.
4. Education and capacity development.
5. Rehabilitation plan.
6. Long-term sustainability plan.
7. Adaptive learning through monitoring and evaluation.

- may also engage in illegal activities for livelihood, such as dynamite fishing or smuggling.
- Rather than being specialised, and therefore vulnerable to a sudden change, many households in coastal communities are well situated to adapt to changing circumstances.
- The net result of this occupational diversity is that many coastal communities are best understood as dependent not on a single resource but on a whole ecosystem, marine and terrestrial.
- If it is deemed appropriate to provide an alternative occupation that is attractive to fishers, it should, at least, have some of the same characteristics as those considered desirable in fishing. A common alternative livelihood considered for fishers is aquaculture.

2.3. Incentives

- The incentive structures that individuals and households face are in part economic and in part related to other external factors, such as property rights, rules governing resource use, and levels of enforcement.
- Many coastal resource users exist at the subsistence level and have a short-run survival strategy of taking care of the daily needs of themselves and their family.
- Cultivating an awareness of the problems of unsustainable resource use is therefore only a small first step. Efforts must build on the array of opportunities and resources at people's disposal so that they become less directly dependent on the local natural resources for their daily subsistence and strengthen their security of tenure (whether private or communal) so that users have a greater stake in a longer term perspective.

2.4. Vulnerability

- The physical isolation of some coastal communities makes them highly resource-dependent and reduces access to alternative livelihoods; this can make them especially vulnerable to any disruptions.
- Some aspects of household vulnerability vary with the seasons. While occupational diversification may allow households to

maintain a level of income throughout the year, there may be periods of high income (as when crops are harvested or fishing is good) and low income (as when fishing is poor or not possible due to storms).

- Other root causes of vulnerability in coastal communities are social and economic power imbalances, lack of participation in decision-making, limited asset ownership, resource dependence, and laws and regulations that influence people's ability to use assets.
- Once the root causes of vulnerability are recognised, interventions can be put in place to address them and to increase the resilience of the community to shocks, seasonal factors, and human and natural changes.
- Building resilience means, in part, reducing reliance upon natural resources for livelihoods, strengthening community institutions, organisations and infrastructure, and diversifying livelihoods.

3. A process for rehabilitating coastal livelihoods

- The rehabilitation of coastal livelihoods after a natural disaster should be seen as an opportunity to strengthen and revitalise coastal communities.
- The focus of rehabilitation efforts should be on rebuilding the economic basis of livelihoods rather than on physical reconstruction, and on giving coastal people the skills and resources for self-recovery.
- The rehabilitation of coastal livelihoods should be undertaken through a process of change that will address the recurrent factors that make them vulnerable.
- Some of these factors can be addressed more immediately, while others, such as equity, power relations, access to resources and markets, asset ownership and sustainability of resource use, are more fundamental and difficult to alter.
- Social development (empowerment, organisations, education and training), economic development (job creation, private sector investment, market access, microfinance) and ecological interventions (rehabilitating coastal habitats, coastal resource

management) that address these fundamental issues must be integrated.
- Organisations engaged in coordinating recovery efforts there have learned that rehabilitation interventions should be phased out over time, not trying to achieve too much too soon.
- A number of different processes can be followed to rehabilitate coastal livelihoods in disaster-affected areas. It is vital that the process be well planned at the operational level and be participatory, involving consultation and collaboration with the community.
- Whatever steps are followed in the process of planning, implementing and evaluating the rehabilitation of livelihoods, it will be necessary to repeatedly evaluate a range of options.
- When the root causes of vulnerability are addressed, they effectively change the boundaries of what is feasible and possible, improving the chances that communities can build resilience to cope with future threats and opportunities.

Key criteria for assessing livelihood options include:

Social feasibility:

- Any process of rehabilitating coastal livelihoods should be 'socially feasible', i.e., that the livelihood options be compatible with the needs and aspirations, existing work ethic and livelihood strategies, organisation, economic and social structure, gender differences, and culture of the affected community and households.
- Various livelihood options need to be made available because individual and household goals, attitudes and preferences will differ.

Technical feasibility:

- Simple livelihoods that require low levels of capitalisation and extensive, rather than intensive, levels of management are often preferable for many coastal communities.
- A labour-intensive operation is better suited for communities where labour is abundant, wage rates low, and capital relatively scarce.
- A livelihood with a low-gain, low-risk strategy is likely be more attractive to the target group than one which offers high gain and high risk.

- The livelihood must be at a level that can be maintained and operated by the target group and that can generate cash flow over the long term.

Institutional sustainability:

- Approaches to rehabilitate coastal livelihoods must be sustained by the beneficiaries after external organisations phase out assistance to the community.
- Successful efforts to enhance existing livelihoods, diversify, or adopt alternative livelihoods typically stem from participatory decision-making, bearing in mind the capacity and incentives for coastal people to engage in the livelihood strategy.
- Approaches to rehabilitating coastal livelihoods must acknowledge that change on the coast is an ongoing process. A viable livelihood today will only be sustainable if it incorporates the capacity to evolve with the changes around it.

Supporting infrastructure and policy environment:

- The sustainability, and therefore choice, of a livelihood option will depend on the availability of supporting infrastructure and the enabling environment, including credit, inputs, markets and technical assistance.
- Policy reform may be needed to create an enabling environment for the livelihood options.
- Effective technical assistance, such as extension services, should be in place to provide specialised training and technical assistance on a continuing basis.
- The strength of property rights will influence the nature of use and management of the resources, and the economic returns from the resource.

4. From disaster to opportunity: Addressing the root causes of vulnerability

To rehabilitate coastal livelihoods in a manner that will be effective and sustainable requires addressing the factors that have led to vulnerability unsustainable livelihoods, high levels of poverty and a declining quality of life in coastal communities.

- Rehabilitation of coastal livelihoods means addressing fundamental social, economic and environmental reforms that affect coastal communities and livelihoods.
- Reforms are challenging because it involves coordination across sectors that are often governed through separate planning and regulatory processes.
- Disasters bring the spotlight of societal attention on affected communities, however, they often open a window of opportunity to address more fundamental problems that at other times would not be possible.

4.1 Securing resource tenure and access

Coastal areas that offer a range of easily accessible livelihood opportunities that are rarely available in inland areas, often attract the poor, who exploit a range of resources from the land and sea with a range of harvesting methods. One reason that the poor are attracted to these resources is that they are typically 'open access' and are easily exploited with minimal capital resources.

- Measures to control access, such as marine protected areas, are being used to conserve and protect these resources.
- The land-tenure status of families and entire communities can often be unclear in modern legal terms; and the poor are often the least able to defend their livelihoods or to establish legal tenure rights over the resource.
- The rights of the poor to security of tenure over the resources upon which they depend for their livelihood need to be established and asserted so that they can make long-term investments in sustainable livelihoods and resource management.

The rehabilitation process is a way of clarifying and correcting past injustices and administrative inefficiencies and of providing equity and new tenure arrangements.

- Property rights over the resources need to be specified and secured to enable resource users to optimise use and ensure conservation of resources.

4.2 Building equitable market access

The free market both provides for and restricts livelihood opportunities for resource users and the community. With specialised traders, resource users often have little, if any, control over marketing outlets and the prices that they receive.

Low incomes create a situation of potential dependence that influences the resource user's decisions about credit sources and marketing decisions.

- New livelihood options, and increases in market power, can be found in the integration of resource users forward in the market chain once they have the skills to undertake the activity supplemented by microfinance.
- The setting up of new microfinance institutions as a post-disaster response has been found to be largely ineffective; it is more effective to build on existing microfinance institutions.
- Microfinance can help poor households multiply income-generating opportunities.
- Actions to restore people's productive capacity and actions to revive market demand for their output and labour need to be complementary.

4.3 Reducing excess capacity

In most coastal fisheries in Asia there is an excessive level of factor inputs (capital and labour) relative to that needed to catch the available fish. Thus, most fisheries can be characterised as having the problem of 'excess capacity', 'overcapitalisation', or simply, too many fishers chasing too few fish. Fishing capacity in the areas affected by the tsunami should generally not exceed the levels that existed before, and in most places should be reduced to help ensure the sustainability of the fisheries. Rehabilitation efforts should ensure that less destructive and more sustainable fishing gears and practices are adopted.

4.4 Protecting ecosystems

The region's pattern of development and, more specifically, the persistence of widespread poverty, rapid and uncontrolled coastal development and environmental degradation have led to an increase in its vulnerability to natural disasters.

- Coastal ecosystems, such as wetlands, mangroves, coral reefs, seagrass beds and sand dunes, should be identified and protected from development and uses that compromise their structural integrity.
- Protection of these coastal ecosystems will enhance their ability to provide long-term economic benefits to coastal communities by way of coastal protection, generation of a diversity of sustainable livelihoods and maintenance of the communities' natural resource capital.

4.5 Reducing vulnerability to natural hazards

Coastal areas are inherently exposed and dangerous. Despite this, people are attracted to the coast for economic and other reasons.

4.6 Empowering coastal communities

The economic and political marginalisation of coastal communities has led to poverty and resource degradation. Addressing marginalisation requires empowerment of community members and the transfer of economic and political power from a few to the impoverished majority. Individual and community empowerment are central to rehabilitating coastal livelihoods.

Empowerment is concerned with the capability building of individuals and the community in order to increase social awareness, to gain greater autonomy over decision-making, to gain greater self-reliance and to establish a balance in community power relations. While decentralisation reforms that shift authority for decision-making to more local levels of government do not guarantee community empowerment, they can increase the opportunities for disadvantaged community members to access information or participate in decision-making.

4.7 Rebuilding community organisations

Disasters often weaken social structures and processes in the community, but rehabilitation efforts provide an opportunity to reinforce the positive strengths of existing social structures. Community organisations facilitate participation in decision-making over rehabilitation efforts and are essential for institutional sustainability.

- Many coastal resource users, due to the nature of their livelihood, act individually or as a family unit. Most are not represented in community organisations that enable them to effectively engage in collective action to take advantage of an opportunity or overcome a threat, or to influence the direction of policies and decision-making.

- Community organisations should be re-established (if lost with the disaster) or newly established. In some cases, traditional community organisations that typically serve one function, such as regulating resource access, may take on new leadership roles in response to a disaster.

- Community organisations must have the legal right to exist and make arrangements related to their needs. They must be recognised as legitimate and credible by the community and be transparent and accountable to their members.

- Community organisations can be networked to further strengthen their ability to serve and represent their members.

4.8 Integrating coastal communities into national economic development

Many rural coastal communities have been left behind as economic development progressed in other parts of the country; this furthered their economic marginalisation. In addition (or as a result), rural coastal communities have often been a low priority in national economic development planning.

- Livelihood development in coastal communities needs to be linked to national economic development plans and to current and future employment needs in the country.

- Greater attention to and investment in social and physical infrastructure that will improve the overall quality of life in coastal communities should be made.

4.9 Investing in education and training

Knowledge and information is power.

- The rebuilding of coastal communities is a good time to address educational and training needs in coastal communities.
- Young men and women from the community can learn new skills that match the immediate needs of rebuilding, such as carpentry, masonry, plumbing, that can also be used after rebuilding is completed.
- Training in disaster preparedness and management, such as safety at sea, can be linked with environmental education to improve the conservation and management of coastal resources.

5. Conclusions

Rehabilitation should look beyond reinstating the problems of the past and seek to address the root causes of vulnerability of coastal people and communities and to build their resilience to future threats and capacity to exploit opportunities.

- Coastal communities should be involved in a dialogue about the future they envision, the steps needed to get there, and the lessons learned along the way. This requires engaging a much broader array of actors across government, civil society and the private sector.

References

Beatley, T. and P. Berke (2006). *After the hurricane: Linking recovery to sustainable development in the Caribbean.* Baltimore (MD): Johns Hopkins University Press.

———. (1998). in R.S. Pomeroy *et al. Marine Policy* 30: 786-793.

CONSRN (2005). *Regional impacts of the tsunami on fisheries and aquaculture.* (updated 3 February). www.apfic.org/modules/wfsection/download.php?fileid=58

FAO (Food and Agriculture Organization) (2001). *Report of the FAO Asia-Pacific conference on early warning, prevention, preparedness and management of disasters in food and agriculture.* Bangkok, Thailand. RAP Publication 2001/14. FAO: Regional office for Asia and the Pacific.

Grunewald, F., V. de Geoffroy, S. Lister (2000). "NGO responses to hurricane mitch: Evaluations for accountability and learning", *HPN Paper* 34. London: Overseas Development Institute.

Houghton, R. (2005). "Tsunami emergency: Lessons from previous natural disasters", *Action learning network for accountability and performance*, London, Overseas Development Institute. www.alnap.org/lessons_tsunami.htm

IMM Ltd. (2001). *Learning lessons from the cyclone: A study of DFID's support for post-cyclone livelihoods rehabilitation in Orissa, India*. University of Exeter, United Kingdom.

———. (a) (n.d.). "The sustainable livelihoods approach in the coastal context", SCL Working Paper 1. University of Exeter, UK.

———. (b) (n.d.). "Sustainably enhancing and diversifying the livelihoods of the coastal poor", *SCL Working Paper 7*. University of Exeter, UK.

International Food Policy Research Institute (IFPRI) (2005). *When disaster strikes*. IFPRI Forum, March. www.ifpri.org/pubs/newsletters/ifpriforum/IF200503.htm

Inter-American Development Bank (IDB) (2000). *Facing the challenge of natural disasters in Latin America and the Caribbean: An IDB action plan*. Environment Division, Washington DC.

Kurien, J. (2003). "The blessing of the commons: Small-scale fisheries, community property rights and coastal natural assets", *Working Paper 349*. India: Center for Development Studies, Thiruvananthapuram.

McDonald F. (1999). "Environmental influences on disaster impacts: An emerging concern for the Caribbean—Environmental issues in disaster prevention, preparedness and response", *Conference Proceedings 18*, March. UK: GREEN Cross. www.ww2bw.org/eforums/mcdonaldthoughts

Messer, N.M. (2003). *The role of local institutions and their interaction in disaster risk mitigation: A literature review*. Rome, UN Food and Agriculture Organization. www.fao.org/documents/show_cdr.asp?url_file-\=/docrep/006/ad710e/ad710e0i.htm

Organization of American States (1997). *The Caribbean disaster mitigation project: Technical assistance for natural disaster management at the regional scale*. Washington, DC. www.oas.org/CDMP/document/papers/cdmptech.htm

Pomeroy, R.S. (2004). "Socioeconomic dimensions of aquaculture as a solution to destructive fishing", *Technical Report FTR4*. Honolulu, Hawaii: Community Conservation Network.

Silvestre, G., L.R. Garces, I. Stobutzki, M. Ahmed, R.A. Valmonte Santos, C. Luna, L. Lachica- Aliño, P. Munro, V. Christensen and D. Pauly (eds.) (2003). "Assessment, management and future directions for coastal fisheries in Asian countries", *WorldFish Center Conference Proceedings 67*. Penang, Malaysia.

Whittingham, E., J. Campbell and P. Townsley (2003). "Poverty and reefs", *Global Overview 1*. UK: DFID- IMM-Ioc/ UNESCO, IMM Ltd, Exeter.

Wisner, B. (2002). "The communities do science! Proactive and contextual assessment of capability and vulnerability in the face of hazards", in G. Bankoff, G. Frerks and T. Hilhorst (eds.), *Vulnerability: Disasters, development and people*. London: Earthscan.

World Bank (2005). *Lessons from natural disasters and emergency reconstruction*. Operations Evaluation Department. 10 January, Washington DC.

2

Ecosystem Threats and Impacts in Coastal Areas

Dramatic population growth, rapid urbanisation, and destructive practices pose a threat to the coastal ecosystem.

Source:

Robert Kay *(Adapted from: Improved coastal zone planning and management. Integrated coastal zone planning in Asian tsunami-affected countries. The regional perspective in proceedings of the workshop on coastal area planning and management in Asian tsunami-affected countries. 27-29 September 2006. Bangkok, Thailand).*

Uses of the coastline are generally considered under four main categories: resource exploitation (e.g., fisheries, forestry, gas and oil, and mining); infrastructure (including transportation, ports, harbours, shoreline protection works and defence); tourism and recreation; and the conservation and protection of biodiversity (Kay and Alder, 2005). Of specific interest here are the major land uses in the coastal zones of tsunami-affected countries, which include agriculture, shrimp and fish farming, forestry and human settlement (UNEP, 2005).

1. Mangroves' use and ecological role

Resource-based industries such as fisheries and tourism are particularly important within the Asia-Pacific region. Fishing provides a basic source

---

Mangroves as effective tsunami buffers

While the ecosystem value of mangroves and coastal forests is well-known and widely accepted, their widespread clearance has subsequently become a common feature of coastal zones in all tsunami-affected countries (IUCN, 2005).

In the islands of Indonesia, Java alone had lost 70 per cent of its mangroves by 1991, while Sumatra had lost 36 per cent (*www.earthisland.org*).

Mangrove reduction has led to a loss of biodiversity and a reduction in food production and cooking fuel, which exacerbates the problem in other areas. In addition, a source of income is eliminated for marginalised communities that are already considered socially and economically worst off (Kay and Alder, 2005).

Anecdotal evidence in the aftermath of the tsunami suggested that mangroves were effective in buffering its impacts (Dahdouh-Guebas *et al.*, 2005; FAO, 2005; and Wetland International, 2005). This was subsequently confirmed by systematic analysis of the effectiveness of mangrove buffering against tsunami waves (Chang *et al.*, 2006). Their preliminary analysis suggested that villages that were behind substantial mangroves suffered relatively little damage in comparison to those not likewise protected. However, assessment of the role of healthy ecosystems in reducing damage to coastal communities is ongoing and remains a matter for debate within the scientific community (Baird, 2006).

of food and income for up to 13 million people, while the extensive tourist industry may directly account for as much as 20 per cent of the GDP (UNDP/IUCN, 2006). Mangroves and the coastal forest play a crucial role in coastline stabilisation and storm-surge protection (Rutinbeek, 1991; 1994). Table 2.1 shows the uses and environmental functions of mangroves. In addition, mangroves act as important pollutant and nutrient sinks (Ibid.).

2. Effects of development activities on the coastal ecosystem

Within the Asia-Pacific region, healthy coastal ecosystems including mangroves, wetlands, estuaries, lagoons, sandy beaches, sand dunes, coral reefs and seagrass communities are fundamentally linked to human well-being (UNDP/IUCN, 2006). However, dramatic population growth has led to increased pressure on ecosystems throughout the coastal zone. Table 2.2 summarises these effects.

Within the Asia-Pacific region, an ad hoc approach, where no coherent planning strategy has been adhered to, has frequently provided greater short-term economic returns, although sometimes at great social and environmental costs, and often, these returns benefit only some sections of society. Single objective, single output land management has resulted in the conversion of land from directly productive purposes; in many cases, leading to degradation or loss as a result of erosion, salinity, inundation and other interventions (FAO, 2005).

It is clear that the enormous contribution of healthy coastal ecosystems in safeguarding production and consumption, reducing vulnerability, strengthening resilience and mitigating disasters has generally been undervalued, poorly understood, and improperly safeguarded within the region (UNDP/IUCN, 2006).

In fact, it has been suggested that unsustainable development activities and their associated degradation of the coastal zone led to exacerbated effects of the 2004 Indian Ocean tsunami in terms of ecosystem destruction and loss of life (Table 2.3).

In addition to the problems brought about by the effects of the tsunami, pressure on coastal land is expected to continue in step with increases

TABLE 2.1
Uses and environmental functions of mangroves

Sustainable production functions	Regulatory or carrier functions	Information functions	Conversion uses
• Timber	• Erosion prevention (shoreline)	• Spiritual & religious information	• Industrial/urban land use
• Fuel wood	• Erosion prevention (riverbanks)	• Cultural & artistic inspiration	• Aquaculture
• Woodchips	• Storage & recycling of human waste/pollutants	• Educational, historical & scientific information	• Salt ponds
• Charcoal	• Biodiversity maintenance	• Potential information	• Rice fields
• Fish	• Migration habitat	• Plantations	
• Crustaceans	• Nursery	• Mining	
• Shell fish	• Breeding grounds	• Dam sites	
• Tannins	• Nutrient supply		
• Nipa	• Nutrient regeneration		
• Medicine	• Habitat for indigeneous people		
• Honey	• Recreation sites		
• Traditional hunting, fishing			
• Genetic resources			

Source: Adapted from Rutinbeek, 1991; 1994.

TABLE 2.2

Threats to coastal ecosystems and key drivers

Threat	Drivers
Habitat loss or conversion	
Coastal development (ports, urbanisation, tourism related development, industrial sites)	Population growth, poor development policies for industry and tourism, environmental refugees and internal migration
Destructive fisheries (dynamite, cyanide, trawling)	Shift to market economies, demand for aquaria fish and bottom live food fish, increasing competition in light of diminishing resources
Coastal deforestation (especially mangrove deforestation)	Lack of alternative materials, poor national policies increased competition
Mining (coral, sand, minerals, dredging)	Lack of alternative materials, global commons' perceptions
Civil engineering works	Transport and energy demands, poor public policy, lack of knowledge about impacts and costs
Environmental change brought about by conflict	Increased competition for scarce resources, political war and instability, inequality in wealth distribution
Aquaculture related habitat conversion	Demand for luxury items, regional food needs, declining wild stocks, loss of property rights in fisheries, inability to compete
Habitat degradation	
Eutrophication from land based sources (agricultural waste, sewage, fertilisers)	Urbanisation, lack of wastewater and sewage treatment systems, poor agricultural practices, loss of wetlands and other natural controls
Pollution: toxins and pathogens from land based sources	Lack of awareness, increasing pesticide and fertiliser use (especially as soil quality diminishes), unregulated industry
Pollution: dumping and dredge spoils	Lack of alternative disposal methods, increasing costs for land disposal, belief in unlimited assimilative capacities, waste as a commodity
Pollution: shipping-related	Substandard shipping regulations, no investment in safety, policies promoting flags of convenience, increases in ship-based trade
Salinisation of estuaries due to decreased freshwater inflow	Demand for electricity and water, territorial disputes

contd...

...contd...

Threat	Drivers
Alien species' invasions	Ballast discharge regulations lacking, increased aquaculture related escapes, lack of international agreements on deliberate introductions
Global warming and sea level rise	Emission controls lacking, poorly planned development (vulnerable development), stressed ecosystems less able to cope
Overexploitation	
Directed take of low value species at high volumes exceeding sustainable levels	Subsistence and market demands (food and medicinal), industrialisation of fisheries, improved fish-finding technology, poor regional agreements, lack of enforcement, breakdown of traditional regulation systems, subsidies
Directed take for luxury markets (high value, low volume)	Demand for specialty foods and medicines, aquarium fish and curios, lack of awareness or concern about impacts, technological advances, commodification
Incidental take or bycatch	Subsidies, bycatch has no cost

Source: Adapted from Kay and Alder, 2005.

TABLE 2.3

Exacerbation of biophysical and socioeconomic impacts as a result of unplanned development in the coastal zone

Biophysical impacts	Socioeconomic impacts
• Increased coastal erosion	• Loss of property and coastal habitats
• Siltation	• Loss of life
• Loss of biodiversity	• Damage to coastal protection works and other infrastructure
• Extensive coastal inundation	• Loss of renewable and subsistence resources
• Higher level flooding	• Loss of amenity value
• Decline in water quality	• Loss of non-monetary cultural resources and values
	• Impacts on agriculture and aquaculture

in the regional population. In light of this, the need for an integrated approach to the management and rehabilitation of the coastal zone has been brought into even starker relief.

References

Baird, A. (2006). "Tsunami viewpoint: the myth of green belts", *Samudra* 44: 14-19.

Chang, S.E., B.J. Adams, J. Alder, P.R. Berke, R. Chuenpagdee, S. Ghosh and C. Wabnitz (2006). "Coastal ecosystems and tsunami protection after the December 2004 Indian Ocean Tsunami", *Report to the National Science Foundation of the USA*.

Dahdouh-Guebas, F., L.P. Jayatissa, D. Di Nitto, J.O. Bosire, D. Lo Seen and N. Koedam (2005). "How effective were mangroves as a defence against the recent tsunami?", *Current Biology* 15(12): R443-47.

FAO (2005). "Building back better livelihoods in tsunami-affected countries". http://ftp.fao.org/FI/DOCUMENT/tsunamis_05/FAO_guiding/FAOBriefSpecialEnvoy130550.pdf

IUCN (2005). "Rapid environmental and socio-economic assessment of tsunami-damage in terrestrial and marine coastal ecosystems of Ampara and Batticaloa districts of Eastern Sri Lanka", http://www.iucn.org/tsunami/docs/rapid-ass-easte-sri-lanka.pdf

Kay, R.C. and J. Alder (2005). *Coastal planning and management*. London: E&F Spon. p. 380.

Rutinbeek, H.J. (1991). *Mangrove management: An economic analysis of management option with a focus on Bintuni Bay, Iran Jaya, Jakarta*. Government of Indonesia, Dalhousie University.

———. (1994). "Modeling economy-ecological linkages in mangroves: Economic evidence for promoting conservation in Bintuni Bay, Indonesia", *Ecological Economics* 10: 233-47.

UNDP/IUCN (2006). *Mangroves for the future: A strategy for promoting investment in coastal ecosystem conservation, 2007-2012*.

UNEP (2005). "After the tsunami: Rapid environmental assessment", February, http://www.unep.org/tsunami/tsunami_rpt.asp

Wetlands International (2005). "Natural mitigation of natural disasters", Assessment report to Ramsar STRP12. Cited 15 June. Available at http://www.wetlands.org/tsunami/

3

Coastal Erosion and its Management

Clearing of mangrove forests makes coastal areas more susceptible to erosion.

Source:

Gegar Prasetya Adapted from: The Role of coastal forests and trees in protecting against coastal erosion. Coastal protection in the aftermath of the Indian Ocean Tsunami: What role for the forests and trees? Proceedings of the Regional Technical Workshop Khao Lak, Thailand, 28-31 August 2006.

1. Introduction

Coastal erosion is a complex process. There is a strong relation between major coastal erosion problems and the protective functions of coastal systems. Erosion is widespread in the coastal zones of Asia. Bilan (1993) reported that the erosion rate in the northern part of Jiangsu Province in China is serious and as high as 85 metres/year; in Hangzhou Bay the rate is 40 metres/year, while in Tianjin it is 16-56 metres/year. Erosion persists even where preventive measures such as sea dykes are constructed. Beach scour has been found along coasts with sea-dyke protection. This erosion is attributable to many factors such as river damming and diversion that leads to less sediment supply to the coast, and the clearing of mangrove forests, which makes coastal areas more susceptible to the hazard. Juxtaposing these phenomena, the intensification of typhoons and storm surges during the 42-year period between 1949 and 1990 has meant that storm surges with increasing tidal levels exceeding one and two metres have occurred 260 and 48 times respectively, thus exacerbating the erosion problem. Most of the sediment taken offshore by the storm waves has been returned in minimal quantities to the coast during normal conditions owing to the frequent storm intensity.

2. Cases of coastal erosion in Asia

2.1 Malaysia

According to Othman (1994), nearly 30 per cent of the Malaysian coastline is undergoing erosion. Many of these areas are coastal mudflats, fringed by mangroves. Behind the mangroves there are usually agricultural fields protected from tidal inundation by bunds (dykes). Locally, mangroves are known to reduce wave energy as waves travel through them; thus, the Department of Irrigation and Drainage has ruled that at least 200 metres of mangrove belts must be kept between the bunds and the sea to protect the bunds from eroding. However, the mangroves themselves are susceptible to erosion when the soil under their root systems is undermined by wave action that mostly occurs during periods of lower water level or low tide. The value of intact mangrove swamps for storm protection and flood control alone in Malaysia is approximately US $300, 000/kilometre (*http://ramsar.org*).

2.2. Vietnam

In Vietnam, most of the coastline in the south that is located in a wide and flat alluvial fan and bordered by tidal rivers fringed by wide mangrove swamps, has been eroded continuously at a rate of approximately 50 metres/year since the early 20th century (Mazda *et al.*, 1997; Cat *et al.*, 2006). This massive erosion—mostly due to wave and current action—and vanishing mangrove vegetation is attributable to the long-term impacts of human activities since the late 19th century and also human-induced change within watersheds (dam construction that has reduced the sediment supply to the shore). Erosion still occurs in the central coastal zone of Vietnam and preventive measures such as sea dykes, revetments, and tree plantations have been implemented in many coastal areas. In the southern coastal zone, however, mangrove plantations have mitigated wave action and prevented further erosion (Cat *et al.*, 2006).

2.3 India

The rapid erosion of the coast of Sagar island in West Bengal, India, is caused by several processes that act in concert. These are natural processes that occur frequently (cyclones, waves and tides that can reach six metres in height) and anthropogenic activities such as human settlement and aquaculture that remove mangroves and other coastal vegetation. The erosion rate from 1996 to 1999 was calculated to be 5.47 square kilometres/year (Gopinath and Seralathan, 2005). The areas that are severely affected by erosion are the northeastern, southwestern, and southeastern faces of the island. Malini and Rao (2004) reported coastal erosion and habitat loss along the Godavari delta front owing to the combination of the dam construction across the Godavari and its tributaries that diminishes sediment supply to the coast and continued coastal land subsidence.

2.4 Sri Lanka

Sri Lanka's experience with coastal erosion dates back to 1920 (Swan, 1974; 1984). It has become more serious because mangroves are being eradicated by encroachment (human settlement), fuel wood cutting and the clearing of coastal areas for intensive shrimp culture. Mangrove forest cover was estimated to be approximately 12,000 hectares in 1986. This dwindled to 8,687 hectares in 1993 and was estimated to be only 6,000 hectares in 2000 (Samrayanke, 2003). Approximately US $30 million has already been spent on breakwaters and other constructions to combat coastal erosion on southern and western coasts (UNEP, 2006); though coastal erosion still persists in some coastal areas.

2.5 Indonesia

In Indonesia, coastal erosion started in the northern coast of Java Island in the 1970s when most of the mangrove forest had been converted to shrimp ponds and other aquaculture activities, and the area was also subjected to unmanaged coastal development, diversion of upland freshwater and river damming. Coastal erosion is prevalent throughout many provinces (Bird and Ongkosongo, 1980; Syamsudin and Riyanduni, 2000; Tjardana, 1995) such as Lampung, Northeast Sumatra, Kalimantan, West Sumatra (Padang), Nusa Tenggara, Papua, South Sulawesi (Nurkin, 1994) and Bali (Prasetya and Black, 2003). US $79.667 million was provided by the Indonesian government to combat coastal erosion from 1996 to 2004, but only for Bali island in order to protect this valuable coastal tourism base (Indonesia water resource donor database: *http://donorair.bappenas.go.id*). A combination of hard structures and engineering approaches (breakwaters/jetties/revetments) of different shapes that fused functional design and aesthetic values, and soft structures and engineering approaches (beach nourishment) was used. They succeeded in stopping coastal erosion on Sanur, Nusa Dua and Tanjung Benoa beaches, but were neither cost-effective nor efficient, because during low tide all of the coastal area was exposed up to 300 metres offshore; thus, these huge structures were revealed and became eyesores.

2.6 Thailand

In Thailand, intensification of coastal erosion came in focus during the past decade (Thampanya *et al.*, 2006). Overall, the net erosion is approximately 1.3 to 1.7 metres per year along the southern Thailand coastline. Total area losses amount to 0.91 square kilometres per year for the Gulf coast and 0.25 square kilometres per year for the western coast. Most of the eroded areas increase with larger areas of shrimp farms and less mangrove forest area. When dams reduce riverine inputs and coastal land subsidence transpires. In areas where erosion has prevailed, the presence of mangroves has reduced erosion rates. Mangroves dominating coastal locations exhibit less erosion than areas with non-vegetated land or former mangrove areas.

Such examples indicate that there is a strong relation between major coastal erosion problems throughout the region and degradation of the protective function of coastal systems such as coastal forest and trees— particularly mangrove forest. Artificial and natural agents that induce mangrove loss and make coastal areas more susceptible to coastal erosion include anthropogenic factors such as excessive logging, direct land reclamation for agriculture, aquaculture, salt ponds, urban development and settlement, and to a lesser extent fires, storms, hurricanes, tidal waves and erosion cycles, owing to changing sea levels (Kovacs, 2000). More scientific investigation and quantification of the physical processes and dynamic interaction of the system is needed to understand how and under what circumstances mangrove forests and other coastal vegetation effectively protect the shoreline from erosion. A number of efforts have focussed on field observations, laboratory and numerical model experiments and theoretical analysis (Wolanski, 1992; Mazda *et al.*, 1997; Massel *et al.*, 1999).

3. The causes of coastal erosion

Coastal erosion and accretion are complex processes that need to be investigated from the angles of sediment motion under wind, wave and tidal current action; beach dynamics within a sediment/littoral cell; and human activities along the coast, within river catchments and watersheds and offshore, both at spatial and temporal scales. In terms of temporal scales, the issue of sea level rise is complex and produces a range of environmental problems. As the sea level rises, the water depth increases and the wave base becomes deeper; waves reaching the coast have more energy and therefore can erode and transport greater quantities of sediment. Thus, the coast starts to adjust to the new sea level to maintain a dynamic equilibrium. Figure 3.1 lists the processes of coastal erosion and accretion, as well as natural factors and human activities.

FIGURE 3.1
The complex processes of coastal erosion and accretion

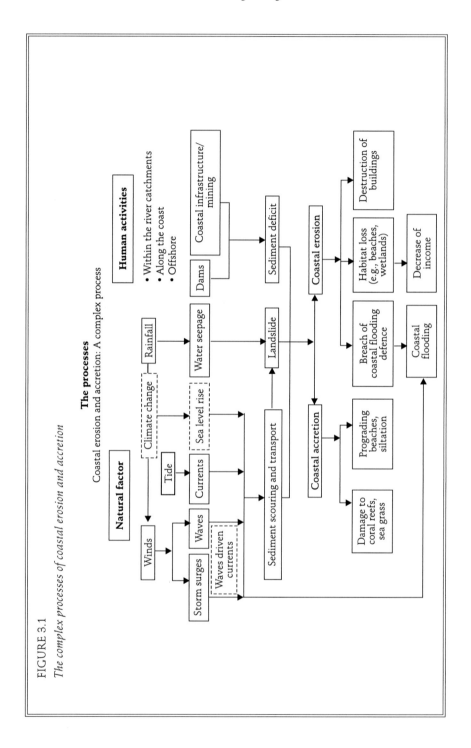

The key physical parameters that need to be understood to identify coastal erosion as a problem in the coastal zone are:

- *Coastal geomorphology*: Coastline type and sensitivity to coastal processes.
- *Wind*: The main force in wave generation. Under the right environmental conditions, wind may transfer sediment from the beach environment landward on all open coastlines.
- Waves: Most important forces for sediment erosion and transport to the coastal zone. They introduce energy to the coast and also a series of currents that move sediment along the shore (longshore drift) and normally to the shore (cross-shore transport). It is important to understand the movement of wave forms as well as water particles and their interaction with seabed material; also how the waves determine whether the coasts are erosive or accretional.
- *Tides*: Influential in beach morphodynamics. They modulate wave action, controlling energy arriving on the coast and drive groundwater fluctuation and tidal currents. The interaction of groundwater with tides in the coastal forest environment is crucial in understanding why coastal forest clearance causes intensive coastal erosion in particular environments.
- *Vegetation*: Important for improving slope stability, consolidating sediments and providing some shoreline protection.

Equally significant human activities that must be considered over the range of spatial and time scales are:

- *Activities along the coast*: Building houses via land reclamation or within sand dune areas and port/harbour development has a long-term impact on shoreline change; protective seawalls lead to erosion at the end of the structures, generate beach scouring at the toe of seawall and shorten the beach face. This can occur in the short term (less than five years) or the long term (more than five years). Other structures such as groynes and jetties typically cause erosion down-drift of the structure within a short period of time (between 5 and 10 years). Removal of dune vegetation and mangroves will expose low energy shorelines to increased energy and reduced sediment stability, causing erosion within 5 to 10 years.

- *Activities within river catchments/watersheds*: Dam construction and river diversion cause reduction of sediment supply to the coast that contributes to coastal erosion. The effects of dam and river diversion in terms of coastal erosion are not straightforward, but there are mid- to long-term impacts (20 to 100 years) with spatial scales approximately from 1 to 100 kilometres.

- *Onshore and offshore activities*: Sand and coral mining and dredging may affect coastal processes in various ways such as contributing to sediment deficit in the coastal system and modifying water depth that leads to altered wave refraction and longshore drift. The impact of these activities will be obvious within a short period of time (1 to 10 years).

Understanding the key processes of coastal dynamics and how the coasts function both in spatial and temporal time scales (short and long term), as well as human activities along the coast, within the river watershed and offshore is essential for managing coastal erosion because it may occur without reason. A quantitative understanding of changes in spatial and short- and long-term time scales is indispensable for the establishment of rational policies to regulate development in the coastal zone (NRC, 1990). Table 3.1 summarises possible natural factors and human activities that affect shoreline change over a range of time scales, leading to coastal erosion.

4. The coastal type and protective function of the coastal system

Coastlines comprise the natural boundary zone between the land and the ocean. Their natural features depend on the type of rocks exposed along the coastline, the action of natural processes and the work of vegetation and animals. The intensity of natural processes formed their origin—either as erosional or depositional. The geological composition of a coastal region determines the stability of the soil, as well as the degree of rocky materials and their breakdown and removal.

TABLE 3.1
Possible natural factors and human activities that affect shoreline change

Factor	Effects	Time scale								
		Hours	Days	Months	Years	10 years	50 years	100 years	1000 years	10,000 years
Natural factors										
Short wave period	Erosion		▓			▓			▓	▓
Wave of small steepness	Accretion	▓								
Large wave height	Erosion			▓						
Storm surge	Erosion		▓							
Along shore current	Accretion, no change or erosion			▓						
Rip currents	Erosion			▓						
Underflow	Erosion		▓							
Overwash	Erosion		▓							
Wind	Erosion			▓						
Sediment supply (source and sink)	Accretion or erosion							▓		
Inlet presence	Net erosion, high instability							▓		
Sea level rise	Erosion								▓	▓
Land subsidence (tectonic)	Accretion or erosion								▓	▓
Human activities										
Dredging	Accretion or erosion					▓				
Coastal defence	Accretion or erosion						▓	▓		
Vegetation clearing	Erosion				▓					
Harbour development	Accretion or erosion							▓		
River damming	Erosion								▓	
Land reclamation	Erosion									

4.1 Clayey bank coast

This type of coast can be classified as a 'semi-hard' coast, consisting of cohesive soils; it is common on estuarine coastlines and often has nearly vertical banks ranging from one to five metres in height. The rate of erosion is relatively high compared with the hard coast because it is composed of weaker and less resistant material. Erosion is mostly due to coastal processes, weathering and loss of vegetation cover (ARC, 2000). For extreme events such as storms and tsunami, vegetation cover plays a significant role in protecting the coast from flooding and inundation by reducing wave height and energy and decelerating tsunami flow speed; hence, erosive forces and inundation distance are decreased.

4.2 Intertidal/muddy coast

This type of coast is characterised by fine-grained sedimentary deposits, predominantly silt and clay that come from rivers; it can be classified as a 'soft' coast. It has a broad gentle seaward slope, known as an intertidal mud flat where mangrove forest, saltmarshes, shrubs and other trees are found. Most erosion is generated by river damming that reduces sediment supply, diminishes vegetation cover (usually mangroves and saltmarshes) and exposes vegetation roots by lowering the mud flat that leads to their final collapse. During storms, healthy and dense vegetation/coastal forest and trees can serve as barriers and reduce storm wave height, as well as afford some protection to the area behind them. In the case of a tsunami, coastal forest and trees can decrease wave height and tsunami flow speed to some extent if the forest is dense and wide enough. Both extreme events can cause severe erosion and scouring on the coast and at the river mouth.

4.3 Sand dune coast

This type of coast consists of unconsolidated material, mainly sand, some pebbles and shells; it can be classified as a soft coast. It has a gentle seaward slope known as dissipative beaches that have broad fine sand and gradually steep slopes at the backshore/foredunes. Its profile depends on wave form, energy and wind direction; hence, profiles can be adjusted to provide the most efficient means of dissipating incoming wave energy. This type of coast experiences short-term fluctuation or cyclic erosion—accretion. Long-term assessment is needed to identify erosion as a problem here. Generally, erosion is a problem when the sand dunes completely lose their vegetation cover. The cover traps wind-borne sediment during rebuilding, improves slope stability and consolidates the sand. During extreme events such as storms and tsunamis, this type of coast can act as a barrier for the area behind the dunes. Sand dunes and their vegetation cover are the best natural protective measures against coastal flooding and tsunami inundation.

4.4 Sandy coast

This type of coast consists of unconsolidated material mainly sand from rivers and eroded headlands, broken coral branches (coralline sand) and shells from the fringing reefs. It can be classified as a soft coast with reef protection offshore. The beach slope varies from gentle to steep slopes depending on the intensity of natural forces (mainly waves) acting on them. Coconut trees, *waru* (*Hibiscus tiliaceu*s), *Casuarina catappa*, *pandanus*, pine trees and other beach woodland trees are common here. Most erosion is caused by loss of: (1) the protective function of the coastal habitat, especially coral reefs (where they are found) that protect the coast from wave action; and (2) coastal trees that protect the coast from strong winds. During extreme events, healthy coral reefs and trees protect coasts to some extent by reducing wave height and energy as well as severe coastal erosion.

5. The protective function of coastal systems

Coastal areas with natural protective features can re-establish themselves after natural traumas or long-term changes such as sea level rise. The protective features of the coastal system vary (Figure 3.2). The role of coral reefs in coastal protection has been studied for some time and recent efforts have focussed on the role of coastal vegetation, especially mangrove forest and salt marshes in this context.

5.1 Scientific findings on the protective functions of coastal forests and trees

Scientific investigations on how coastal vegetation provides a measure of shoreline protection have been conducted (Sale, 1985; Kobayashi *et al.*, 1993; Mazda *et al.*, 1997; Massel *et al.*, 1999; French, 2001; Blasco *et al.*, 1994; Moller *et al.*, 1999; Wu *et al.*, 2001; Baas, 2002; Jarvela, 2002; Mendez and Losada, 2004; Lee, 2005; Dean and Bender, 2006; Daidu *et al.*, 2006; Moller, 2006; Turker *et al.*, 2006). These field, laboratory and numerical studies show that mangrove forests and other coastal vegetations of certain density can reduce wave height considerably and protect the coast from erosion, as well as effectively prevent coastal sand dune movement during strong winds. Healthy coastal forests such as mangroves and saltmarshes can serve as a coastal defence system where they grow in equilibrium with erosion and accretion processes generated by waves, winds, and other natural actions.

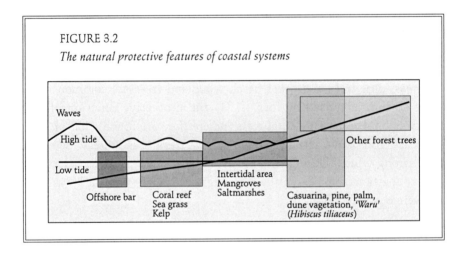

FIGURE 3.2
The natural protective features of coastal systems

6. Combinations of options

As mentioned already, combining hard and soft solutions is sometimes necessary to improve the efficiency of the options and provide an environmentally and economically acceptable coastal protection system. Hard solutions are known to:

- cause erosion and unnecessary accretion;
- be expensive and often further aggravate the problem; and
- spoil the aesthetic aspect of the beaches or coastlines they seek to protect, hence decreasing their economic value, especially for tourism purposes.

Meanwhile, many soft solutions can:

- take time to become effective (not overnight or quick-fix solutions), which generates negative public response; and
- be effective solutions only in medium- to long-term perspectives (5 to 10 years).

A planned retreat where the coast is left to erode can be expensive, unnecessary, and sometimes impossible, especially in highly modified environments such as tourism areas and waterfront cities.

To optimise the long-term positive impact of soft solutions, many combinations with hard solutions can be selected; combining beach nourishment and artificial headlands/groynes and revegetation and temporary offshore breakwaters/artificial reefs that act as interim hard structures is the most common approach.

6.1 Revegetation and temporary offshore breakwaters/artificial reefs

In some cases, revegetation in a low energy environment is required because deforestation of the coastal forest has led to direct exposure to wave action. There is also a need to establish offshore breakwaters/ artificial reefs as temporary wave protection structures for mangroves and saltmarshes; otherwise, seawalls/revetments for vegetation that grows above the highest water mark such as *waru*, Casuarina, pine and palm trees can be built. Once the plants have established themselves, the structures may be removed.

7. General guidelines on managing coastal erosion and their options

Understanding the key processes of coastal dynamics and how coasts developed in the past and present, as well as over the short and long term, is very important for managing coastal erosion problems because coastal erosion may occur without cause for concern. This can be very complex and possibly controversial where many conflicts of interests exist within the coastal environment. The main underlying principles for coastal erosion management are as follows (NRC, 1990; ARC, 2000):

- Identify and confirm coastal erosion as a problem.
- Identify, confirm and quantify the cause of the problem and ensure that any management option is well thought out before implementing coastal erosion measures.
- Understand the key processes and characteristics of coastal dynamics and system boundaries that reflect the natural processes of the erosion problem.
- Determine the coastal erosion measure options and implement them using proper design, construction and maintenance with careful evaluation of the effects on adjacent shores.
- Consider the balance of the options' costs and their associated benefits.

7.1 Setting up a green belt/buffer zone

The purposes of setting up coastal green belts must not be solely for preventing coastal erosion and mitigating other natural hazards, but also for addressing the socioeconomic status of the local communities as well as ecological sustainability.

The purposes of coastal green belts/buffer zones are:

- control and stabilise the shoreline by holding and trapping sediments and consolidate land for areas such as intertidal mudflats with mangrove green belts and sandy coasts with Casuarina, pine trees or coconuts and palm trees;
- attenuate the force of devastating storm surges and waves that accompany cyclones and tsunamis;
- provide an amenity and a source of food, materials and income for local communities; and
- benefit biodiversity and create habitat corridors for wildlife that can be used for conservation activities and ecotourism development.

In general, the underlying concepts of setting up green belt/buffer zones are:

- *Social forestry*: This should not be considered as a source of government or private sector revenue, but to support sustainable livelihood development among the coastal community.
- *Ecodevelopment*: This is beneficial for conservation activities, educational and recreational opportunities.
- *Participatory planning, implementation and monitoring*: The indigeneous knowledge of local communities should be used in decision-making so that they receive benefits directly.

Selection of the vegetation for setting up the green belt/buffer zone should take into account the natural protective function of the coastal system:

- Start with vegetation at the water's edge and gradually proceed to hydric species inland.
- Select water-edge vegetation that is found locally on each type of coast. In most cases, the width of the buffer zone for the intertidal delta ranges between 300 and 500 metres, depending on the slope of the region.
- Select beach vegetation that is found locally on each type of coast. The width of the buffer zone should be a minimum of 100 metres for the flat area, even with sand dunes or coastal embankments.

8. Conclusions

Coastal erosion and accretion are natural processes; however, they may become a problem when exacerbated by human activities or natural disasters. They are widespread in the coastal zone of Asia and other countries in the Indian Ocean owing to a combination of various natural forces, population growth and unmanaged economic development along the coast, within river catchments and offshore. This has led to major efforts to manage the situation and to restore the ability of the coast to accommodate short- and long-term changes induced by human activities, extreme events and sea level rise.

Understanding the key processes of coastal dynamics and how coasts are functioning both in spatial and temporal time scales (short and long term), in juxtaposition with human activities along the coast, within river watersheds and offshore is crucial for managing coastal erosion problems.

Three main conclusions can be drawn on the roles that coastal forest and trees can play in combating coastal erosion:

1) There is evidence that they provide some coastal protection and their clearance has increased the vulnerability of coasts to erosion. Based on scientific findings, the presence of vegetation in coastal areas will improve slope stability, consolidate sediment and diminish the amount of wave energy moving onshore, therefore protecting the shoreline from erosion.

2) Increased interest in soft options (in this case the use of coastal forest and trees) for coastal protection is becoming predominant and is in line with advanced knowledge on coastal processes and the natural protective function of the coastal system. This is because hard options are mostly satisfactory in the short term, while soft options are effective in medium-to long-term perspectives (5 to 10 years).

3) A combination of hard and soft solutions is sometimes necessary to improve the efficiency of the options and to provide an environmentally and economically acceptable coastal protection system.

References

Auckland Regional Council (ARC) (2000). "Technical Publication", *Coastal Erosion Management Manual* 130.

Baas, A.C.W. (2002). "Chaos, fractals and self-organization in coastal geomorphology: Simulating dune landscapes in vegetated environments", *Geomorphology* 48: 309-28.

Bilan, D. (1993). "The preliminary vulnerability assessment of the Chinese coastal zone due to sea level rise", Proceedings of the IPCC eastern hemisphere workshop, Tsukuba, Japan 3-6 August.

Bird, E.C.F. and O.S.R. Ongkosongo (1980). "Environmental changes on the coast of Indonesia", NRTS-12/UNUP-197. Tokyo-Japan, the United Nation University. p. 55.

Blasco, F., E. Janodet and M.F. Bellan (1994). "Natural hazards and mangroves in the Bay of Bengal", *Journal of Coastal Research* (Special Issue) 12: 277-288.

Cat, N.N., P.H. Tien, D.D. Sam and N.N. Bien (2006). *Status of coastal erosion of Viet Nam and proposed measures for protection*.

Clark, J.R. (1995). *Coastal zone management handbook*. New York: Lewis Publishers. p. 695.

Daidu, F., Y. Guo, P. Wang and J.Z. Shi (2006). "Cross-shore variations in morphodynamic processes of an open-coast mudflat in the Changjiang Delta, China: With an emphasis on storm impacts", *Continental Shelf Research* 26: 517-38.

Dean, R.G. and C.J. Bender (2006). "Static wave setup with emphasis on damping effects by vegetation and bottom friction", *Coastal engineering* 53: 149-56.

French, P.W. (2001). *Coastal defenses: processes, problems & solutions*. Florence, KY, USA: Routledge. http://site.ebrary.com/

Gopinath, G. and P. Seralathan (2005). "Rapid erosion of the coast of Sagar island, West Bengal - India", *Environment Geology* 48: 1058-67.

Jarvela, J. (2002). "Flow resistance of flexible and stiff vegetation: A flume study with natural plants", *Journal of Hydrology* 269: 44-54.

Kobayashi, N., A.W Raichle and T. Asano (1993). "Wave attenuation by vegetation", *Journal of Waterway, Port, Coastal, and Ocean Engineering* 119(1): 30-48.

Kovacs, J.M. (2000). "Perceptions of environmental change in a tropical coastal wetland", *Land Degradation & Development* 11: 209-20.

Lee, T.M. (2005). "Monitoring the dynamics of coastal vegetation in Southwestern Taiwan", *Environmental Monitoring and Assessment* 111: 307-23.

Malini, B.H. and K.N. Rao (2004). "Coastal erosion and habitat loss along the Godavari delta front: A fallout of dam construction (?)", *Current Science* 87(9): 1232-126.

Massel, S.R., K. Furukawa and R.M. Brinkman (1999). "Surface wave propagation in mangrove forests", *Fluid Dynamics Research* 24: 219-49.

Mazda, Y., E. Wolanski, B. King, A. Sase, D. Ohtsuka and M. Magi (1997a). "Drag force due to vegetation in mangrove swamps", *Mangroves and Salt Marshes* 1: 1993-99.

Mazda, Y., M. Magi, M. Kogo and N.P. Hong. (1997b). "Mangroves as a coastal protection from waves in the Tong King delta, Viet Nam", *Mangroves and Salt Marshes* 1: 127-35.

Mendez, F.J. and I.J. Losada (2004). "An empirical model to estimate the propagation of random breaking and nonbreaking waves over vegetation fields", *Coastal Engineering* 51:103-18.

Moller, I., T. Spencer, J.R. French, D.J. Leggett and M. Dixon (1999). "Wave transformation over salt marshes: A field and numerical modeling study from North Norfolk, England", *Estuarine, Coastal and Shelf Science* 49: 411-26.

Moller, I. (2006). "Quantifying saltmarsh vegetation and its effect on wave height dissipation: Results from a UK east coast saltmarsh", *Estuarine, Coastal and Shelf Science* XX: 1-15.

National Research Council (NRC) (1990). "Managing coastal erosion", Committee on Coastal Erosion Zone Management, Water Science and Technology Board, Marine Board, Commission on Engineering and Technical System, National Research Council. Washington, DC: National Academy Press. p. 182.

Nurkin, B. (1994). "Degradation of mangroves forest in South Sulawesi, Indonesia", *Hydrobiologia* 285: 271-76.

Othman, M.A. (1994). "Value of mangroves in coastal protection", *Hydrobiologia* 285: 277-282.

Prasetya, G.S. and K.P. Black (2003). "Sanur and Kuta Beaches in Bali—case studies for replacing traditional coastal protection with offshore reef", Proceedings of Artificial Surfing Reef, Raglan, New Zealand.

Sale, E.V. (1985). *Forest on sand, the story of Aupouri state of forest*. The New Zealand Forest Service.

Samarayanke, R.A.D.B. (2003). "Review of national fisheries situation in Sri Lanka", in G. Silvestre, L. Garces, I. Stobutzki, M. Ahed, R.A. Valmonte-Santos, C. Luna, L. Lachica-Alino, P. Munro, V. Christense and D. Pauly (eds.), *Assessment, management and future direction of coastal fisheries in Asian countries*. pp. 987-1012; WorldFish Center Conference Proceedings 67. p. 1120.

Swan, S.B.Stc. (1974). *The coast erosion hazards southwest Sri Lanka: an introductory survey*. University of New England. p. 182.

———. (1984). "The coastal geomorphology of Sri Lanka", *New England Research in Applied Geography* 40: 125.

Syamsudin, K. and F. Riandini (2000). "Coastline evolution monitoring at up drift and downdrift of some coastal structure in Indonesia", Proceedings Institut Tekonologi Bandung on Seminar on *Sediment Transport*. Supplement 32 (3): 45-54.

Thampanya, U., J.E. Vermaat, S. Sinsakul and N. Panapitukkul (2006). "Coastal erosion and mangrove progradation of Southern Thailand", *Estuarine, Coastal and Shelf Science* 68: 75-85.

Tjardana, P. (1995). *Indonesian mangroves forest*. Duta Rimba, Jakarta.

Turker, U., O. Yagci and M.S. Kabdasl (2006). "Analysis of coastal damage of beach profile under the protection of emergent vegetation", *Ocean Engineering* 33: 810-28.

UNEP-WCMC (2006). *In the front line: Shoreline protection and other ecosystem services from mangroves and coral reefs*. Cambridge, UK: UNEP-WCMC. p. 33.

Wolanski, E. (1992). "Hydrodynamics of mangroves swamp and their coastal waters", *Hydrobiologia* 247: 141-61.

Wu, Y., R.A. Falconer and J. Struve (2001). "Mathematical modeling of tidal currents in mangroves forests", *Environmental Modeling & Software* 16: 19-29.

4

Tsunami Impacts and Coastal Land Use Issues

A 100-metre coastal buffer zone allows for dynamic coastal fluctuations and ensures safe public access to waterfront.

Source:

R.A.D.B. Samaranayake *Adapted from: Pre- and post-tsunami coastal planning and land-use policies and issues in Sri Lanka. The regional perspective in proceedings of the workshop on coastal area planning and management in Asian tsunami-affected countries. 27-29 September 2006. Bangkok, Thailand.*

Sri Lanka's coastal zone contributes 40 per cent to the national gross domestic product (GDP) and is pivotal for economic development. Coastal resources such as coral reefs, mangrove forests, seagrass beds, salt marshes, sand dunes and coastal wetlands have been exploited by development activities; consequently, coastal erosion, degradation and transformation of coastal habitats, as well as water quality deterioration, are major threats. Damage generated by the tsunami in December 2004 has exacerbated socioeconomic, cultural and environmental issues. The Sri Lankan government has had to provide immediate relief, restore livelihoods and rebuild housing facilities and other infrastructure, including hard and soft defence mechanisms.

The tsunami affected about one million people initially and devastated two-thirds of Sri Lanka's coastline. The tsunami caused extensive damage and disruption to human life, livelihood, infrastructure, private and public property, and economic and social activities. The official death toll was 35,322; 21,441 people were injured, and over 5,000 are still missing. Approximately 516,150 people have been internally displaced.

The tsunami caused severe damage to fisheries assets, residential houses, schools, hotels, railways, roads, electricity supply, ports, fishery harbours, anchorages, health institutions and private commercial buildings. The fisheries sector *per se* suffered enormous losses along two-thirds of the coastline.

The total cost of the required relief, rehabilitation and reconstruction efforts is estimated to be around US$2 billion. The government had projected that it will take three to five years to rebuild destroyed structures and achieve full recovery.

The government created the Task Force to Rebuild the Nation (TAFREN) as the primary institutional mechanism for recovery and reconstruction, to coordinate, facilitate and assist implementing organisations, to coordinate donor assistance and fund-raising activities, to expedite the procurement process and to enable implementing agencies through capacity building. TAFREN focussed on four thematic areas:

- Getting people back to their homes;
- Restoring livelihoods;

- Health, education and protection for everyone; and
- Upgrading national infrastructure.

Implementing agencies established several innovative mechanisms at the sectoral level such as permanent housing and education.

Displaced families initially found shelter in emergency accommodations such as tents, public buildings and religious institutions or sought refuge with friends and relatives. It was recognised in the early stages that emergency accommodation would be only a temporary solution because the reconstruction of approximately 98,000 houses would take time. Therefore, transitional shelters were required to bridge the gap between emergency accommodation and permanent housing.

As a result of the combined effort of the government, private sector, civil society, and international community, the achievements in restoring livelihoods have been remarkable. The available information shows that between 70 and 85 per cent of households affected by the tsunami had regained their main sources of income by November 2005. On the other hand, at least 15 per cent of the tsunami-affected population is still living off an income of temporary relief work, not necessarily an income earned through regular work. The percentage of persons living off other sources is presumably higher in areas with lower economic activity.

1. Land issues

Twenty-four per cent of Sri Lanka is under the Coastal Divisional Secretariat, but along the coast, beaches and coastal reservations are state lands. Prior to the tsunami, most of the crown lands in urban areas were encroached by the coastal communities. In these populated areas it was very difficult to maintain the set-back zones stipulated under coastal zone management plans owing to unauthorised construction and demolition orders to eject them. A set-back (buffer) zone is defined as an area left free of any physical modification. It is a good planning practice to leave minimum set-back from the mean sea level line. A set-back is desirable to allow for the dynamics of seasonal and long-term fluctuations of the coastline, and to ensure public and visual access to the waterfront. Therefore, a set-back area belongs to the owner of the land and it benefits the owner to protect the property from hazards. However, the enforcement of these regulations was not successful in some areas owing to socio-economic conditions and political interferences. The encroachers of the set-back areas were devastated by the tsunami and had to be provided with emergency accommodation.

Vis-à-vis vulnerability factors such as exposure to extreme natural events, geomorphological characteristics, current development densities and availability of free space, the cabinet decided to implement set-back zones of 100 and 200 metres in the west and south coasts and north and east coasts respectively. A senior committee was then appointed to prepare interim guidelines for all the development activities within the coastal zone. These will be used until formal guidelines are formulated based on the findings of studies on vulnerability assessments, coastal bathymetry and coastal mapping. The buffer zones for development activities within the coastal zone and interim guidelines as declared by the government, are described hereunder.

2. Interim guidelines

'A 100-meter buffer zone should be delineated for any new constructions in the west and south coast from Kala Oya river mouth (Ganga Wadiya) to Kirindi Oya river mouth and a 200-metre buffer zone should be delineated for any new construction in the East and the North coast from Kirindi Oya river mouth to Kala Oya (Ganga Wadiya)' (*sic*).

- In compliance with the government decision, the declared buffer zones (south and west: 100 metres, north and east: 200 metres) will be applied in issuing development permits within the coastal zone. No developments will be permitted within the declared set-back zones.

- Exemptions may be granted only if the Coast Conservation Advisory Council (CCAC) determines that there are compelling reasons for allowing an exemption.

- Set-back exemptions are determined by the CCAC for the following development activities:
    - nationally important projects;

- fisheries related buildings and infrastructure; and
- tourism related developments.

The elevation of the development site, soil and geomorphological conditions, coastal hazard-proof building plans, evacuation and safety plans and national or sectoral importance of the projects will be assessed in granting exemptions.

- Similar set-backs will also be applied for developments within coastal islands.
- With the concurrence of the CCAC, approval will be granted for a set-back (25 metres, structure-free) from the edge of the cliff, when developments are carried out in areas 5 metres from mean sea level in the identified 13 high grounds, characterised by rock outcrops or hard soil.
- Within the coastal zone, a 100-metre set-back should be maintained either on the banks of rivers, streams, lagoons and any other waterbodies that are permanently or temporarily connected to the sea.
- A structure-free set-back should be maintained from the seaward toe of the fore dunes up to the landward toe of the back dunes, where the total width is greater than 100 metres on the west and south coasts and 100 metres on the east and north coasts.
- Approvals will not be granted for any development activity within any national parks or sanctuaries that lie in the coastal zone.

Due to limited or non-availability of land to relocate displaced families under donor-built housing programmes and homeowner-driven programmes, the Secretary of the Ministry of Urban Development and Water Supply and the district secretaries requested the CCAC to reduce the set-back zone size.

The CCAC considered the request and approval was granted for construction, following the set-back standards stipulated in the national Coastal Zone Management Plan (CZMP), 1997.

Now the land in the set-back zone is vacant. Prior to relaxation of the set-back zone size, the market value of other land coming within the set-back areas declined, but subsequently it has gradually increased despite tsunami risk. Incidences of land capture by powerful players following the tsunami have not been reported. However, people who own coastal lands have already started reconstructing their businesses, mainly in the tourism and fisheries sectors.

2.1 Rehabilitation and reconstruction of coastal protective structures

After the tsunami, rapid surveys were carried out by government officers along the coastline of the country, including the north and east quadrants. It was observed that the tidal wave had penetrated from 500 metres to 1 kilometre on the west coast and about 1-2 kilometres on the east coast. More damage to buildings and structures was incurred within a 200-metre range on the north and east coasts. Further, erosion of the coast has been aggravated and approximately 20-30 metres of land width have been lost on the western and eastern coasts. Significant morphological changes to the coastline have been observed at Bentota beach barrier, Gin Oya barrier, north of Mahaoya, dunes in Hambantota and Pottuvil (beaches) and some coastal stretches in the east.

A team of experts from the Coastal Resources Management Project (CRMP) carried out a rapid survey on damage to harbour breakwaters and coastal structures and submitted a report to the Coastal Conservation Department (CCD), the Ministry of Fisheries and the Asian Development Bank (ADB).

Following recommendations from the report, most of the coastal protective structures damaged by the tsunami were to be rehabilitated before the onset of the monsoon. Failure of the revetments and groynes to work would result in submergence of the low-lying areas of coastal stretches.

In early 2005, the Ministry of Environment and Natural Resources also carried out a country-wide field analysis of tsunami impacts on the coastal belt of the country. This report highlighted the key issues in the aftermath of the disaster, including the pollution of the coastal strip by solid waste and debris and its contamination by seawater, as well as impacts on coastal ecosystems.

In compliance with UNEP's guiding principles formulated at Cairo in February 2005, the CCD is constructing set-backs, reconstructing protective structures, rehabilitating sand dunes and establishing green belts along the coastline.

Under the Coastal Protective Structure Rehabilitation Programme, the CCD identified priority areas to be rehabilitated before the onset of monsoons in 2005. The CCD completed rehabilitation and reconstruction of revetments, groynes and offshore breakwaters in Galle, Matara, and Kalutara districts. These structures are approximately 3,000 metres in length. Under the ADB-funded CRMP, rehabilitation of Bentota sand spit, Hikkaduwa groyne and Tangalle revetment was carried out. The total cost of the work was around Rs.300 million. The cost of the work at Galle marine drive and the Kinniya coastal stretch in the east is estimated to be around Rs.200 million.

2.2 Rehabilitation of sand dunes

The CCD has initiated the rehabilitation of damaged sand dunes. A survey was carried out in Hambantota district to identify the damaged areas of sand dunes with the assistance of the National Aquatic Resources Research and Development Agency (NARA). Designs and plans to carry out this pilot project have cost Rs.25 million. This programme will be extended to the eastern coast with the assistance of a proposed project on restoration of coastal ecosystems funded by Global Environment Facility (GEF)/International Fund for Agriculture Development (IFAD).

2.3 Establishment of a greenbelt

The Forest Conservation Department (FCD) commenced the rehabilitation of coastal forests and establishment of shelterbelts along the coastline by following major landscape types. The CCD, as the custodian of the coastal zone, coordinated the programme to restore and rehabilitate vegetation to create a greenbelt that may be resilient and stable enough to prevent or mitigate the devastating effects of natural disasters such as cyclones, storm surges and tsunamis; this was done in collaboration with the FCD, the Urban Development Authority, the Coconut Development Authority, local authorities, NGOs and coastal communities in the area.

This programme commenced in Galle district and was extended to other parts of the island; guidelines for establishment of the green belt were developed with the assistance of International Union for Conservation of Nature (IUCN) and stakeholders. This programme is mainly implemented by NGOs and coastal communities under the guidance of CCD extension staff and the FCD.

Coastal communities learned that areas where coastal protection structures were in place were secure and their properties and lives were protected to some extent from tsunami waves. Therefore, requests came from the communities to reconstitute and reconstruct the damaged structures before the onset of the monsoon following the tsunami. Breakwaters, piers and fishery harbour groynes, anchorages and landing sites have to be rehabilitated and reconstructed as requested by fisherfolk and other coastal inhabitants. Biodefensive methods of ecosystem restoration programmes such as the establishment of greenbelts and sand dune restoration are being implemented and are well accepted.

However, in different locations such as beach seining areas and landing sites, conflicts have arisen with fisherfolk apropos the execution of coastal protection structures; remedial action, such as the implementation of beach nourishment programmes in areas for beach seining and inclusion of alternative anchorage facilities in the planned coast protection structures, have been taken to mitigate issues in consultation with the coastal communities.

Restoration and reconstruction are major challenges because many of the affected shorelines are densely populated and multiple developments are taking place. Hence, any reconstruction effort must ensure efficiency, sustainability, and equality, and be guided by principles that are scientific, participatory and in total harmony with natural systems.

3. Lessons learned from tsunami reconstruction and rehabilitation

Tsunami reconstruction and rehabilitation programmes have entered into the latter part of their second year. Now it is possible to learn lessons from past achievements and shortcomings.

- *Equity and gaps*: The first and primary guiding principle of the tsunami programme was equity. While there is no evidence of intentional unequal treatment, some communities (notably those in Ampara) are much less advanced than others; this is mainly attributable to differing local capacity and the impact of the set-back zone restrictions. As work progresses, it must be ensured that these factors are addressed and no one is left behind.

- *Lack of coordination*: The number of actors and the size of the reconstruction needs have made coordination a huge challenge, especially between state and non-state actors. An intensified effort will be needed during this major reconstruction period among agencies, partners and government authorities at the Central level, and among officials, development partners and communities at the local level, including in particular representatives of all affected ethnic groups. Strengthening of weak capacity at local and district government levels has to be continued, building on existing programmes.

- *Monitoring and measuring impact*: Another emerging lesson from the first year is that it is a huge challenge to monitor inputs—such as all the different sources of funding—and to measure outputs and the results on the ground.

- *Improved links to conflict-prone communities*: Much of the tsunami-affected area were in locations that were only recently conflict-prone. In the first year of the tsunami recovery effort, it was clear that the impact on the conflict-affected areas of this massive tsunami effort needed to be carefully considered and monitored, and equity of treatment, at least with respect to standards, needed to be ensured.

- *Integration of disaster risk concerns in coastal management and development*: The 2004 tsunami forced scientists to reconsider the guidelines for development in the coastal zone. The following components were intrinsic for developing guidelines on a scientific platform:
    - Assessment of coastal hazard, vulnerability and risk; analysis of the elevation model for both the hinterland and nearshore;

coastal hydrodynamics and sediment processes; mitigation measures, including defensive shoreline structures and buffers; appropriate guidelines.
- Criteria for design and construction of buildings and other structures in the coastal environment.
- Assessment of socioeconomic impacts on set-back standards.

Coastal zone management began in the early 1960s when coastal issues received greater attention from the government. A coast protection unit was established in the Colombo Port Commission in 1963, with the realisation that a comprehensive approach to coastal zone management was required. The need for a more management-oriented approach was first recommended in 1971 by the unit. In 1978, the issue of coast conservation was transferred to the Ministry of Fisheries, which created a separate Coast Conservation Division (CCD), basically directed towards maintenance of coastal stability by planned engineering works combined with a sound planning and management approach. This division was upgraded to the status of government department in 1984 under the Ministry of Fisheries and Aquatic Resources.

The mandate of the CCD is: The sustainable development of coastal resources and the management of coastal processes to optimize social, economic, and environmental status of Sri Lanka (sic).

When these assessments and studies have been completed, new guidelines including the new set-back zones for different coastal stretches will be developed for the next revision of the national CZMP.

- *Reduction of future vulnerabilities*: Sri Lanka was not considered to be a country prone to large-scale natural disasters and did not have appropriate institutional capacity to deal with the impact of a tsunami. In light of this experience, establishment of a suitable early warning system and disaster management and response capabilities quickly emerged as important priorities. An interinstitutional committee for early warning was established in the first quarter of 2005. With an interim mechanism for an early warning system in place, the government is working with other stakeholders for a permanent early warning system.

Realising the need for prompt and organised action, the Parliamentary Select Committee on Disaster Management was appointed. Subsequently, the Disaster Management Act, No. 13 of 2005 was passed by parliament on 13 May 2005. As a result, a Disaster Management Centre (DMC) was established in September 2005. This centre will promote community based disaster management programmes. The Ministry for Disaster Management was established in November 2005 to take the lead role in directing the strategic planning for disaster response, risk mitigation, preparedness planning and risk reduction.

The CCD is playing an important role in disaster risk management of the coastal environment. Therefore, the mission of the DMC will assure the sustainability of future coastal area management in Sri Lanka.

4. Projected needs and recommended actions

Measures are required to stop further degradation of the coastal resource base and, where possible, reverse the trend by implementing activities designed to regenerate and improve the quality of the natural resources of the coastal zone.

For this purpose, it is necessary to undertake various activities to be planned and implemented at different levels. These activities should be complementary and mutually reinforcing. Some activities may be planned at the national level or at the regional level, while other activities are more site-specific.

4.1 Survey and research needs

A research agenda should be set up to narrow down information gaps and to strengthen the scientific basis for coastal area management. At present, the database for coastal resources is very poor. A nationwide programme of data gathering and research is required to improve access to information needed for sound resource management and decision-making.

4.2 Strengthening of enforcement

Amendments to the Coast Conservation Act are required for the proper implementation of the revised CZMP. Proposed amendments are to be discussed with government policy-makers and the CCAC and will be submitted to the Legal Draftsman's Department for finalisation of the procedure.

4.3 Develop CZM capacity

The capacity of stakeholders should be strengthened to effectively contribute to the CZMP. A major constraint is the lack of technical and management capacities at the local level. Education and public awareness on coastal area management should be strengthened at national and local levels.

4.4 Establish mechanisms for integration and coordination

Institutional mechanisms should be developed to facilitate integration and coordination of the CZMP. Integration brings about the harmonisation of policies and legislation among national, provincial and local governments. Coordination plays a central role in fostering understanding and cooperation among all stakeholders.

4.5 Monitoring the coastal area management programme

Monitoring provides a powerful tool for assessing the performance of projects and programmes and gives early warning of adverse effects so that corrective action can be taken to modify design and management to avoid irreversible impacts. Due to the complexity and dynamic nature of coastal systems, it is not always feasible to accurately predict economic effectiveness and environmental performance.

5

Coastal Planning and Regulatory Issues: Post-Tsunami

Fortifying the coast by installing sea walls, shelterbelts and mangrove plantations to protect coastal communities from future tsunami attacks.

Source:

R.A.D.B. Samaranayake *Adapted from: Pre- and post-tsunami coastal planning and land-use policies and issues in Sri Lanka. The regional perspective in proceedings of the workshop on coastal area planning and management in Asian tsunami-affected countries. 27-29 September 2006. Bangkok, Thailand.*

1. Introduction

The 2004 Indian Ocean tsunami-affected the states of Tamil Nadu, Andhra Pradesh, Kerala and the Union Territory of Pondicherry on the mainland Indian coast. It also had a major impact on the Andaman and Nicobar islands (ANI). The tsunami caused a water level rise all over the Indian coast, with inundation of coastal lands ranging between 300 metres to 3 kilometres inland. Destruction was serious and incurred loss of life as well as damage to property in the Andaman and Nicobar islands, the southern Bay of Bengal coast from southern Andhra Pradesh to halfway down Tamil Nadu (particularly from Chennai to Kodikkarai), Kanyakumari district on the southern extremity of the Arabian Sea and a small stretch of coastline bordering the Quilon-Alleppey districts in Kerala on the Arabian sea coast. Of the aforesaid four areas, the areas that were impacted hardest are clearly the first three (TRINet[1], 2005). Other than the Nicobar group of islands and several parts of the Andaman islands, the badly affected districts were on the mainland—Nagapattinam, Cuddalore and Kanyakumari, in the order mentioned.

Several significant changes followed the tsunami, influencing many facets of coastal planning and governance. In retrospect, the most striking is the scale and intensity of attention the subject of coastal management and development has since received in India. Coastal habitats and coastal communities continue to be recipients of post-tsunami local, national and international consideration. The philosophy, information and history of coastal resource use and planning in India largely determine the nature and direction of the present focus on post-tsunami policies and interventions. The tsunami has been described as a 'wake-up call' (Anon, 2005a), but for India, and perhaps for other areas in the Indian Ocean region, the destruction it caused was almost inevitable, given the history of resource exploitation in the affected states.

---

1. TRINet is a network set up to address the information requirements of the ongoing tsunami rehabilitation and reconstruction activities in different parts of South India. Some of its members include the Nagapattinam NGO Coordination and Resource Centre (NCRC), Kanyakumari Rehabilitation Resource Centre (KRRC) and the Auroville Information Centre.

2. Post-tsunami land tenure and use issues

2.1 Tsunami damage estimates

Early in January 2005, the Government of India estimated reconstruction costs in the four Indian states devastated by the tsunami to be around 70 billion rupees (US $1.6 billion). The report was based on an assessment by federal government teams in the southeastern coastal regions of the states of Tamil Nadu, Andhra Pradesh, Kerala and Pondicherry and did not include costs of reconstruction in the ANI. The nationwide death toll at this stage stood at 9,995, with 5,689 people registered as missing. Of these people, 5,592 were from the Andamans archipelago alone, according to a Home Ministry statement (ReliefWeb[2], 2005a).

A Joint Assessment Mission undertaken by the World Bank, the United Nations and the Asian Development Bank, at the behest of the Government of India, made a preliminary estimation of damage and losses from its survey between the 1 and 15 February 2005 (Table 5.1).

TABLE 5.1
Preliminary estimation of damage/losses between 1 and 15 February 2005

In US $million	Damage	Losses	Total	Effect on livelihoods†
Andhra Pradesh	29.7	15.0	44.7	21.2
Kerala	61.7	39.1	100.8	36.3
Tamil Nadu	437.8	377.2	815.0	358.3
Pondicherry	45.3	6.5	51.8	5.9
Total (by sector)	**574.5**	**448.3**	**1,022.8**	**421.7**
Relief‡		200.7	200.7	

Note: †Effect on livelihoods refers to the impacts on agriculture and livestock, fisheries and micro-enterprises and other livelihood sectors, measured in income terms.

‡Relief provided by local, state and national governments (not included in the total [by sectors]).

Source: JAM estimates on the basis of information made available by the governments and direct observation.

---

2. ReliefWeb is the world's leading online gateway to information (documents and maps) on humanitarian emergencies and disasters. ReliefWeb was launched in October 1996 and is administered by the UN Office for the Coordination of Humanitarian Affairs (OCHA).

According to the Government of India's Report to the Nation in June 2005, as many as 12,405 lives were lost: 8,009 in Tamil Nadu, 3,513 in ANI, 599 in Pondicherry, 177 in Kerala and 107 in Andhra Pradesh. Non-governmental organisations say the death toll was much higher, and there is no clear proof of the exact number of deaths. Official estimates say the tragedy affected 2,792,000 people in 1,089 villages, including 43,000 people in Pondicherry; 196,000 in Andhra Pradesh; 130,000 in Kerala; 356,000 in ANI and 897,000 in Tamil Nadu. The tsunami destroyed over 235,000 homes, damaged 83,788 boats and rendered 39,035 hectares of cropped area unusable. The social infrastructure—schools, primary health centres, drinking water supply, anganwadis (child care centres) and other community assets in these areas were totally destroyed. The Prime Minister's National Relief Fund - the agency that requests citizens' donations in the wake of any national tragedy, collected Rs.8.29 billion. According to the Prime Minister's Office, over 91,000 contributions from organisations and individuals were made for tsunami relief (Iype, 2005).

The Kerala government assessed the damage to be about Rs.1,358 crore (1 crore = 10 million). In its request for aid to the Government of India, the Kerala Government stated that 172 persons were declared dead. The report also stated that the communities of Alappad and Arattupuzha were completely devastated—219 villages and a population of 2,578,000 were affected. The government offered monetary compensation of Rs.100,000[3] for dependents of dead adults, Rs.50,000 to families with dead children and Rs.25,000 to the injured. The document states that a total of 13,044 houses were affected by the tsunami, of which 2,919 houses were badly damaged, while 10,125 houses were partially damaged. There were 3,059 houses requiring minor repairs, while 7,066 needed major repairs (Anon, 2005b).

In the ANI, in the 38 affected islands, 3,513 persons were reported dead or missing. Fifty thousand (50,000) persons were reported as being affected by the tsunami, 10,000 households were badly damaged, about 10,000 hectares of agricultural land were affected and 354 kilometres of roads were destroyed (Anon, 2006).

---

3. Rs.100,000 = US $2,251.69 (December 2006).

The tsunami destroyed over 235,000 homes.

For apropos fisheries only, in 2005, the ANI administration released information on the immediate impact of the tsunami on fisherfolk (Equations, 2006). Besides several being rendered homeless, many fisherfolk lost their fishing inputs such as their craft, gear, engines and iceboxes. The ice plants and cold storages of the Department of Fisheries were also severely affected. The department assessed losses to government property, departmental staff and losses to fisherfolk as detailed below:

1. Loss to government property was estimated at Rs.82,000,000.
2. Two technical officials from Katchal islands were reported missing.

3. 69 fisherfolk were reported missing/dead.
4. A total of 2,323 fisherfolk were directly affected.
5. 622 locally made dongies (boats without engines) were badly damaged.
6. 471 locally made dongies were partially damaged.
7. 316 engine-fitted boats were badly damaged/lost.
8. 294 engine-fitted boats were partially damaged.
9. Several fisherfolk lost their nets, fishing implements, marketing assets, etc.

In addition, the department also received about 1,600 additional claims for losses concerning craft and gear, which were being processed and at the time of writing this report. These claims were to be approved by a committee consisting of the Assistant Director of Fisheries of the Zone (convener), a representative of the Revenue department of the area and Panchayati Raj Institution (PRI) representatives.

2.2 Displaced persons

In the report Tsunami: one year on—India, ReliefWeb reported that the death toll for the entire country was 10,881, about 5,792 people were reported missing and 6,913 were injured. More than three million livelihoods were described as having been affected by the tsunami (ReliefWeb, 2005b). There are no aggregated and reliable statistics to date about the nationwide number of displaced persons. Various regional centres have been collecting data on deaths, houses damaged and so forth, and these are currently being compiled into databases by various efforts such as the United Nations Development Programme (UNDP) supported Post-tsunami Environment Initiative (*www.ptei-india.org*). The United Nations Tsunami Response System (UNTRS) supported a beneficiary tracking system developed by Pricewaterhouse Coopers. These data were not yet available for this paper. The amount of money allocated or spent on tsunami rehabilitation so far is not really an indication of the number of displaced persons or the damage that occurred. The state governments have provided rehabilitation cost estimates to the Government of India to include the estimated cost of disaster mitigation work as well, so the costs do not reflect actual damage.

Coastal Planning and Regulatory Issues: Post-Tsunami **91**

In Nagapattinam, Tamil Nadu, NGOs initiated the construction of permanent shelters using materials such as metal sheets and particler boards.

2.3 Temporary shelters

Despite the immediate surge of concern and aid for the victims of the tsunami, the relief phase saw a number of shortfalls in coordination and planning; consequently temporary shelters were inadequate and could not fully meet the needs of the affected persons. Shelters are categorised as temporary, intermediate or permanent, depending on the material utilised. Several hundred persons are still housed in temporary shelters nearly two years after the tsunami. In the ANI, many families still live in temporary shelters made from tin sheets or tar-coated sheets and corrugated metal with tarpaulins. There are reports of human rights violations in the temporary and intermediate shelters in the ANI (Chaudhry *et al.*, 2006).

In many instances, NGOs began permanent shelter construction. Reliable estimates of the exact number of temporary shelters and intermediate shelters are not available, except where NGO coordination centres were established, such as in Nagapattinam in Tamil Nadu. It is known that the following materials have been used:

- bitumen-coated sheets for walls and roofs;
- roofing of either asbestos or metal sheets;
- cement-impregnated particle boards (for walls); and
- corrugated FRP (fibre-reinforced plastic).

Available information on temporary shelters built with this material and their disposal has caused some concern. In many places, the bitumen sheets have completely disintegrated. Toilets were constructed (at a 1:20 toilet/people ratio) in the temporary shelters and many of the septic tanks were not properly built, which has led to groundwater contamination. In terms of the environmental implications of these temporary shelter measures, the following information is needed:

- the volume of different types of material being utilised for temporary shelters;
- the total number of shelters, material used, septic tanks per shelter;
- the total number of shelters currently in use;
- information on reuse/recycling of the material;
- evaluation of the health and environmental concerns generated by the material;
- guidelines for dismantling of the shelters; and
- guidelines for possible reuse/recycling and disposal of the material.

It is still not clear who is responsible for dismantling the temporary structures, which were abandoned after the occupants moved to intermediate or permanent housing. In Nagapattinam, it is reported that the government has undertaken the responsibility of clearing this additional debris[4] (Rodriguez, undated).

In Tamil Nadu, the government has initiated a massive Tsunami Housing Reconstruction Programme, which envisages the construction of about 130,000 concrete houses. The State Relief Commissioner communicated

---

4. Information on temporary shelters was gathered from Sudarshan Rodriguez of the Post Tsunami Environment Initiative (PTEI) project.

a model memorandum of understanding (MoU) to be entered into with NGOs and other rehabilitation agencies; designs and specifications of permanent houses developed by experts were to be approved by the district collectors.

The Government of Tamil Nadu estimated that 54,000 houses were damaged in Tamil Nadu[5] (2006). However, the construction of new homes had already been begun by the NGOs. The World Bank has revised its assistance programme and is now extending another loan for the construction of approximately 65,000 houses for coastal areas not affected by the tsunami.

2.4 Shelter related issues

The Government of Tamil Nadu introduced Government Order (GO) 172 on 30 March 2005 declaring that all new government-sponsored houses would only be constructed at least 200 metres from the high tide line. With the stated objective of providing built houses in safe locations to the tsunami-affected families, the government pledged assistance only to those who agreed to be relocated beyond 200 metres of the high tide line. Those who intended to construct buildings within 200 metres would not be eligible for government assistance. The government also extended assistance to those people whose homes were not damaged, but who wanted to relocate nonetheless. The ambiguity of the Coastal Regulation Zone (CRZ) notification led to the above interpretation, and it was deemed that no new constructions would be permitted within 200 metres for all categories of the CRZ. Sridhar (2005) stated that several ambiguities lay within the CRZ notification and in the post-tsunami context, and that they required urgent clarification by the Ministry of Environment and Forests. There was strong resentment towards GO 172 in various quarters. Several fishworker groups and NGOs termed it a discriminatory order. They put forth various arguments against the GO, stating that fishing communities have a right to stay close to the shoreline and visibility of the coastal waters is important for their fishing activities. The lack of consultation with, and participation of fishing communities

---

5. Compiled from NGO presentations at the Workshop on *Disaster Preparedness in Agriculture*, NCRC.

in these shelter guidelines has been severely criticised. Others argue that the GO effectively tries to remove fishing communities from the coast, and this makes it easy for the tourism industry and other real estate interests to exploit the coast.

In the early stages of relief and temporary construction, NGOs were fairly ignorant of CRZ laws and were mining sand dunes on the coast for building activities, which is strictly prohibited by law. Fishing communities do not have *pattas* or land rights and title deeds in most cases. However, the CRZ only allows authorised construction on the coast. The dichotomy has not been addressed yet.

Tamil Nadu has developed guidelines for construction on the coast. In the Andaman islands, the Ministry of Urban Development and the Disaster Authority of the Home Ministry have set guidelines for housing. There are several matters that still remain unaddressed in the ANI. As noted earlier, there were several encroachments in the islands by settlers who constructed illegal houses. Many of these settlers have incurred losses from the tsunami. A debate rages as to whether these families are entitled to compensation and housing on humanitarian grounds or not. Whether permanent housing would mean security of tenure is not clear particularly for those who may have been classified as 'encroachers'. It has also been noted that only one house is being issued per *patta* holder, although houses and people have multiplied since the last housing estimates were collated by the government. There have also been several debates on the matter of building design for the ANI. The Ministry of Urban Development and Housing has finalised a prefabricated house design for the Nicobar. However, the appropriateness of these designs is under debate since Nicobaris traditionally live in locally designed *machans* (made principally of bamboo and *dhani* leaves).

2.5 Agriculture

The tsunami damaged large tracts of agricultural areas, creating the following problems: salinised soils, eroded topsoil damaged standing crops, silt and sand casting and siltation of ponds, irrigation and drainage channels. In Tamil Nadu, the Nagapattinam Tsunami Resource Centre estimated damage to 8,460 hectares of agricultural land in Tamil Nadu and about 5,000 hectares in Nagapattinam district itself. Many initiatives are underway in Tamil Nadu to restore agriculture related livelihoods. These include removal of mud, clearing of drainage and irrigation channels, deep ploughing of fields and excavating trenches around fields. Twenty-three NGOs are involved in Nagapattinam district alone. Short-term measures include green manuring of fields and growing salt-resistant crops, while long-term measures are aimed at overall improvements in agriculture.

Limitations to the rehabilitation efforts for agriculture appear to be process related. Coordination, unrealistic community demands and expectations were listed at a recent workshop in Tamil Nadu on the tsunami and agricultural impact. Timely interventions appear to have been lacking in this sector.

2.6 Post-tsunami coastline stabilisation

2.6.1 Bioshields

The immediate reactions of the state governments to the tsunami were to fortify the coast by constructing sea walls. A Tamil Nadu government press release, dated 4 January 2005, quoted the Chief Minister as having stated: 'To ensure the Tamil Nadu coast is not ravaged by the tsunami in future, protection works such as construction of sea walls, groynes, beach protection measures will have to be taken up. It is also proposed to take up shelter belts, mangrove plantations along the coastline to protect the coastal areas from the tsunami attack in future' (Anon, 2005c). The first month after the tsunami witnessed many news reports quoting the Tamil Nadu government's demands for a 1,000-kilometre sea wall for tsunami protection (Das, 2005). Recently, it has been reported that a 3.2-kilometre sea wall will be built at Kalpakkam township near the Kalpakkam nuclear power plant. While the penchant for constructing sea walls has not completely waned, it has gradually given way to another slogan—that of 'bioshields' or coastal plantation defences. There are various arguments about the appropriateness of these plantations, their impacts on coastal ecosystems and also conflicts arising out of land use in coastal commons (Kerr et al., 2006).

The MSSRF has been spearheading the promotion of mangrove plantations and non-mangrove bioshields, which it describes as 'shelterbelts'. These shelterbelts are strips of vegetation composed of trees and shrubs grown along the coasts to protect coastal areas from high velocity winds, and also presumably from devastation like that caused by the recent tsunami. They are stated to act as sand binders and to inhibit sand erosion. Shelterbelts are promoted as a means to reduce windspeed and ameliorate the local microclimate. The toolkit for establishing coastal bioshields states that well-placed and well-managed shelterbelts or bioshields can be used to increase agricultural productivity (Selvam *et al.*, 2005). The document states that in order to make bioshields effective at proposed sites, the choice and mix of species should be decided based on the height and depth of the bioshield required. The perception is that these plantations will augment incomes in the medium to long-term.

Prior to the tsunami, the Tamil Nadu forest department was involved with the plantation of Casuarina along the coast, although largely on revenue lands. Post-tsunami, the World Bank-funded Emergency Tsunami Reconstruction Project (ETRP) is supporting the plantation of mangroves and shelterbelts along the Tamil Nadu coast. The data from various coastal forest divisions along Tamil Nadu show that only Casuarina is being planted along the coast and the entire exercise appears to be devoid of any science. The plantation exercise does not currently follow any guidelines on how these bioshields should be raised.

The Swaminathan Committee report on revised coastal management legislation strongly recommended the use of bioshields. The authors of the Review of the Swaminathan report have been critical of this recommendation (Sridhar *et al.*, 2006). They assert: 'The use of exotics in the putative "Bio-shield" is strongly advised against.' There should be some concern about the unmitigated zeal with which the "bioshield" concept is being promulgated as a win-win solution in the wake of the tsunami. While it certainly has some benefits for local communities in the short term, one is uncertain about how much protection it affords the coast in actual terms. The last thing required is the further transformation of the coasts into groves of fast-growing exotic species. It is also potentially quixotic to invest large amounts of energy and funds in the regeneration of mangroves in habitats where the primary conditions that led to their decline still exist. It may be instead important to more completely understand what those conditions are before large-scale eco-engineering operations like this are undertaken. Where possible, the regeneration of mangrove, beach and dune vegetation, and coastal forests should definitely be considered, but the conditions under which they will be warranted and successful would be limited when compared with the much more important task of understanding and protecting coastal processes against the primary influences affecting it.'

The reviewers also state that access and visibility of the seashore and sea are crucial for fisherfolk's daily decision-making and the bioshields could be a hindrance in this respect. Furthermore, there have been cases of conflicts between the forest department (who promoted and implemented afforestation projects) and local communities. They advise against carbon sequestration as a goal or major benefit from the creation of bioshields. They state: 'Viewing bio-shields from a climate change/carbon sequestration angle may encourage a forestry paradigm on coastal systems, which is not desirable.'

References

Anonymous (2005a). *Report of the committee to review the Coastal Regulation Zone Notification 1991*. New Delhi: Ministry of Environment and Forests. p. 116.

———. (2005b). "Revised proposal presented to the Govt. of India seeking assistance from NCCF/other schemes for tsunami relief rehabilitation and reconstruction measures", Government of Kerala. p.109.

———. (2005c). "Central Inter-Ministerial Team for assessment of the situation in the state", Press release 4 January. Government Information Cell, Disaster Management and Mitigation Department, Chennai. *http://www.tn.gov.in/tsunami/pressreleases/pr040105/pr04012005-3.htm*

———. (2006). *State development report of Andaman and Nicobar Islands, 2005*. New Delhi: National Institute of Public Finance and Policy.

———. (2006). "Post-tsunami relief, rehabilitation and reconstruction measures in Andaman and Nicobar Islands", Note circulated during the Prime Minister's Press Meeting on 4 January 2006, Port Blair.

Chaudhry, S. and E.G. Thukral (2006). *Battered islands: Report of a fact finding mission to tsunami-affected areas of the Andaman and Nicobar Islands*. New Delhi, South Asia Regional Programme Housing and Land Rights Network.

Equations (2006). *Coastal area assessments - A post-tsunami study on coastal conservation and regulation: Andamans*. Bangalore.

Government of Tamil Nadu (2006). *Tiding over tsunami*. Chennai: Government of Tamil Nadu.

Iype, G. (2005). "The tsunami: Putting the record straight", *http://www.rediff.com/news/2005/dec/19spec.htm*

Kerr, A.M., A.H. Baird and S.J. Campbell (2006). "Comments on 'coastal mangrove forests mitigated tsunami", by K. Kathiresan and N. Rajendran. *Estuarine, Coastal and Shelf Science* 65(2005): 601-06; 67: 539-41.

ReliefWeb (2005a). "India estimates tsunami damage at over 1.6 billion dollars", *http://www.reliefweb. int/rw/rwb.nsf/db900SID/VBOL-68EHFP?OpenDocument*

———. (2005). "Tsunami: one year on - India", December. *http://www.reliefweb.int/rw/rwb.nsf/db900SID/HMYT-6KGN5Z?OpenDocument &rc=3&cc=ind*

Rodriguez, S. (n.d.). personal communication.

Selvam, V., T. Ravishankar, V.M. Karunagaran, R. Ramasubramaniyan, P. Eaganathan and A.K. Parida (2005). *Toolkit for establishing coastal bioshields*. Chennai, M.S. Swaminathan Research Foundation. p.117.

Sridhar, A. (2005). "Statement on the CRZ Notification and post tsunami rehabilitation in Tamil Nadu", *UNDP Discussion Paper*. New Delhi.

Sridhar, A., R. Arthur, D. Goenka, B. Jairaj, T. Mohan, S. Rodriguez and K. Shanker (2006). *Review of the Swaminathan committee report on the CRZ notification*. New Delhi: UNDP.

TRINet (2005). *http://www.tsunami2004-india.org/modules/xoopsfaq/index.php?cat_id=4#q4*

6

Trends in Land and Resource Use and Land Tenure

Traditional coastal land uses may be harmonised within carefully designed land use zoning systems.

Source:

Arcadis Euroconsult *Adapted from: Land tenure and land-use change in relation to poverty, livelihoods, the environment and integrated coastal management in Asian tsunami-affected countries. The Regional Perspective in Proceedings of the workshop on coastal area planning and management in Asian tsunami-affected countries 27-29 September 2006, Bangkok, Thailand.*

The overall pattern in land and resource use in Asian countries has intensified compared to previous decades (even then needing management) (Ambio, 1988). The Asian tsunami-affected countries, perhaps with the exception of Myanmar, chose economic growth at the cost of the environment in a world dominated by trade liberalisation (Time International, 2006). Worldwide economic growth that existed before the 1990s accelerated as free trade expanded. Globalisation today is driving most of the land uses on a scale that did not exist in the 1980s. Globalisation can harm the environment for the following reasons (Harford, 2006):

- Companies adopted the 'race to the bottom system' rush overseas to produce goods under cheaper, lenient environmental laws, additionally influenced by corruption.
- Physically moving goods inevitably consume resources and cause pollution.
- Economic growth produces externalities that harm the planet.

The purpose of this article is to indicatively trace the manner in which selected land use changes between 1995 and 2005 relate to poverty and the environment; and implications for Integrated Coastal Management (ICM). The complexity of the relationships is illustrated by Figure 6.1.

Conflicts characterise the relation among economically disparate coastal land uses. Those backed by global finance are capable of influencing the highest level of government (Stiglitz, 2002). Money is the ultimate determinant of power and influence (Todd, 2004). The implication for the coastal poor is whether governments may or may not regulate access by the rich and powerful to the same resources that they demand. The dominant political elites in all the Asian tsunami-affected countries, whether or not backed by global capital, have an exploitative relationship with coastal residents engaged in traditional livelihoods. The future appears bleak for the poor until they acquire countervailing power by way of becoming organised (as in Kerala) in order to negotiate with governments (Kurien, 2005). Land reforms in some countries, like in Bangladesh and India have provided apparent benefits to coastal farmers during early stages. However, these benefits appear to have dissipated in

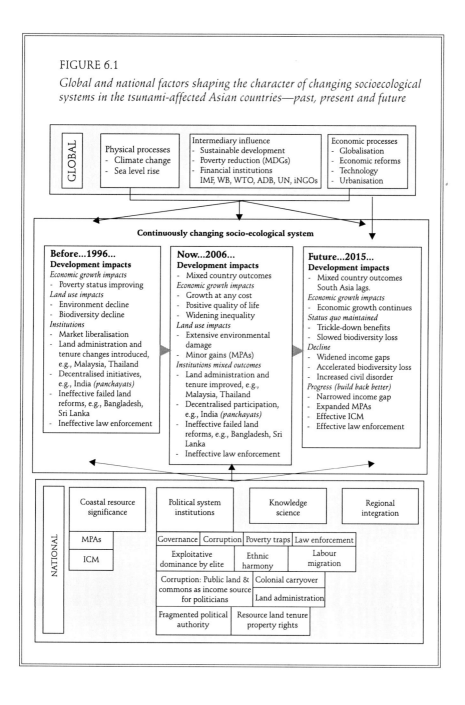

FIGURE 6.1
Global and national factors shaping the character of changing socioecological systems in the tsunami-affected Asian countries—past, present and future

the face of competition from capital-intensive and commercial land uses such as shrimp farming, tourism, industry and port development.

Agricultural production is the mainstay of economic sustenance in most Asian tsunami-affected countries. Possession of land rights also typically ensures a baseline of shelter and food supply and allows people to turn latent assets into live capital through entrepreneurial activity. Once secure in their land rights, rural households invest to increase productivity. Moreover, the use of land as a primary investment vehicle allows households to accumulate and transfer wealth between generations. The ability to use land rights as collateral for credit helps create a stronger investment climate and land rights are thus, at the level of the economy, a precondition for the emergence and operation of financial markets (De Soto, 2000; Stern, 2005; World Bank, 2003).

1. Land uses and trends that influence coastal communities, livelihoods and the environment

1.1 Land uses

- *Impacts of pollution*: Inadequately regulated land use results in discharges and emissions that cause air and water pollution. The consequences of pollution for small-scale fisherfolk were recognised in the late 1980s (Ambio, 1988). New Delhi, Calcutta, Kanpur and Jakarta rank among the 10 most polluted cities in the world. Their discharges eventually reach the coast and impact on coastal resources. Asia's coral reefs are already partially destroyed by climate change, destructive fishing, and pollution. Air and water pollution impose severe stresses on health, mainly on the poor, particularly in South Asia, where many people lack safe drinking water and sanitation (Time International, 2006). Pollution in the Asian tsunami-affected countries is likely to become worse unless strict environmental safeguards are enforced (GPA, 2005).

- *Small-scale artisanal fisheries*: Marine fisheries and coastal aquaculture production in the Asian tsunami-affected countries expanded substantially during the past decade with the exception of Thailand, where it declined. Coastal aquaculture in the Maldives is insignificant. A feature of marine fisheries is competition and the juxtaposition of technologically modernised fishing fleets and traditional small-scale crafts. The former is frequently supported by government policy and by development financing, the latter relatively marginalised and dependent on informal financing.

The significance of small-scale artisanal fishery to national economies is high. The statistics provided by India indicate an associated population of about 14.4 million living mainly in about 3,937 coastal villages. The vast majority of the fisherfolk engage in subsistence fishing, consuming a part of their daily catch and selling the remainder for consumption at the local level. It contributes about US $6 billion to national income, which is about 1 per cent of the total GDP. This production elevated India to the position of fourth largest fish producer in the world with total production of about 6.3 million tonnes in 2003 and 2004. This production constitutes about 90 per cent of domestic fish consumption. Fisherfolk have customary use rights to the marine and terrestrial resources that they depend on (Salagrama, 2005).

The livelihoods of small-scale artisanal fisherfolk are now clashing with commercialised fisheries and other uses of coastal resources. Non-fishery activities are supported by government policies, and allocated investment and property rights. Where conflict resolution is attempted through litigation, small-scale artisanal fisherfolk are highly vulnerable in the face of statutory and common law based upon written legal procedure. These procedures are alien and incomprehensible to them, as in agriculture, because they are accustomed to customary rights (Wily, 2006). Comprehensive literature exists on poverty in fishing communities (Macfadyen and Corcoran, 2002). The clash of small-scale traditional enterprises and sometimes multinational business ventures is not confined to developing countries.

- *Aquaculture (shrimp)*: Shrimp farming is not a uniform activity. Both small-scale operators and more powerful larger scale operators exist side by side. In Indonesia tambak (fish pond) aquaculture dates back many centuries. Commercial shrimp farming has achieved massive growth during recent decades. The industry

> In Indonesia, in 1998, the area under shrimp cultivation approximated 305,000 hectares. The impetus for expansion was provided mainly by ADB and the World Bank. The Government of Indonesia asserts that about 800,000 hectares, about one-third of the remaining mangroves, are available for shrimp ponds (Down To Earth No. 58, August 2003).

has been hailed as being capable of producing large volumes of food without impacting on marine stocks and increasing the availability of food for the hungry (FAO, 2004). The sustainability of aquaculture as the 'blue' counterpart of the agricultural green revolution is questionable (Wolowicz, 2005).

Shrimp farming has increased land values and led to conflict over land rights and access to natural resources. Resulting social problems include increased poverty, landlessness and food insecurity, displacement of communities, pollution of drinking water, poor working conditions and impacts on health and education. Large tracts of agricultural land have been inundated with saline water to create shrimp ponds. Shrimp farming physically invades farmland and saltwater intrusion can change soil composition and pollute water supplies. Shrimp aquaculture has had direct impacts on crop productivity and on the health and livelihoods of rural farming communities (EJF, 2003).

- *Agriculture and forestry*: In Asian tsunami-affected countries, agriculture and forestry are in the same land use equation. The destruction of rain forests which generated a loss of 33 per cent in Asia, particularly in Malaysia and Indonesia, between 1960 and 1990, continues at even a greater rate today. The direct causes are agriculture, cattle raising, damming and megaprojects, logging, plantations, shrimp farming, slash-and-burn activities and mining (BBC, 2005). The World Rainforest Movement, WRM (*http://www.*

wrm.org.uy/forests/future.html) identifies several underlying causes including:

a. Unequal land tenure.
b. Discrimination against indigeneous people, subsistence farmers and poor people.
c. Absence of participatory democracy.
d. Military influence.
e. Exploitation of rural areas by the urban elite.

> Since 1996, Indonesian forest loss has accelerated to 2 million hectares per year. Forests have been almost entirely cleared in Sulawesi while they are predicted to disappear in Sumatra and Kalimantan in the coming decade if the existing trend persists (Global Forest Watch *http://www.globalforestwatch.org/english/indonesia/forests.htm*).

Tropical deforestation is most serious in Indonesia and Malaysia. It is likely that China, like Japan, will export deforestation activities to Indonesia and Malaysia while conserving its own forests (Diamond, 2005).

In all Asian tsunami-affected countries where the natural forests are owned by the state, more extensive support is required for community based agroforestry, with appropriate tenure rights, to reduce the pressure on primary forests for subsistence products (WCFSD, 1999 and Ostrom *et al.*, 1999). Customary tenure rights are claimed by indigeneous populations who live in the forests of India, Indonesia, Malaysia and Myanmar. Privately owned forests are rare although rubber plantations are a major source of wood for industries in Malaysia and Thailand.

The forests in the coastal zones include mangroves and peat swamps. Many coastal communities have traditionally harvested forest products in mangrove forests in Bangladesh, India, Indonesia, Malaysia, Myanmar and Thailand. Loss of mangroves has been compensated in some countries—

142,000 hectares of mangroves have been planted in Bangladesh in shallower sections of the delta. In Indonesia and Malaysia, mangrove cultivation has been incorporated into coastal aquaculture (Primavera, 2000).

The WWF provides regular updates on serious problems in Southeast Asia with regard to forestry management, biodiversity, agriculture and poverty (http://www.livingplanet.com/about_wwf/where_we_work/asia_pacific/where/indonesia/index.cfm).

- *Ports and navigation*: Ports and navigation facilities have expanded during the past decade. The changes have been driven by the emergence of China as a global economic power. China's continued growth is crucially linked to energy supplies, primarily oil and gas. The Malacca Straits, which in places narrows to about half a nautical mile and is only 25 metres deep, is today the busiest sea lane in the world; this has serious implications for coastal communities. Some 50,000 vessels, carrying roughly a quarter of the world's maritime trade pass through it annually. So do about half of all seaborne oil shipments on which the economies of Japan, China, and Republic of Korea depend. (The Economist, 2004). Port expansion at diverse coastal locations in Indonesia, Malaysia and Thailand may be anticipated.

- *Industry: steel production and ship breaking*: Worldwide demand for steel is increasing. Chinese and Indian steel producers have become influential players in the world market. India's recent acquisition of major European production facilities suggests expansion in future. One source of raw material is steel scrap including discarded ships. Bangladesh and India are leaders in the ship-breaking industry which supplies scrap iron as raw material. In India, annual consumption of scrap iron is 8 million tonnes while supply is

6 million tonnes (OECD, 2006). It is likely that the production of scrap iron from various sources, including ship breaking, will increase in the future. Ship breaking is a coastal activity. Recycling ship construction materials is a highly desirable activity that can be environmentally beneficial. However, the manner in which it is being carried out has no social and environmental safeguards. (*http://www.greenpeace.org/india/campaigns/toxics-free-future/ship-breaking*).

- *Tourism and recreation*: Tourism is the world's primary export earner, ahead of automotive products, chemicals, petroleum and food. It was valued in the region of US $500 billion in 2000 (*http://padh.gpa.unep.org/page.cfm? theme=1*). In ASEAN countries, regional tourism has grown sharply because of relaxed travel and currency regulations. Coastal tourism has been a leading contributor of foreign exchange to the national economies of all the countries, perhaps with the exception of Myanmar.

- *Urbanisation and settlements*: Settlement growth in ASEAN countries has been driven by a strategy of cooperation and complementarity, which has combined planning with ongoing developments in industry and regional economic cooperation—so-called 'growth triangles' (Kakazu *et al.*, 1998). In South Asia urbanisation has been mainly a consequence of increasing poverty in rural areas, driving urban migration. The consequences are significant when urban migration occurs without planning for the key labour population.

- *Marine protected areas (MPAs)*: Many MPAs have been established in Asian tsunami-affected countries (*http://www.mpaglobal.org/index.php?action =aboutus*). They are a mechanism for safeguarding habitats critical for biodiversity, marine productivity, and small-scale fishing income. The recent findings on the status of the world's marine fishery stocks suggest that MPAs are an important management intervention for arresting the collapse of overfished fishery stocks (Worm *et al.*, 2006). The near future is likely to witness a major increase in the area under MPAs, with countries like Indonesia setting ambitious targets (Salm *et al.*, 2000). Increasing the area under MPAs is likely to have major repercussions for small-scale fishing communities. They could be positive if traditional rights of access and use of resources are strengthened. They could be negative if implemented in non-participatory and exclusionary ways.

Summary: The challenge for integrated coastal management (ICM)

Meaningful contribution from ICM to the problem of land use conflicts could be substantial. ICM may facilitate integrated land uses which provide win-win solutions to both traditional coastal resource users and to commercial enterprises. How ICM confronts predation by powerful commercial interests—the extraction of benefits from the capture of assets that rightfully belong to coastal communities—will be the major challenge. Traditional coastal land uses on which millions of households depend for their livelihoods and national and commercial interests may be harmonised within carefully designed land use zoning systems.

References

Ambio (1988). "East Asian Seas", *Ambio* 17(3).
BBC News (2005). "Logging threatens Borneo species", 11 June, http://news.bbc.co.uk/1/hi/world/asia-pacific/4083016.stm
De Soto, H. (2000). *The mystery of capital: Why capitalism triumphs in the west and fails everywhere else*. Bantam Press.
Diamond, J. (2005). *Collapse: How societies choose to fail or survive*. London: Penguin Books Ltd. p. 575.
Environmental Justice Foundation (EJF) (2003). "Smash & grab: Conflict, corruption and human rights abuses in the shrimp farming industry", www.ejfoundation.org
FAO (2004). *The state of world fisheries and aquaculture*. Rome: FAO.
GPA (2005). "Ecosystem based management: Markers for progress", UNEP Global Programme of Action for the Protection of the Marine Environment from Land-based Activities (GPA). http://www.gpa.unep.org/documents/ecosystem-based_management_english.pdf
Harford, T. (2006). *The undercover economist*. UK: Little Brown. p.278.
Kakazu, H., M. Tang and M. Thant (eds.) (1998). *Growth triangles in Asia: A new approach to regional economic cooperation*. (Second edition).
Kurien, J. (2005). "Evolving toward unsustainability: A personal statement on Kerala's marine fishery spanning three decades", *International journal of rural management* 1:73. http://irm.sagepub.com/cgi/reprint/1/1/73
Macfadyen, G. and E. Corcoran (2002). "Literature review of studies on poverty in fishing communities and of lessons learned in using the sustainable livelihoods approach

in poverty alleviation strategies and projects", *FAO Fisheries Circular No. 979* FIPP/C979. Rome.

OECD (2006). *India: Steel-making raw materials, issues, policy and outlook.* DSTI/SU/SC(2006)31, Directorate for Science, Technology and Industry Steel Committee. 2006,12 May.

Ostrom, E., J. Burger, C.B. Field, R.B. Norgaard and P. Policansky (1999). "Revisiting the commons: Local lessons, global challenges", *www.sciencemag.org*

Primavera, J.H. (2000). "Integrated mangrove-aquaculture systems in Asia", *Integrated Coastal Zone Management* : 121-128. Autumn.

Salagrama, V. (2005). "Fish out of water: The story of globalization, modernization and artisanal fisheries in India", *http://www.onefish.org/servlet/CDSServlet?status=NDOyNDM3MzMu MjQzODI1JjY9ZW4mMzM9ZG9jdW1lbnRzJjM3P WluZm8~#koin fo1111124640817_Fish_out_o... 08457.pdf*

Salm, R.V., J.R. Clark and E. Siirila (2000). *Marine and coastal protected areas - A guide for planners and managers*, (Third edition). Gland, Switzerland: IUCN.

Stern, N. (2005). "Making development happen: growth and empowerment", *Eighteenth Presidential Lecture Series*, 27 April. Washington DC: World Bank. *http://www.info. worldbank.org/etools/docs/library/138945 Transcript%20Stern%20042705.doc*

Stiglitz, J. (2002). *Globalization and its discontents*. London: Penguin Books. p. 282.

The Economist (2004). "Going for the jugular", 10 June.

Times International (2006). *Time Asia* 168(6), October 9.

Todd, E. (2005). "The spectre of a Soviet-style crisis. Truthout interview on Hurricane Katrina", 12 September. *www.truthout. org/docs_2005/091205H.shtml*

WCFSD (1999). "Our forests our future - Summary report", Canada: World Commission on Forests and Sustainable Development. *http://iisd.ca/wcfsd*

Wily, L.A. (2006). *Land rights reform and governance in Africa: How to make it work in the 21st century?*. Nairobi: UNDP Drylands Development Centre.

Wolowicz, K. (2005). "The fishprint of aquaculture. Can the blue revolution be sustainable? Redefining progress". *www.rprogress.org/publicatons/2005/The_Fishprint_of_Aquaculture_1205.pdf*

World Bank (2003). *Land policies for growth and poverty reduction*. Washington DC: World Bank. p. 288.

Worm, B., E.B. Barbier, N. Beaumont, J. Emmett Duffy, C. Folke, B.S. Halpern, J.B.C. Jackson, H.K. Lotze, F. Micheli, S.R. Palumbi, E. Sala, K.A. Selkoe, J.J. Stachowicz and R. Watson (2006). "Impacts of diversity loss on ocean ecosystem services", *Science* 314 (5800): 787-90.

7

Management Challenges and Approaches for Sustainable ICM

Pressure to revive tourism in tsunami-affected areas is one of the barriers to effective implementation of Integrated Coastal Management (ICM).

Source:

Robert Kay Adapted from: *Improved coastal zone planning and management. Integrated coastal zone planning in Asian tsunami-affected countries. The regional perspective in proceedings of the workshop on coastal area planning and management in Asian tsunami-affected countries. 27-29 September 2006. Bangkok, Thailand.*

1. Introduction

A major challenge in the post-tsunami rehabilitation process was how to avoid a knee-jerk reaction and control the investor-led need for rapid, uncontrolled redevelopment (Teerakul and Renshaw, 2005). Pre-existing forms of coastal development were generally unplanned and often inefficient, inequitable and unsustainable, pushing the poor into the most unhealthy and hazardous corners of the coast (UNEP/GPA, 2005). The need to undertake some kind of coastal management planning process within the post-tsunami rehabilitation effort was widely acknowledged throughout the region in light of the significant potential long-term benefits afforded by an integrated and interdisciplinary approach.

The goals of an integrated approach to coastal planning and management are to:

- minimise possible future tsunami and other coastal hazard impacts;
- maximise coastal resource use in a sustainable manner; and
- improve the living conditions of the coastal inhabitants.

There are numerous well-known examples in the Asia-Pacific region of long-term ICM programmes and extensive analysis of the factors that lead to and enhance the effectiveness of these programmes, as well as the barriers to their success.

The study by Christie *et al.* (2005) provided an overview of important considerations towards the improvement of ICM project design with a view to fostering long-term sustainability.

2. Framework of a long-term and sustainable ICM

Successful, sustainable ICM in the wake of the Indian Ocean tsunami should strive to:

- Highlight the importance of the participatory management processes, specifically those based at a community level over a long-term perspective (Pollnac *et al.*, 2005);

Chang et al. (2006) found that there was considerable variation in the performance of local conservation efforts across villages. Performance appeared to be influenced by the design of external aid programmes, with higher levels of performance attributed to bottom-up programmes. In this respect, practitioners involved in coastal management should view the disaster as providing a window of opportunity to enable and build relationships with local people that allows them to become active participants in the process.

'A renewed, and expanded, long-term commitment to ICM, and other forms of long-term planning and environmental management which will require careful implementation by cooperative efforts of national governments, NGOs, and civil society who have growing experience with ICM.'

- Share the benefits from ICM within and between constituency groups with specific reference to potentially contentious relations between different groups (Thiele et al., 2005; Sievanen et al., 2005).

  Pollnac and Pomeroy (2005) found that early involvement and participation in ICM are influenced by initial project benefits and perceptions of benefits. This involvement enhances the chances that the ultimate benefits will be those desired for the target population. Achievement of these benefits also stimulates continuing involvement in the activities, sustaining the ICM process. Their findings indicate that both community involvement and achievement of desired benefits will be necessary to impact ICM sustainability in the post-tsunami era (Christie et al., 2005).

- Carry out evaluative and adaptive processes to ensure that ICM projects result in tangible benefits to the stakeholders involved (Olsen and Christie, 2000); and

- Monitor which is a significant predictor of ICM sustainability. Pollnac (2004) provides quantitative evidence that community monitoring is one of the principal predictors of both compliance and biological success of ICM.

3. Barriers to effective ICM

While the objectives provide a sound framework for the long-term sustainability of ICM projects in the region, there were a number of barriers to effective employment of ICM principles in the immediate aftermath of the tsunami.

These included the following (updated from Kay and Alder, 2005):

- Pressure to rebuild tsunami-affected areas as quickly as possible—both to rehouse local people and to encourage tourists to return.
- Competing donor agendas to promote rapid redevelopment on the one hand, while promoting sustainable development on the other (combined with the sheer complexity and scale of the international relief effort).
- Perception that CZM is about ecosystem management and not about land use planning, tourism management, hazard management, urban development or sustainable livelihood promotion.
- Difficulties faced by donors and national governments in obtaining international CZM expertise quickly and effectively.
- Procurement processes faced by tsunami-affected countries are tied to many different needs and systems of releasing funds by donors.
- Problems in engaging with local-level coastal managers charged with making on-the-ground land use planning decisions.

Almost two years after the disaster, there is still a large percentage of regional coastal populations that is highly vulnerable and ill-equipped to deal with future disasters (UNDP/IUCN, 2006). The critical factor in this context is that ICM-planning strategies are designed to be long-term proactive approaches, and that in the face of sudden catastrophic disasters their ability to respond rapidly to such events is problematic. Fundamentally, because they are long-term processes, they thus take a long time to develop, with consensus building among the key stakeholder communities proving to be a painstaking process. While clearly valid for the long term, ICM processes require modification to cope with the requirement for immediate responses on short-time frames resulting from sudden events.

References

Chang, S.E., B.J. Adams, J. Alder, P.R. Berke, R. Chuenpagdee, S. Ghosh and C. Wabnitz. (2006). "Coastal ecosystems and tsunami protection", *Report to the National Science Foundation of the USA*.

Christie, P., K. Lowry, A.T. White, E.G. Oracion, L. Sievenen, R.S. Pomeroy, R.B. Pollnac, J.M. Patlis and R.V. Eisma (2005). "Key findings from a multidisciplinary examination of integrated coastal management process sustainability", *Ocean and Coastal Management* 48: 408-83.

IUCN (2006).

Kay, R.C. and J. Alder (2005). *Coastal planning and management*. London: E&F Spon. p. 380.

Olsen, S.B. and P. Christie (2000). "What are we learning from tropical coastal management experiences?", *Coastal Management* 28: 5-18.

Pollnac, R.B. (2004). "Multimethod analysis of factors contributing to the sustainability of community based marine protected (no-take) areas in the Philippines", Paper presentation at *American Association for the Advancement of Science (AAAS) Annual Meeting*. Seattle.

Pollnac, R.B. and R. Pomeroy (2005). "Factors influencing the sustainability of integrated coastal management projects in the Philippines and Indonesia", *Ocean and Coastal Management* 48: 233-51.

Sievanen, L., R.B. Pollnac, C. Lowe and B. Crawford (2005). "Weeding through assumptions of livelihood approaches in ICM: Eucheuma farming in the Philippines and Indonesia", *Ocean and Coastal Management* 48: 297-313.

Teerakul, B. and I. Renshaw (2005). *Draft comments: December 2004 tsunami event. Thailand*. Thailand Office: Wetlands International.

Thiele, M., R.B. Pollnac and P. Christie (2005). "Relationships between coastal tourism and ICM sustainability in the Central Visyas Region of the Philippines", *Ocean Costal Management* 48: 378-92.

UNEP/GPA (2005). "After the tsunami: Rapid environmental assessment", http://www.unep.org/tsunami/reports/TSUNAMI_report_ complete.pdf#search=%22 UNEP%2FGPA%2C%202005%20tsunami%20%22

8

Integrated Coastal Zone Planning

Integrated Coastal Management and Planning allows rational decisions concerning conservation and sustainable use of resources and space.

Source:

Robert Kay *Adapted from: Improved coastal zone planning and management. Integrated coastal zone planning in Asian tsunami-affected countries. The regional perspective in proceedings of the workshop on coastal area planning and management in Asian tsunami-affected countries. 27-29 September 2006. Bangkok, Thailand.*

1. Introduction

The 2004 Indian Ocean tsunami resulted in the widespread devastation of coastal communities throughout the region from a social, environmental and economic perspective. The huge loss of life generated by the tsunami waves was coupled with catastrophic destruction of the coastal zone in areas of India, Indonesia, Sri Lanka and Thailand. Areas most affected were generally those where marginalised members of the population were concentrated and unsustainable management practices were endemic. In order for the principles of 'build back better' to be achieved, a focussed and sustained coastal planning and management effort is required. In this context, the internationally recognised principles of Integrated Coastal Management (ICM) will play a key role in the rehabilitation and reconstruction process. With this in mind, the aim of this paper is to provide a perspective on ICM and its role in the post-tsunami era within the Asia-Pacific region.

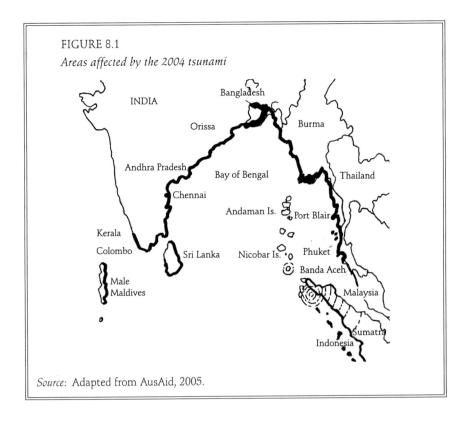

FIGURE 8.1
Areas affected by the 2004 tsunami

Source: Adapted from AusAid, 2005.

The contents draw on perspectives from the country papers formulated for the workshop on 'Coastal area planning and management in Asian tsunami-affected countries', presented in this proceedings. The workshop was held in Bangkok, Thailand, in September 2006, and forms one of a series of FAO workshops in the region that address the range of forestry, fisheries, aquaculture and agriculture problems faced by Asian countries in the aftermath of the 2004 tsunami.

Coastal planning and management policies and issues are considered in focus for each of the affected countries in the region, with the exception of Myanmar, for which limited information was available. In addition, a review of initiatives taken in the post-tsunami period is carried out in conjunction with a critical analysis of the broader principles of coastal management practice.

2. Integrated coastal zone planning and management: Concept and principles

2.1 Coastal management terms

The transitional region between the land and the ocean is commonly referred to as the coastal zone or coastal area (Kay and Alder, 2005). The variety of terms used internationally to refer to efforts to manage this coastal space include: Integrated Coastal Zone Management (ICZM); Integrated Coastal and Marine Area Management (ICMAM); Integrated Coastal Area Management (ICAM); Integrated Marine and Coastal Area Management (IMCAM), or even Integrated Management of Maritime Affairs (IMMA). Despite the variety of labels applied, it is generally accepted that they refer to the same over-riding set of principles (Cicin-Sain and Knecht, 1998).

2.2 Coastal management concept and approach

The unique nature of coastal zones is widely understood to require a concerted, well-considered and holistic coordinated integrated management and planning approach (Kay and Alder, 2005). This is most often referred to as an 'integrated' management approach. Indeed, 'integrated' as a prefix to coastal management is now so widely adopted as a concept critical in striving for sustainable coastal environments that it has effectively been adopted into the daily language of decision-makers and policymakers in many coastal nations. ICM explicitly defines its goal in terms of progress towards more sustainable forms of development seeking a balance between:

- economic development and use of the coastal region;
- protection and preservation of the coastal areas;
- minimisation of losses of human life and property; and
- public access to the coastal zones.

Examples of principles and objectives advocated for successful ICM are listed in Table 8.1.

ICM is now widely viewed as a mainstream activity. At the international level, all key institutions with involvement in the management and planning of coastal zones have embraced the concept including FAO, UNEP, UNDP, World Bank, IOC/UNESCO and IUCN. In addition, an estimated 142 national governments and semi-sovereign states are assessed to be actively engaged in ICM (Sorensen, 2002).

ICM is based on long-term consensus building and must be supported with a range of methods and techniques for the provision of sound information to aid the decision-making process. Many coastal managers

TABLE 8.1
Principles and objectives of ICM

Principles of ICM	Objectives of ICM
- A long-term view	- Strengthening sectoral management by improving training, legislation, and staffing
- A broad holistic approach	
- Adaptive management	- Preserve the biological diversity of coastal ecosystems by preventing habitat destruction, pollution and over expliotation
- Working with natural processes	
- Support and involvement of all relevant administrative bodies	- Promote the rational development and sustainable use of coastal resource
- Use of a combination of instruments	
- Participatory planning	
- Reflecting local characteristics	

Source: After Clarke, 1992.

seek to adopt this approach through planning activities that are either stand alone ICM plans or include ICM approaches into other planning and policy instruments that influence or have an impact on coastal resource management. Plans used in the management of the coast can be classified according to a number of methods, the most common of which are shown in Table 8.2.

2.3 Principles of ICM

In broad terms, all ICM programmes are focussed on seeking operational mechanisms for achieving sustainable development goals for the use and management of coastal resources. As with all such multidisciplinary integrated approaches, the resulting objectives for management must be locally appropriate to ensure maximum uptake and, therefore, implementation effectiveness. An overarching principle of ICM is that it is able to be flexibly implemented according to both the coastal issues it is being focussed on and the unique circumstances of a particular coastal nation. For example, the European Union (2002) outlines a clear set of objectives for member states, to be adhered to when formulating national strategies for coastal management, which highlights the need to consider the following:

> ...local specificity and the great diversity of European coastal zones, which will make it possible to respond to their practical needs with specific solutions and flexible measures.

2.4 Implementing ICM

Given the widespread adoption of ICM, there is a significant pool of expertise to facilitate the design of tailored approaches to employing ICM concepts. Indeed, many coastal managers describe this as a 'toolkit' approach. For example, in Australia a recent initiative explicitly developed a set of capacity-building fact sheets to provide such tools to local governments attempting to improve its coastal management efforts (Coastal CRC, 2006).

Once individual projects are evaluated and decisions are taken, implementation begins. Implementation of a project must be supported with policy tools. These may be supplemented with voluntary agreements

TABLE 8.2
Coastal management plan classification methods and plan types

Classification elements	Plan types			
Scope				
Geographic coverage	International	Whole of jurisdiction	Regional	Local	Site
Focus	Operational	Strategic			
Degree of integration	Subject	Integrated			
Statutory basis	Statutory	Non-statutory			
Reason for plan	Required for funding	Required to clear statutory works	Legislation which requires management plans	Direct response to management problem	Create business value
Process				
Participation	Expert	Participatory			
Flexibility	Fixed goals	Adaptive learning			
Worldview	Rationalist	Value-based			
Acceptance	Consensus	Directed			
Goal setting	Single goal	Scenario-based			
Context				
Cycle	New plan	Building on previous planning cycle			
Plan/programme	Stand-alone plan	Plan within programme			

Source: Adapted from Kay and Alder, 2005.

between various parties to achieve environmental or conservation objectives (UNEP, 1995).

The policy cycle as applied to ICM is illustrated in Figure 8.2 below, and is returned to later in the paper within the specific context of the Indian Ocean tsunami.

In general, increased population pressure has resulted in unplanned and unregulated development throughout the coastal zones within the region. Here, 'unplanned' development is understood to mean development that has not occurred under the umbrella of community—or government-led plans, or through an incomplete framework of plans and policies. The philosophy underpinning ICM and many of its tools and approaches focusses on planning for the long-term sustainability of coastal resources and the people that depend on these resources for their livelihood. The fundamental mandate of planning (and plan production) is to harness and focus community desires in the broader sense, i.e., local communities, communities of users and communities of elected representatives to develop a shared vision and practices for the use and allocation of coastal resources. This approach is most often articulated in some form of planning strategy. While there are numerous challenges in effectively implementing such strategies in whatever form they take (Figure 8.2), they are invariably better than either no plan at all or plans that have excluded key communities, either through a poorly conceived process or deliberate (and often malicious) choice.

3. Competing uses of the coastal zone

3.1 Costs and benefits of unplanned development

Uses of the coastline are generally considered under four main categories: resource exploitation (e.g., fisheries, forestry, gas and oil, and mining); infrastructure (including transportation, ports, harbours, shoreline protection works and defence); tourism and recreation; and the conservation and protection of biodiversity (Kay and Alder, 2005). Of specific interest here are the major land uses in the coastal zones of tsunami-affected countries, which include agriculture, shrimp and fish farming, forestry and human settlement (UNEP, 2005).

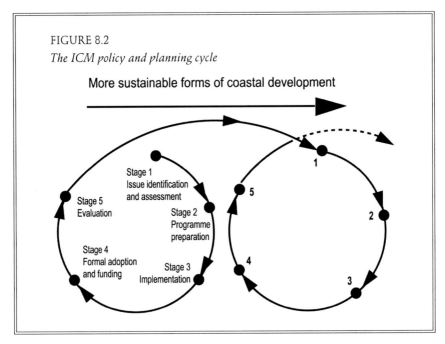

FIGURE 8.2
The ICM policy and planning cycle

Source: GESAMP, 1996.

Resource based industries such as fisheries and tourism are particularly important within the Asia-Pacific region. Fishing provides a basic source of food and income for up to 13 million people, while the extensive tourist industry may directly account for as much as 20 per cent of the GDP (UNDP/IUCN, 2006). Mangroves and the coastal forest play a crucial role in coastline stabilisation and storm-surge protection (Rutinbeek, 1991; 1994). In addition, mangroves act as important pollutant and nutrient sinks (Ibid).

The link between fisheries production and mangrove forestry are also widely reported in the literature (Janssen and Padilla, 1999; Mumby *et al.*, 2004; Murphy, 2004; Thampanva, 2006), as is the important income provided by traditional use of mangrove products to many of the poorest households within the region (Sathirathai, 1998).

Within the Asia-Pacific region, healthy coastal ecosystems including mangroves, wetlands, estuaries, lagoons, sandy beaches, sand dunes, coral reefs and seagrass communities are fundamentally linked to human well-being (UNDP/IUCN, 2006). However, dramatic population growth has led to increased pressure on ecosystems throughout the coastal zone.

3.2 Effects of economic development on coastal zones

Within the Asia-Pacific region, an ad hoc approach, where no coherent planning strategy has been adhered to, has frequently provided greater short-term economic returns, although sometimes at great social and environmental costs—and often these returns benefit only some sections of society.

Single objective and single output land management has resulted in the conversion of land from directly productive purposes; in many cases, leading to degradation or loss as a result of erosion, salinity, inundation and other interventions (FAO, 2005). For example, over the last 30 years the significance of the shrimp sector in particular has grown rapidly in tandem with increased demand and high prices for shrimp in international markets (Phillips and Budhiman, 2005). The increase in shrimp production has resulted in huge pressure on land in the coastal zone with agriculture and mangrove areas being converted for shrimp farming (IUCN, 2005).

While the ecosystem value of mangroves and coastal forests is well-known and widely accepted, their widespread clearance has subsequently become a common feature of coastal zones in all tsunami-affected countries (IUCN, 2005). In the islands of Indonesia, Java alone had lost 70 per cent of its mangroves by 1991, while Sumatra had lost 36 per cent (*www.earthisland.org*).

3.3 Role of mangroves in reducing vulnerability

Mangrove reduction has led to a loss of biodiversity and a reduction in food production and cooking fuel, which exacerbates the problem in other areas. In addition, a source of income is eliminated for marginalised communities that are already considered socially and economically worst-off (Kay and Alder, 2005).

In this context, it is clear that the enormous contribution of healthy coastal ecosystems in safeguarding production and consumption, reducing vulnerability, strengthening resilience and mitigating disasters has generally been undervalued, poorly understood and improperly safeguarded within the region (UNDP/IUCN, 2006). In fact, it has been suggested that unsustainable development activities and their associated degradation of the coastal zone led to exacerbated effects of the 2004 Indian Ocean tsunami in terms of ecosystem destruction and loss of life (Table 8.3). For example, anecdotal evidence in the aftermath of the tsunami suggested that mangroves were effective in buffering its impacts (Dahdouh-Guebas et al., 2005; FAO, 2005; Wetlands International, 2005). This was subsequently confirmed by systematic analysis of the effectiveness of mangrove buffering against tsunami waves (Chang et al., 2006). Preliminary analysis suggested that villages that were behind substantial mangroves suffered relatively little damage in comparison to those not likewise protected. However, assessment of the role of healthy ecosystems in reducing damage to coastal communities is ongoing and remains a matter for debate within the scientific community (Baird, 2006).

TABLE 8.3
Exacerbation of biophysical and socioeconomic impacts as a result of unplanned development in the coastal zone

Biophysical impacts	Socioeconomic impacts
• Increased coastal erosion	• Loss of property and coastal habitats
• Siltation	• Loss of life
• Loss of biodiversity	• Damage to coastal protection works and other infrastructure
• Extensive coastal inundation	• Loss of renewable and subsistence resources
• Higher level flooding	• Loss of amenity value
• Decline in water quality	• Loss of non-monetary cultural resources and values
	• Impacts on agriculture and aquaculture

In addition to the problems brought about by the effects of the tsunami, pressure on coastal land is expected to continue in step with increases in the regional population. In light of this, the need for an integrated approach to the management and rehabilitation of the coastal zone has been brought into even starker relief.

4. Coastal planning and management: A synopsis of regional policy and adherence

It is now widely understood that tsunami waves undergo a complex set of interactions in conjunction with the varying landscape of the continental shelf and a suite of other physical variables (Synolakis, 2002). However, the Indian Ocean tsunami in 2004 also interacted with another varying landscape, that is, the policy and practices under which coasts are managed. As with variations to the overall ICM landscape, there are significant differences between countries regarding their coastal hazard management landscape in line with those provided in Table 8.4 below.

TABLE 8.4
Options and measures for coastal erosion hazard management

Options	Measures
Event protection	• Hard engineering (e.g., sea walls, groynes)
	• Soft engineering (e.g., beach nourishment, dune enhancement)
Damage prevention	• Avoidance (e.g. prevent development)
	• Mitigation (e.g., relocatable or flood-proofed buildings, building codes)
Loss distribution (transfer)	• Individual measures (e.g. insurance)
	• Community measures (e.g. insurance, cost pooling, disaster relief and rehabilitation
Risk acceptance	• Do nothing

Source: Kay et al., 1994.

4.1 Problems in implementation of coastal zone laws

Overall, while several countries instituted legal and planning instruments for the management of the coastal zone, these generally lacked adherence. This non-implementation of coastal laws often gave way to unregulated development along many stretches of the coast ultimately affected by the tsunami. A consideration of the information identifies a set of common problems leading to this poor implementation and adherence at a regional scale. Key issues include:

- low capacity to participate in ICM programmes;
- no focal agency or coordinating body;

- lack of integration—persistence of fragmentation and sectoral isolation;
- overlapping of jurisdiction and misunderstandings over responsibilities; and
- inadequate legal framework to oversee implementation.

4.2 Factors hindering effective coastal management

Fundamentally, managing coastal zones is a people-centered exercise and as a result effective management revolves around the foibles of the individual and the institutions and practices built by the individuals. Practitioners understand that corruption, cronyism, individual egos, power and factors such as politics and familial relationships all play a part in daily life and, consequently, also play an important role in managing the coast. The global spotlight brought to the region by the tsunami highlighted these issues and while both the government and donor community have made stringent efforts to mitigate these factors, they will always play a role.

5. Post-tsunami management challenges and approaches for sustainable ICM

A major challenge in the post-tsunami rehabilitation process was how to avoid a knee-jerk reaction and control the investor-led need for rapid, uncontrolled redevelopment (Teerakul and Renshaw, 2005). Pre-existing forms of coastal development were generally unplanned and often inefficient, inequitable and unsustainable, pushing the poor into the most unhealthy and hazardous corners of the coast (UNEP/GPA, 2005).

'A renewed, and expanded, long-term commitment to ICM, and other forms of long-term planning and environmental management which will require careful implementation by cooperative efforts of national governments, NGOs, and civil society who have growing experience with ICM.'

The need to undertake some kind of coastal management planning process within the post-tsunami rehabilitation effort was widely acknowledged throughout the region in light of the significant potential long-term benefits afforded by an integrated and interdisciplinary approach. The goals of an integrated approach to coastal planning and management are to:

- minimise possible future tsunami and other coastal hazard impacts;
- maximise coastal resource use in a sustainable manner; and
- improve the living conditions of the coastal inhabitants.

There are numerous well-known examples in the Asia-Pacific region of long-term ICM programmes and extensive analysis of the factors that lead to and enhance the effectiveness of these programmes, as well as the barriers to their success.

The study by Christie et al. (2005) provided an overview of important considerations towards the improvement of ICM project design with a view to fostering long-term sustainability.

5.1 Guide to a successful ICM

Successful, sustainable ICM in the wake of the Indian Ocean tsunami should strive to:

- Highlight the importance of the participatory management processes, specifically those based at a community level over a long-term perspective (Pollnac and Pomeroy, 2005). Chang et al. (2006), found that there was a considerable variation in the performance of local conservation efforts across villages. Performance appeared to be influenced by the design of external aid programmes, with higher levels of performance attributed to bottom-up programmes. Practitioners involved in coastal management should view the disaster as providing a window of opportunity to build relationships with local people and become active participants in the process.
- Share the benefits from ICM within and between constituency groups with specific reference to potentially contentious relations

between different groups (Thiele et al., 2005; Sievanen et al., 2005). Pollnac and Pomeroy (2005) found that early involvement and participation in ICM are influenced by initial project benefits and perceptions of benefits. This involvement enhances the chances that the ultimate benefits will be those desired by the target population. Achievement of these benefits also stimulates continuing involvement in the activities, sustaining the ICM process. Their findings indicate that both community involvement and achievement of desired benefits will be necessary to impact ICM sustainability in the post-tsunami era (Christie et al., 2005).

- Carry out evaluative and adaptive processes to ensure that ICM projects result in tangible benefits to the stakeholders involved (Olsen and Christie, 2000).

- Monitor is a significant predictor of ICM sustainability. Pollnac (2004) provides quantitative evidence that community monitoring is a principal predictor of compliance and biological success of community based no-take areas in Visayas, the Philippines.

5.2 Barriers to effective ICM

While the aforementioned objectives provide a sound framework for the long-term sustainability of ICM projects in the region, there were a number of barriers encountered in employing ICM principles in the immediate aftermath of the tsunami. These included the following (updated from Kay, 2005):

- Pressure to rebuild tsunami-affected areas as quickly as possible—both to rehouse local people and to encourage tourists to return.

- There were often competing donor agendas to promote rapid redevelopment on the one hand, while promoting sustainable development on the other (combined with the sheer complexity and scale of the international relief effort).

- There was the perception that CZM is about ecosystem management and not about land-use planning, tourism management, hazard management, urban development or sustainable livelihood promotion.

- There were difficulties faced by donors and national governments in obtaining international CZM expertise quickly and effectively.
- The procurement processes faced by tsunami-affected countries are tied to many different needs and systems of releasing funds by donors.
- There were problems in engaging with local-level coastal managers charged with making on-the-ground land use planning decisions.

5.3 Consequences of inefficient ICM

Requirements for quick action resulted, in many instances, in duplication and overlap of resources and effort (Khazai et al., 2006). While the rebuilding process has strived to avoid needlessly repeating the mistakes of the past, uncoordinated initiatives have meant that rehabilitation and reconstruction in areas affected by the tsunami were not always economically or environmentally sound (FAO, 2005). This was due, in large part, to the immediate needs of post-tsunami reconstruction that resulted in restoration and rehabilitation activities guided by short-term planning perspectives (UNDP/IUCN, 2006). Currently, many of these short-sighted projects remain incomplete, unfinished or have failed to achieve their intended impact (Ibid).

It is important to recognise the trade-off between the need for rapid inputs to restore livelihoods in communities affected by the tsunami *versus* good governance and sustainable management (UNESCO, 2005). In some instances, the lack of good governance that builds on supporting institutions and policies has hindered recovery and even resulted in a return to some of the undesirable pre-tsunami situations. That said, it is also necessary to acknowledge that economic development has been, and still remains, an important component of the coastal zone.

The potential impacts of unplanned coastal development may only be mitigated through established democratic and mature legislative processes. In the absence, or poor functioning, of these legislative processes it is inevitable that there will be wide variations in management outcomes. These are clearly sensitive issues which are known by practitioners but not widely discussed.

Almost two years after the disaster there is still a large percentage of regional coastal populations that is highly vulnerable and ill-equipped to deal with future disasters (UNDP/IUCN, 2006). The critical factor in this context is that ICM-planning strategies are designed to be long-term proactive approaches, and that in the face of sudden catastrophic disasters their ability to respond rapidly to such events is problematic. Fundamentally, because they are long-term processes, they thus take a long time to develop, with consensus building among the key stakeholder communities proving to be a painstaking process. While clearly valid for the long term, ICM processes require modification to cope with the requirement for immediate responses on short-time frames resulting from sudden events.

Kay (2005), in reflecting on the role of ICM/CZM to date in tsunami response stated:

> By-and-large the potential central role of CZM in the regional tsunami response has not eventuated. This is not to say that CZM won't play a critical role in long-term recovery efforts—it is that the potentially integrative role that CZM could have played has not occurred. Rather, the response and recovery efforts have used mainstream disaster management approaches. The tsunami appears to have been viewed as a disaster that occurred on the coast, rather than a critical CZM issue that was a disaster.

The fundamental question then is: Are the underlying principles of ICM at fault here? In response to this question, Prof. Wong (2005), in an eloquent summary of ICM effectiveness wrote:

> ... if integration in CZM is to be carried out in the post-tsunami phase, it must first be an integration of livelihood restoration and habitat restoration. This would require a paradigm change or a change in the mindset of those implementing CZM. Too many of the principles expounded in various post-tsunami recovery programs may not offer practical solutions. To start, there should be a list of immediate tasks with which local communities can become involved, both to earn a livelihood and at the same time to restore coastal habitats. These are challenging tasks for coastal managers in the affected areas and they need to understand the fundamental change in CZM wrought by the 26 December 2004 event. At the very least, tropical CZM in the post-tsunami phase will never be the same.

The author agrees with Prof. Wong's assessment. ICM must be adapted to both manage the long-term response of the tsunami-affected coastlines and also to be able to deal more effectively with future coastal disasters.

5.4 What can be done?

Examples following the Asian tsunami illustrated the complete absence of methodologies and the outstanding need to formulate a potential disaster coastal response action plan. This highlighted the imperative for quick and robust ICM tools and techniques which are accepted as an approach by governments and donors alike. In this context, recent collaboration among national, provincial and local emergency management agencies and local communities under the USAID-funded US Indian Ocean Tsunami Warning System (IOTWS) Program, has led to the development of the concept of Coastal Community Resilience (CCR) (USAID, 2006).

5.5 Coastal Community Resilience (CCR)

The concept of CCR blends elements of disaster management and ICM. It provides a promising framework that has been tested through local scale workshops and bodes well for future efforts to minimise social disruption and mitigate the effects of events and impacts.

CCR promotes tsunami and other hazard readiness through better and more consistent tsunami awareness and mitigation efforts among communities at risk. The main goal is to improve public safety during tsunami emergencies and to build resilience against recurring coastal events (USAID, 2006). To meet this goal, the following objectives need to be met:

- Create minimum standard guidelines for a community to follow to become tsunami resilient.
- Encourage consistency in educational materials and response among communities and national emergency systems.
- Recognise communities that have adopted CCR guidelines.
- Increase public awareness and understanding of tsunamis and other hazards; and

- Improve community pre-planning for tsunami and other disaster impacts.

While CCR is relevant to ICM, it should not be viewed as a replacement. Rather, CCR should be viewed as a useful component to be adopted into a broader ICM framework i.e., protecting against and preparing for coastal disasters is one component of the holistic planning process. Indeed, the CCR concept may become the conduit through which to address integration shortfalls that have occurred to date in the ICM response to the tsunami as outlined in the next section.

6. Key findings and lessons learned relevant to post-tsunami reconstruction

The 2004 Indian Ocean tsunami has brought into sharp relief the challenges faced by those responsible for managing Indian Ocean coasts and has highlighted the need for improved and more systematic coastal management efforts. The lack of, or poor implementation or adherence to policy and legislative frameworks that support coastal area management within the region have contributed to the unplanned development of coastal zones and the suite of problems with which it is associated.

Impacts of the tsunami served to compound and exacerbate many of these pre-existing problems. Damage was reduced in areas with healthy coral reefs, mangrove forests, and coastal vegetation (Chang et al., 2006). Likewise, had designated set-back areas been enforced along the coastal zone, it is likely that the mortality rate of marginalised fisherfolk and their families throughout the region would have been significantly reduced. Although it is important to stress that while this is both intuitively correct and supported by anecdotal evidence, rigorous examination of this issue is still ongoing and remains a matter for debate.

Equitable and sustainable reconstruction efforts have been an important focus of the post-tsunami response. However, effective ICM requires long-term planning and implementation adherence. This may only be achieved through setting realistic objectives and time frames for implementation and working steadily toward that goal. In this context, it is evident that the aftermath of a disaster is not the appropriate time to attempt to use traditional ICM approaches alone, owing to their reliance on long-term consensus building, generally through community-driven participatory techniques. The CCR approach, by contrast, combines ICZM philosophies with contingency planning for the next disaster.

In the aftermath of the Indian Ocean tsunami, it has become clear that we must ensure that the long-term sustainable goals of ICM do not become compromised by short-term responses. Although efforts to rebuild and reconstruct have been well-intentioned, there is still a large percentage of regional coastal populations that is highly vulnerable and ill-equipped to deal with future disasters. In this climate, the marriage of coastal disaster response plans, CCR principles and traditional ICM tools and techniques may represent the way forward towards a holistic planning process. This will protect against and prepare for future coastal disasters, while at the same time add considerable benefit to sustainable responses to the 2004 tsunami disaster.

References

Baird, A. (2006). "Tsunami viewpoint: The myth of green belts", *Samudra* 44: 14-19.

Chang, S.E., B.J. Adams, J. Alder, P.R. Berke, R. Chuenpagdee, S. Ghosh and C. Wabnitz (2006). "Coastal ecosystems and tsunami protection", Report to the National Science Foundation of the USA.

Christie, P., K. Lowry, A.T. White, E.G. Oracion, L. Sievenen, R.S. Pomeroy, R.B. Pollnac, J.M. Patlis and R.V. Eisma (2005). "Key findings from a multidisciplinary examination of integrated coastal management process sustainability", *Ocean and Coastal Management* 48: 408-83.

Cicin-Sain, B. and R.W. Knecht (1998). *Integrated coastal zone and ocean management concepts and practices*. Washington DC: Island Press.

Clark, J.R. (1992). "Integrated management of coastal zones", *FAO Fisheries Technical Paper* No. 327. Rome: FAO. p. 167.

Coastal (CRC) (2006). "Cooperative research centre for coastal zone, estuary and waterway management", Local Government Authority Information Factsheets, Australia. http://www.coastal.crc.org.au/lg/index.html

Dahdouh-Guebas, F., L.P. Jayatissa, D. Di Nitto, J.O. Bosire, D. Lo Seen and N. Koedam (2005). "How effective were mangroves as a defence against the recent tsunami?", *Current Biology* 15(12): R443-47.

FAO (2005). "Building back better livelihoods in tsunami-affected countries", http://ftp.fao.org/FI/DOCUMENT/tsunamis_05/FAO_guiding/FAOBriefSpecialEnvoy130550.pdf

GESAMP (1996). *The contributions of science to integrated coastal management*.

IUCN (2005). "Rapid environmental and socio-economic assessment of tsunami damage in terrestrial and marine coastal ecosystems of Ampara and Batticaloa districts of Eastern Sri Lanka", http://www.iucn.org/tsunami/docs/rapid-ass-easte-sri-lanka.pdf

Janssen, R. and J.E. Padilla (1999). "Preservation or conversion?Valuation and evaluation of a mangrove forest in the Philippines", *Environmental and Resource Economics* 14(3): 297-321.

Kay, R.C. and J. Alder (2005). *Coastal planning and management*. London: E&F Spon. p. 380.

Khazai, B., G. Franco, J.C. Ingram, C. Rumbaitis del Rio, P. Dias, R. Dissanayake, R. Chandratilake and S.J. Kanna (2006). "Post-December 2004 tsunami reconstruction in Sri Lanka and its potential impacts on future vulnerability", *Earthquake Spectra* 22: S3, S829-44.

Mumby, P.J., A.J. Edwards, E.E. Arias-Gonzales, K.C. Lindeman, P.G. Blackwell, A. Gall, M.I. Gorczynska, A.R. Harbourne, C.L. Pescod, H. Renken, C.C.C. Wabnitz and G. Llewellyn. (2004). "Mangroves enhance the biomass of coral reef fish communities in the Caribbean", *Nature* 427 (6974): 533-6.

Murphy, P.J. (2004). "Mangroves enhance the biomass of coral reef fish communities in the Caribbean", *Nature* 427: 6974, 533.

Olsen, S.B. and P. Christie (2000). "What are we learning from tropical coastal management experiences?", *Coastal Management* 28: 5-18.

Phillips, M. and A. Budhiman (2005). *An assessment of the impacts of the 26th December 2004 earthquake and tsunami on aquaculture in the provinces of Aceh and North Sumatra*. Indonesia: FAO.

Pollnac, R.B. (2004). "Multimethod analysis of factors contributing to the sustainability of community based marine protected (no-take) areas in the Philippines", Paper presentation at *American Association for the Advancement of Science (AAAS) Annual Meeting*. Seattle.

Pollnac, R.B. and R. Pomeroy (2005). "Factors influencing the sustainability of integrated coastal management projects in the Philippines and Indonesia", *Ocean and Coastal Management* 48: 233-51.

Rutinbeek, H.J. (1991). *Mangrove management: An economic analysis of management option with a focus on Bintuni Bay, Iran Jaya, Jakarta*. Government of Indonesia, Dalhousie University.

———. (1994). "Modeling economy-ecological linkages in mangroves: Economic evidence for promoting conservation in Bintuni Bay, Indonesia", *Ecological Economics* 10: 233-47.

Sievanen, L., R.B. Pollnac, C. Lowe and B. Crawford (2005). "Weeding through assumptions of livelihood approaches in ICM: Eucheuma farming in the Philippines and Indonesia", *Ocean and Coastal Management* 48: 297-313.

Synolakis, C. (2002). "Tsunami and seiche", in W.F. Chen and C. Scawthorn (eds.), *Earthquake engineering handbook*. Boca Raton, FL: CRC Press. pp. 9-1-9-90.

Sorensen, J. (2002). *Baseline 2000 background report: Second iteration*. Boston, Urban Harbours Institute, University of Massachusetts.

Teerakul, B. and I. Renshaw (2005). *Draft comments: December 2004 tsunami event, Thailand*. Wetlands International, Thailand Office.

Thampanva, U. (2006). "Coastal erosion and mangrove propagation in Southern Thailand", *Estuarine, Coastal and Shelf Science* 68(2): 75-85.

Thiele, M., R.B. Pollnac and P. Christie (2005). "Relationships between coastal tourism and ICM sustainability in the Central Visayas region of the Philippines", *Ocean and Coastal Management* 48: 378-92.

UNDP/IUCN (2006). *Mangroves for the future: A strategy for promoting investment in coastal ecosystem conservation, 2007-2012.*
UNEP (1995). "Guidelines for integrated management of coastal and marine areas", UNEP Regional Seas Reports and Studies No. 161. p. 80.
———. (2005). "After the tsunami: Rapid environmental assessment", February. *http://www.unep.org/tsunami/tsunami_rpt.asp*
UNEP/GPA (2005). "After the tsunami: Rapid environmental assessment", *http://www.unep.org/tsunami/reports/TSUNAMI_report_ complete.pdf#sear ch=%22UNEP%2FGPA% 2C% 202005%20tsunami%20%22*
UNESCO (2005). "The tsunami disaster and disaster preparedness one year later", Summary prepared by Stefano Belfiore, Intergovernmental Oceanographic Commission, UNESCO.
USAID (2006). "Coastal Community Resilience (CCR)", Fact sheet available from: *http://www.iotws.org/ev_en.php?ID=2142_201 &ID2= DO_TOPIC*
Wetlands International (2005). "Assessment report to Ramsar STRP12. Natural mitigation of natural disasters", Cited 15 June 2005. Available at *http://www.wetlands.org/tsunami*

9

Establishing Resilient Coastal Ecosystems for Sustainable Livelihoods

In Seychelles, coastal and marine biodiversity contributes nearly a quarter of all employment opportunities.

Source:

Ritesh Kumar Wetlands International, South Asia, India.

1. Introduction

Coastal ecosystems, comprising mangroves, coral reefs, estuaries, lagoons, sandy beaches, sand dunes, sea grasses and other forms, play an important role in supporting livelihoods and human well-being through a range of ecosystem services. They are regions of remarkable biological productivity and high accessibility making them centres of human activity (Burke *et al.*, 2001). They are also known to provide protection against extreme weather events, reduce floods, trap and retain sediments, and function as nurseries for several species.

The numerous benefits that could be derived from the coastal ecosystems are palpable (Box 9.1). Yet, ecosystem services are often ignored in developmental planning leading to lopsided policy decisions. Moreover, these ecosystems continue to be under pressure from unsustainable development and climate change. This trend is likely to enhance vulnerability of coastal livelihoods in a number of ways, particularly through decline in resources, loss of employment opportunities, and reduced protection against natural disasters.

BOX 9.1
Services, livelihoods, and economic gains from the coastal ecosystem

- In Seychelles, coastal and marine biodiversity contributes nearly a quarter of all employment opportunities, one-third of government revenues, and two-thirds of foreign exchange earnings (CNPS, 1997).
- Marine and coastal tourism is the largest industry in the Maldives and contributes nearly one-fifth of its GDP.
- Nearly two-thirds of world's fisheries depend on coastal wetlands and over 90 per cent of the marine fisheries production is sourced from coastal zones (Hinrichsen, 1998).
- Reefs contribute up to 25 per cent of the annual fish catch providing food to a billion of people (Cesar *et al.*, 2003).
- The mangroves of Sunderbans within the Gangetic Delta in India and Bangladesh provide livelihoods to more than 0.3 million people, protects them from cyclones and tidal waves and is an important source of revenue through tourism, timber and forest non-wood products.

contd...

...contd...

The coastal ecosystem brings in concrete economic benefits that are comparable with alternate economic opportunities with defined cost and benefits.

- In American Samoa, mangroves with an extent of just 0.5 sq km have an estimated annual value of US $50 million (Spurgeon and Roxburgh, 2005). In Thailand, high values of US $2.7 - 3.5 million per sq km have been reported for the mangroves (Santhirathai and Barbier, 2001). Coral reefs have been associated with a range of ecosystem service values—for natural hazard management (upto 189,000 US$/ha/year), tourism (up to 1 million US$/hectare/year), genetic materials and bio-prospecting (up to US $57,000 US$/ha/year) and fisheries (up to US $3,818 US$/ha/year) (TEEB, 2009).

- Burke *et al.* (2002) have assessed a net economic loss of US $5.1 billion for Indonesia and the Philippines due to degradation of coral reefs over a period of 20 years. Similar assessments in the Caribbean have projected a net loss of US $350 - 870 million over 2050 (Burke and Maidens, 2004).

- An economic assessment study by Badola and Hussain (2005) indicated highest loss in village that was not sheltered by mangroves but by embankments, with the least per capita damage in village with mangrove as barrier. Das and Vincent (2009) validated the storm protection function of mangroves and established that villages with wider mangroves between them and the coast experienced significantly fewer deaths than ones with narrower or no mangroves. The opportunity cost of saving a life by retaining mangroves was assessed to be Rs. 11.7 million per life saved.

2. Drivers and pressures

Threats on the coastal ecosystems emanates largely from the following factors:

- *High concentration of population*: Nearly 39 per cent of the global population lived within 100 kilometres of the coastline (CIESIN, 2000). The high concentration of population puts immense pressure on the coastal resources by way of overexploitation, destructive fishing, and converting habitats for alternate use (e.g., shrimp culture, land use conversion, etc.).

- *Catchments*: A mangrove ecosystem requires a distinct range of salinity gradients for survival, which is dependent on appropriate mixing of saline and freshwater. Intensifying usage of freshwater on the upstream reaches can lead to reduced availability for coastal ecosystems. Selvam (2003) highlights the changes in mangrove floristics, especially reduction in low saline-tolerant species on the east coast of India due to reduction in periodicity and quantity of freshwater flowing to these ecosystems. Catchments are also a source of pollutants entering the coastal and marine environment. Cicin-Sain *et al.* (2002) suggest that land based sources account for more than 70 per cent of pollution in marine systems.

- *Climate change*: Increasing frequency of storms and the rising sea levels are some of the major threats to coastal wetlands as mangroves and coral reefs. The rising sea levels is projected to change the distribution and extent of coastal wetlands. Around 30 per cent of Asia's coral reefs are likely to be lost in the next 30 years due to multiple stresses and climate change (Cruz *et al.*, 2007).

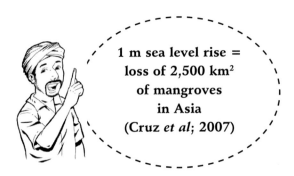

1 m sea level rise = loss of 2,500 km² of mangroves in Asia (Cruz *et al*; 2007)

Establishing Resilient Coastal Ecosystems for Sustainable Livelihoods   143

- *Other forms of localised threats*: include oil and chemical spills, introduction of alien invasive species, and high boat movement. Box 9.2 presents 2000 to 2006 condition status of the global coastal ecosystem, as well as trends in coastal degradation.

3. Ensuring resilient coastal ecosystems for resilient coastal livelihoods

The role of Integrated Coastal Zone Management (ICZM) (Box 9.3), is important in ensuring the conservation and management of coastal ecosystems. Addressing the multiple direct and indirect drivers which influence the resilience of coastal ecosystem requires linking ICZM with river basin and catchment management and oceans and fisheries management so as to secure their conservation and sustainable use. The linkages of the hydrological and ecological processes of coastal ecosystems and river basins can be best addressed through Integrated Coastal Area and River Basin Management (ICARM).

The basic principles of ICARM are same as those for ICZM, but applied simultaneously to the two linked systems. ICARM especially assumes relevance in addressing land-based threats to coastal ecosystems (e.g., regulation of hydrological regimes, pollution, etc.). In promoting adoption of sustainable practices, for example, FAO's Code of Conduct for Responsible Fishing is an important step in this direction.

BOX 9.2
Status and trends

Though the information base on coastal ecosystems in general is poor relative to other terrestrial systems, some findings could be derived as regards the condition of the coastal ecosystem.

- A study by the World Resources Institute (2001) indicated that mangroves lined 8 per cent of the global coastline with their principal distribution concentrated within the tropics.
- Coral reefs covered just 1.2 per cent of the world's continental shelf area (Spalding *et al.*, 2001). The narrow range of distribution of these ecosystems is indicated by that fact that of the 177 countries in the world, only 44 per cent have coral reefs and about half have mangroves (UNEP-WCMC, 2006).
- The Millennium Ecosystem Assessment (2005) confirmed that coastal ecosystems are experiencing rapid loss and degradation with 35 per cent of the original mangrove cover lost within the last two decades. Similarly, 20 per cent of the coral reefs had been lost, with an additional 20 per cent in the 20th century. Estuaries and coral reefs were observed to be the most threatened of all coastal ecosystems due to several direct and indirect impacts.
- Valiela *et al.* (2001) indicate that mangrove losses were as high as 80 per cent of original cover in some countries.
- About 30 per cent of world's reefs are seriously damaged (Wilkinson, 2004).
- The analysis of World Resources Institute (2000) indicated that 88 per cent of coral reefs of Southeast Asia were threatened by human induced pressures (Burke *et al.*, 2002).
- Several reef areas are undergoing shifts from coral-dominated to algal dominated states.
- In general, coral reef degradation is a far stronger trend as compared to decrease in its area. The Millennium Assessment synthesis also reported major loss of seagrass habitats in the Mediterranean, Florida Bay and parts of Australia, Southeast Asia and the Caribbean (Millennium Ecosystem Assessment, 2005).

BOX 9.3
What is ICZM?

Integrated coastal zone management (ICZM) is broadly defined as 'a mechanism to bring together the multiplicity of users, stakeholders, and decision makers in the coastal zone in order to secure more effective ecosystem management while achieving economic development and intra and inter-generational equity through application of sustainability principles.

The purposes of ICZM are to:

a) guide the level of coastal uses or interventions so as to not exceed the carrying capacity of the resource base;

b) respect natural dynamic processes, encouraging beneficial processes and preventing adverse interactions;

c) reduce risks to vulnerable resources;

d) ensure coastal ecosystem biodiversity;

e) encourage complimentarity in activities rather than competition;

f) ensure that environmental, social and economic objectives are achieved at an acceptable costs to society;

g) protect traditional uses and rights and equitable access to resources; and

h) resolve sectoral issues and conflicts (Ramsar Convention Secretariat, 2007).

While conservation and sustainable management of coastal ecosystems remains an embedded principle within ICZM and ICARM, a critical challenge is to maintain a balance between the various ecosystem services. Ecosystem restoration is often constrained by the fact that at several instances, degradation or unsustainable resource use is often an outcome of short-term livelihood need. This in turn deteriorates critical life-supporting ecosystem services. There is a need for approaches that can incentivise safeguarding coastal resources through transfer of benefits that emanate from healthy coastal ecosystems to coastal communities.

In particular, financial mechanisms have a tremendous role in promoting conservation or providing alternate livelihood opportunities to reduce resource dependence within carrying capacities (Box 9.4).

BOX 9.4
Biorights: Linking ecosystem restoration and livelihoods

Biorights is an innovative financial mechanism that addresses poverty trap by integrating sustainable development and environmental conservation. In return for provision of microcredits, a local community is involved in ecosystem protection and restoration. Upon successful delivery of conservation services, these microcredits are converted in definitive payments. Thus, the approach enables community involvement in conservation while providing sustainable alternatives to harmful development practices.

A pilot site for this approach was Central Java in Indonesia. Mangroves of the Pemalang District in Central Java used to abound until the extensive development of shrimp culture in the 80s. However, spread of disease led to the decline in productivity and ultimately, investors abandoned shrimp

contd...

Establishing Resilient Coastal Ecosystems for Sustainable Livelihoods   147

>...contd...
>
> ponds, leaving behind completely devastated mangrove belts. Recognising these challenges, a local NGO Mitra Bahiri, facilitated by Wetlands International and District Forest Agency initiated a mangrove restoration programme in late 90s.
>
> In 1998, Biorights was introduced as an implementation approach. Community groups were formed and mangrove restoration plans were developed by these groups, which included identification of appropriate species and locations. The communities selected appropriate livelihood programmes based on development needs and local feasibility. Bioright contracts were developed with respective community groups. Intensive implementation support was provided in the form of technical expertise, capacity building, and adaptation based on local conditions.
>
> Community-led monitoring demonstrated survival rates well above 75 per cent upon termination of the contract period, leading to conversion of all grants to definitive payments. Approximately 10 years after planting of the first seedling, the mangroves have grown into impressive heights of 4 - 8 metres. The number of commercial fish species has increased from 2 to 6. Slow land accretion is being observed. Increase in household incomes and decrease in vulnerability have been recorded. Access to physical and financial capital has also increased.
>
> Although Biorights contracts have ended in 2005, communities have continued with rehabilitation since then. Convinced of the benefits of restoration, the groups still meet to discuss restoration plans and have undertaken more ponds for rehabilitation.
>
> Further information on Biorights can be found in Eijk and Kumar (2009).

Conventional responses to natural disasters have focussed on reducing damages and eliminating changes. However, approaches that promote conditions enhancing human and ecosystem resilience are required to build resilient livelihoods. There is a need to promote local community-led coastal ecosystem restoration and management approaches as a step in this direction (Box 9.5).

BOX 9.5
Green Coast: A reef to ridge approach

The Indian Ocean Tsunami of 2004 created a massive chain of destruction along Indian Ocean coastline. It destroyed thousands of lives and livelihoods and cost of damages was estimated at almost US $10 billion (Telford and Cosgrave, 2006). One of the responses to the felt gap in post tsunami restoration is Green Coast—an initiative developed by Wetlands International in partnership with International Union for Conservation of Nature (IUCN), World Wildlife Fund (WWF) and Both ENDS. The initiative aims to restore and manage coastal ecosystems through a community-led approach, to improve biodiversity and livelihoods of coastal communities. It works through a three-pronged approach: a) science- and community based assessments identifying ecological damage and priority options for coastal restoration; b) community based restoration of coastal ecosystems and livelihoods through 'biorights' approach' and c) policy guidance and targeted communications aimed towards 'green reconstruction', to influence coastal resource management policies of district and national governments and to increase general awareness on value of coastal ecosystems (Figure 9.1). Within the first phase, the initiative focussed on tsunami affected coastal areas in Aceh, Indonesia, Sri Lanka, South India, South Thailand and Malaysia. A total of 91,000 tsunami-affected people in these coastal areas have benefited from rehabilitated coastal ecosystems; more than 1,100 hectares of mangrove and coastal forests, 2,500 kilometres of sand dunes and 100 hectare of damaged coral reef and sea grass beds were restored and protected. In the second phase, the initiative focusses on Aceh and Nias in Indonesia and Trang Province, Thailand. More than 600 hectares of degraded mangroves at these sites would be restored adopting participatory approaches.

The concept of a well reconstructed coastal zone, coastal protection as well as long-term security of food and income, is enhanced when sustainable tourism and fisheries management are combined with the rehabilitation and management of natural resources.

contd...

...contd...

FIGURE 9.1
Transect map of a well reconstructed coastal zone by the Green Coast

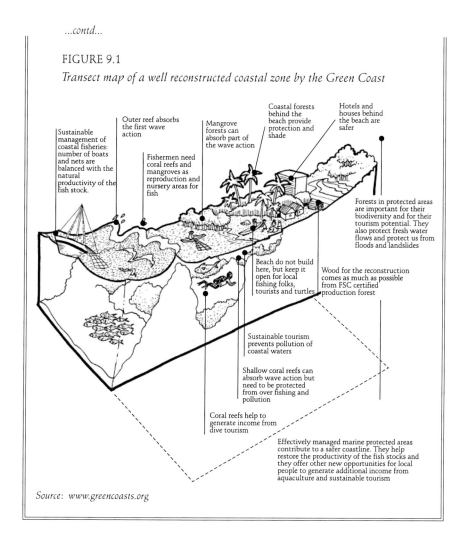

Source: www.greencoasts.org

References

Badola, R. and S.A. Hussain (2005). "Valuing ecosystem functions: An empirical study on the storm protection functions of Bhitarkanika mangrove ecosystem, India", *Ecosystem Conservation* 32(1): 85-92.

Barbier, E.B. (2007). "Valuing Ecosystem Service as Productive inputs", *Economic Policy* 22 (49): 177-29.

Burke, L. and J. Maidens (2004). *Reefs at Risk in the Caribbean*. Washington DC: World Resources Institute. p.80.

Burke, L., E. Selig and M. Spalding (2002). *Reefs at Risk in Southeast Asia*. Washington DC: World Resources Institute.

Burke, L., Y. Kura, K. Kassem, C. Revenga, M. Spalding and D. McAllister (2002). *Coastal Ecosystems: Pilot Analysis of Global Ecosystems*. Washington DC: World Resources Institute.

Cesar, H., L. Burke and L. Pet-Soede (2003). *The Economics of Worldwide Coral Reef Degradation*. Cesar Environmental Economics Consulting, ICRAN/WWF p.23.

Cicin-Sain, B., P. Bernal, V. Vanderweerd, S. Belfiore and K. Goldstein (2002). "Oceans, Coasts and Islands", at the World Summit on *Sustainable Development and Beyond: Integrated Management from Hilltops to Oceans*. Newark, Delaware: Center for the Study of Marine Policy.

Center for International Earth Science Information Network (CIESIN) (2000). "International Food Policy Research Institute and World Resource Institute", *Gridded Population of the World* (Version 2). Palisades, NY: CIESIN, Columbia University.

CNPS (1997). "Seychelles Biodiversity: Economic Assessment", Report prepared for the National Biodiversity Strategy and Action Plan, Conservation and National Parks Section, Division of Environment, Victoria.

Cruz, R.V., H. Harasawa, M. Lal, S. Wu, Y. Anokhin, B. Punsalmaa, Y. Honda, M. Jafari, C. Li and N. Huu Ninh (2007). "Asia Climate Change 2007: Impacts, Adaptation and Vulnerability", Contribution of Working Group II to the *Fourth Assessment Report of the Intergovernmental Panel on Climate Change*. Cambridge, UK: Cambridge University Press, pp.469-06.

Das, S. and R.V. Vincent (2009). "Mangroves protected villages and reduced death toll during Indian super cyclone", *Proceedings of National Academy of Sciences* 106 (18): 7357-60.

Eijk, Peiter van and Ritesh Kumar (2009). *Biorights in theory and practice - A financing mechanism linking poverty alleviation and environmental conservation*. Wetlands International, January.

Hinrichsen, D. (1998). *Coastal Waters of the World: Trends, Threats and Strategies*. Washington DC: Island Press.

Millennium Ecosystem Assessment (2005). *Ecosystems and Human Well-being: Wetlands and Water Synthesis*. Washington DC: World Resource Institute.

Ramsar Convention Secretariat (2007). "Coastal management: Wetland issues in integrated coastal zone management", *Ramsar handbooks for the wise use of wetlands*, (3rd edition) 10. Gland, Switzerland: Ramsar Convention Secretariat.

Sathirathai, S. and E.B. Barbier (2001). "Valuing mangrove conservation in Southern Thailand", *Contemporary Economic Policy* 19(2): 109-22.

Selvam, V. (2003). "Environmental classification of mangrove wetlands of India", *Current Science* 84 (6): 757-65.

Spalding, M.D., C. Ravilious and E.P. Green (2001). *World Atlas of Coral Reefs*. UNEP-WCMC, University California Press, Berkeley, USA.

Spurgeon, J. and T. Roxburgh (2005). "A Blueprint for Maximising Sustainable Coastal Benefits: the American Samoa case study", Proceedings of *10th International Coral Reef Symposium*, Okinawa, Japan.

TEEB (2009). *The Economics of Ecosystem and Biodiversity for National and International Policy Makers*. Bonn.

Telford, J. and J. Cosgrove (2006). *Tsunami Evaluation Coalition*. London. Synthesis Report.
UNEP-WCMC (2006). *In the front line: Shoreline protection and other ecosystem services from mangroves and coral reefs*. Cambridge, UK: UNEP-WCMC.
Valiela, I., J.L. Bowen and J.K. York (2001). "Mangrove forests: One of the world's threatened major tropical environments", *Bioscience* 51(10): 807-15.
Wilkie, M. and S. Fortuna (2003). *Status and Trends in Mangrove Area Extent Worldwide*. Forest Resources Development Service. Rome: FAO.
Wilkinson, C. (ed.) (2004). *Status of Coral Reefs of the World: 2004*. GCRMN/Australian Institute of Marine Science, Australia.
World Resources Institute (2001).

Part 2

Disaster Risk Reduction: Key to Adaptation

The second part of the source book focusses on disaster risk reduction. The fact that poverty increases the level of exposure of local people to disasters is presented as a basic premise. Exposure to natural disasters also further aggravates poverty. Vulnerability is a key feature of poverty and often inseparable from disasters. The reduction of disaster risk is therefore a key element of development efforts. A good understanding of risks, vulnerability and resilience are essential in disaster prevention and preparedness. Climate change has also emerged as a major threat to development with extreme weather events threatening to undo the benefits of past development. Understanding the links between climate change adaptation and disaster risk management is not just desirable but often essential. Disasters have drawn attention to the role of healthy ecosystems as integrating platforms for linking disaster risk reduction and climate change dimensions.

10
Reducing Disaster Risk:
A Challenge for Development

Rural communities, though isolated from the wider market, are not necessarily less at risk from globalisation, an underlying issue in disaster risk reduction.

Source:

Adapted from: Executive Summary in Reducing Disaster Risk: A Challenge for Development. A Global Report. United Nations Development Programme Bureau for Crisis Prevention and Recovery.

Natural disaster risk is intimately connected to processes of human development. Disasters put development at risk. At the same time, the development choices made by individuals, communities, and nations can generate new disaster risk. But this need not be the case. Human development can also contribute to a serious reduction in disaster risk.

1. Development at risk

Meeting the Millennium Development Goals (MDGs) is severely challenged in many countries by losses from disasters. The destruction of infrastructure and the erosion of livelihoods are direct outcomes of disaster. But disaster losses interact with and can also aggravate other financial, political, health and environmental shocks. Disaster losses may set back social investments to ameliorate poverty and hunger, provide access to education, health services, safe housing, drinking water, and sanitation, or to protect the environment, as well as the economic investments that provide employment and income.

A considerable incentive for rethinking disaster risk comes from the goals laid out in the Millennium Declaration. The MDGs direct development planning towards priority goals. Each of these goals interacts with disaster risk, potentially contributing to a reduction of human vulnerability to natural hazard. But it is the processes undertaken in meeting each goal that will determine the extent to which disaster risk is reduced. This implies a two-way relationship between the kind of development planning that can help achieve the MDGs and the development processes that are associated with an accumulation of disaster risk.

How can development increase disaster risk? Rapid urbanisation is an example of how the drive for economic growth generates new disaster risks. The growth of informal settlements and inner city slums has led to the growth of unstable living environments in ravines, on steep slopes, along flood plains or adjacent to noxious or dangerous industrial or transport facilities.

1.1 Can development planning incorporate disaster risk management?

The frequency with which some countries experience natural disaster should place disaster risk at the forefront of development planners' minds. This agenda differentiates between two types of disaster risk management. Prospective disaster risk management should be integrated into sustainable development planning. Development programmes and projects need to be reviewed for their potential to reduce or aggravate vulnerability and hazard. Compensatory disaster risk management (such as disaster preparedness and response) stands alongside development planning and is focussed on the amelioration of existing vulnerability and reduction of natural hazard that has accumulated through past development pathways. Compensatory policy is necessary to reduce contemporary risk, but prospective policy is required for medium- to long-term disaster risk reduction.

Bringing disaster risk reduction and development concerns closer together requires three steps:

- collection of basic data on disaster risk and development of planning tools to track the relationship between development policy and disaster risk;
- collection and dissemination of best practices in development planning and policy that reduce disaster risk; and
- galvanising political will to reorient the development and disaster management sectors.

2. International patterns of risk

United Nations Development Programme (UNDP) has begun the development of a disaster risk index (DRI) in order to improve the understanding of the relationship between development and disaster risk. It examined four natural hazard types (earthquake, tropical cyclone, flood and drought) responsible for 94 per cent of deaths triggered by natural disasters and the populations exposed, and calculated the relative vulnerability of countries to each.

In the last two decades, more than 1.5 million people have been killed by natural disasters. But human deaths, the most reliable measure of human loss, reveal only the tip of the iceberg in terms of development losses and human suffering. Worldwide, for every person killed, about 3,000 people are exposed to natural hazards.

Disaster risk was found to be considerably lower in high-income countries than in medium- and low-income countries. Countries classified as high human development countries represent 15 per cent of the exposed population, but only 1.8 per cent of the deaths (Figure 10.1).

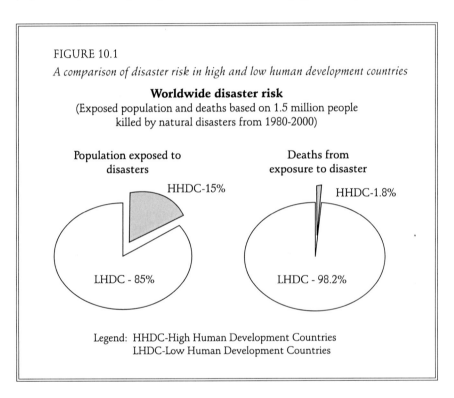

FIGURE 10.1
A comparison of disaster risk in high and low human development countries
Worldwide disaster risk
(Exposed population and deaths based on 1.5 million people killed by natural disasters from 1980-2000)

Legend: HHDC-High Human Development Countries
LHDC-Low Human Development Countries

What are the development factors and underlying processes that configure disaster risk? Analysis of socioeconomic variables and recorded disaster impacts (for earthquake, tropical cyclone, and flood hazard only) enabled some initial associations between specific development conditions and processes with disaster risk (Table 10.1). These findings have very high degrees of statistical significance and highlight the importance of urbanisation and rural livelihoods as development contexts that shape disaster risk.

TABLE 10.1
Patterns of disaster risk, developmental factors and underlying processes that configure disaster risk

Disaster hazard	Average no. exposed (Estimated)	Some countries with high relative vulnerability	Development factors and underlying processes that configure disaster risk
Earthquake	130 million	Iran, Afghanistan, India, Turkey, Russia, Armenia, Guinea	High urban growth rates and high physical exposure
Tropical cyclone	119 million	Bangladesh, Honduras, Nicaragua, India, the Philippines, Vietnam, the small island developing states (SIDS)	High percentage of arable land and high physical exposure
Flood	196 million	Venezuela, Morocco, Somalia, Yemen (190 countries exposed on an average per year to catastrophic floods; climate change to aggravate situation)	Low gross domestic product (GDP) per capita, low local density of population and high physical exposure
Drought	220 million	African states (evidence show translation of drought into famine, mediated by armed conflict, internal displacement, HIV/AIDS, poor governance, and economic crisis)	No data

3. Development: Working to reduce risk?

Two key variables are associated with disaster risk in the DRI: urbanisation and rural livelihoods. For each, a critical dynamic pressure likely to shape the future characteristics of these variables was also examined. For urbanisation, economic globalisation; and for rural livelihoods, global climate change.

3.1 Urbanisation

Urbanisation does not necessarily have to lead to increased disaster risk and can actually, if managed properly, help reduce it. A number of factors contribute to the configuration of risk caused by urbanisation. The first is history. For example, when cities have been founded in or expanded into hazardous locations. Second, the urbanisation process leads to the concentration of populations in risk-prone locations. Third, with

With urbanisation, residential areas have increasingly encroached the landslide-prone areas.

transient or migrant populations, social and economic networks tend to be loose, causing social exclusion and political marginalisation, leading to lack of access to resources and increased vulnerability.

Urbanisation can also modify hazard patterns. For instance, the urbanisation of watersheds can modify hydraulic regimes and destabilise slopes, increasing flood and landslide hazard. Urbanisation also has the power to radically shape disaster risks. Major investments in infrastructure and productive facilities, the development of new urban areas and trade corridors, and the unplanned urbanisation of new regions are all examples of modalities through which urbanisation can shape risk in broad territorial areas.

3.2 Urbanisation affected by dynamic pressures: Economic globalisation

Globalisation and the growing interconnectedness of the global society mean that catastrophic events in one place have the potential to affect lives and public policies in distant locations. At the same time, globalisation has the power to shape new local economic relationships and subsequent geographies of risk. Given that the decisions that generate such conditions (i.e., free trade agreements) are made at the international level and without detailed knowledge, it is uncommon that existing risk patterns are taken into account.

Economic globalisation can provide opportunities to enhance the livelihoods and the quality of life for people and places benefiting from new investments. To prevent these investments from creating large inequalities and further polarising the world into those who are at risk and those who are not, the opportunities and benefits of globalisation need to be shared much more widely. This requires strong international, national, and local governance.

3.3 Rural livelihoods

About 70 per cent of the world's poor live in rural areas. Rural poverty is one of the key factors that shape risk to hazards such as flooding or drought. Often the poorest in rural areas occupy the most marginal lands, forcing them to rely on precarious and highly vulnerable livelihoods in areas prone to drought, flooding, and other hazards. Local ecological and environmental changes as a consequence of agricultural practices can itself create risk. Deforestation to make way for agricultural production leads to soil erosion, loss of nutrients, and the marginality of agriculture. These processes can generate new patterns of hazard.

For the majority of rural communities connected to the global economy, livelihoods are vulnerable to fluctuations in world commodity prices. When low commodity prices coincide with natural hazards, rural livelihoods come under high stress. However, rural communities that are isolated from the wider market are not necessarily any less at risk.

3.4 Rural livelihoods affected by dynamic pressures: Global climate change

Global climate change brings long-term shifts in mean weather conditions and the possibility of increasing frequency and severity of extreme weather events. The impact of climate change will be more severe on developing countries because they are dependent on agricultural production and are less equipped to deal with extreme weather events.

The lack of capacity to manage and adapt to climate related risks is already a central development issue in many developing countries, particularly in small island development states (SIDS). The lack of capacity to manage risks associated with current climate variability will likely also inhibit countries from adapting to the future complexity and uncertainty of global climate change.

If development is to be advanced in countries affected by climate risks and not to aggravate climate change risk, an integrated approach to local climate risk reduction needs to be promoted. Successful risk reduction approaches already practiced by the disaster risk community should be mainstreamed into national strategies and programmes.

Climate change has caused drought in many countries that are dependent on agricultural production.

4. Factors of risk

4.1 Armed conflict and natural disasters

Armed conflict and governance are factors that can turn low rainfall episodes, for example, into famine events. At the turn of the 21st century, some countries suffered episodes of drought, earthquake, or volcanic eruption on top of years of armed conflict, causing a particularly acute humanitarian crisis. Obviously, disaster management is a potential tool for conflict prevention.

4.2 Diseases

Epidemic diseases can be seen as disasters in their own right. They also interact with human vulnerability and natural disasters. There is a great deal of variation in the relationships between disease, disaster and development. Hazard events such as flooding or temperature increase in highland areas can extend the range of vector-borne diseases, such as malaria. HIV/AIDS and other diseases can exacerbate the disaster risks brought on by climate change, urbanisation, marginalisation, and war.

4.3 Governance

Mainstreaming disaster risk reduction with development policy is a key challenge. The need for strong intervention following a disaster is recognised. The challenge now is to increase the focus on disaster risk reduction as a central element of ongoing development policy. This approach requires decentralised disaster risk planning strategies that can empower communities and open the window for local participation. The most vulnerable in society are also often those most excluded from community decision-making and in many cases this includes women. Enabling participation in these circumstances requires a long-term commitment to social development as part of vulnerability reduction programmes.

Within reforms, legislation often remains a critical element in ensuring a solid ground for other focal areas, such as institutional systems, sound planning and coordination, local participation and effective policy implementation. But legislation on its own cannot induce people to follow these rules. Monitoring and enforcement are needed.

> Elements of better governance for disaster risk reduction:
> - Economic governance—the decision-making process that affects a country's economic activities and relationships with other economies.
> - Political governance—the policymaking process including national disaster reduction policy and planning.
> - Administrative governance—the system of policy implementation; disaster risk reduction requires effective enforcement of building codes, land use planning, environmental risk and human vulnerability monitoring and safety standards.

4.4 Social capital

In recent years the concept of social capital has provided additional insights into the ways in which individuals, communities and groups mobilise to deal with disasters. Social capital refers to those stocks of social trust, norms, and networks that people derive from membership in different types of social collectives. Social capital, measured by levels of trust, cooperation, and reciprocity in a social group, plays the most important role in shaping actual resilience to disaster shocks and stress. Local level community response remains the most important factor enabling people to reduce and cope with the risks associated with disaster.

The appropriateness of policies for enhancing the positive contribution of civil society depends on developmental context. For many countries in Africa, Latin America and Asia that have undergone structural adjustment and participatory development, the challenge may not be so much the creation of a non-governmental sector as is coordination.

5. Conclusions and recommendations

The following are the six emerging agendas in disaster risk reduction:

1. *Appropriate governance*: This is fundamental if risk considerations are to be factored into development planning and if risks are to

Legislation has to run parallel with a strong monitoring and enforcement of rules on the use of water and other resources.

be successfully mitigated. Development needs to be regulated in terms of its impact on disaster risk. Perhaps the greatest challenges to mainstreaming disaster risk into development planning are political will and geographical equity.

2. *Factoring risk into disaster recovery and reconstruction*: The argument made for mainstreaming disaster risk management is doubly important during reconstruction following disaster events.

3. *Integrated climate risk management*: Building on capacities that deal with existing disaster risk is an effective way to generate capacity to deal with future climate change risk.

4. *Managing the multifaceted nature of risk*: People and communities most vulnerable to natural hazards are also vulnerable to other sources of hazard. For many, livelihood strategies are all about the playing off of risks from multiple hazard sources—economic, social, political and environmental.

5. *Compensatory risk management*: In addition to reworking the disaster development relationship, a legacy of risk accumulation exists and there is a need to improve disaster preparedness and response.

6. *Addressing gaps in knowledge for disaster risk assessment*: A first step towards more concerted and coordinated global action on disaster risk reduction must be a clear understanding of the depth and extent of hazard, vulnerability and disaster loss.

Specific recommendations towards this end are to:

- enhance global indexing of risk and vulnerability, enabling more and better intercountry and interregional comparisons;
- support national and subregional risk indexing to enable the production of information for national decision-makers;
- develop a multi-tiered system of disaster reporting; and
- support context-driven risk assessment.

11

Linking Poverty, Vulnerability, and Disaster Risk

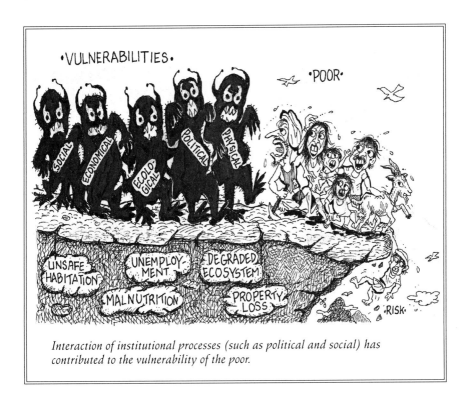

Interaction of institutional processes (such as political and social) has contributed to the vulnerability of the poor.

Source:

Ranasinghe Perera *Practical Action, Sri Lanka.*

1. Introduction

Poverty is the oldest, largest, and most oppressive socioeconomic problem of the world. It is a cause as well as an effect in the development process. It becomes a cause as it determines the economic and social destiny of an individual or a nation. It becomes an effect, because it is a product of certain dynamics and structures of the system. Poverty is predominantly a rural phenomenon in the world. In South Asia, 70 per cent of the population is poor and 75 per cent of them are from rural sector (World Bank, 2008). The rural poor mainly comprises tenant farmers, agricultural labourers, small livestock keepers, people living in marginal lands, tribal groups, and socially excluded groups such as widows, disabled, and other backward communities.

2. Causes of poverty

Groups situated in resource poor locations experience economic and social exclusion and inequitable distribution of income. As a whole, poverty has become the end result of physical, economic, political, social and institutional processes that interact with each other in ways that make worse the deprivation in which poor people live (Perera, 1998). Although poverty is predominantly an end product of the above factors, sudden physical or economic calamities aggravates poverty. There are also cases where natural disasters, epidemics, political calamities, and global, regional, or local economic recessions, likewise, contribute to increasing poverty.

Calamities aggravate poverty:

- The steady declining trend of poverty over a period of two decades in Indonesia was drastically upturned during the latter part of 1990s due to a sudden calamity.
- Sudden calamities or economic shocks have led to increased absolute poverty in Thailand by 6 per cent, Poland by 14 percent, Brazil by 5 per cent and in Costa Rica by 7 per cent at different time periods (The World Bank, 1995; Republic of Indonesia, 1998; Perera, 1998; ITDG, 2005).

3. Vulnerability

In the context of disaster management, vulnerability is defined as the degree of exposure of people to the damages of disasters. It has been perceived as the counterpart of resilience. People's exposure to natural disasters, epidemics, political calamities and global, regional or local economic recessions expands poverty incidence and aggravates poverty conditions among the poorest communities. Poverty is increased under these circumstances basically due to vulnerability of people to face the adverse impacts of disasters or economic shocks. In comparison with rich people, poor people are more vulnerable to the negative impacts of disasters which are mainly due to three different causative factors; physical, economic and social.

3.1 Causative factors of poor's vulnerability

3.1.1 Physical vulnerability

- Generally, poor people have limited or no access to land. Usually, they live in marginal, unfertile, or disaster-prone lands. No matter where they live, they remain physically marginalised from many means. But they have no other option as the lands in safer areas are owned by rich and powerful people and not accessible to them.

- The poor content themselves in living in calamity prone areas and simply cope with the disaster when they occur. Despite experiencing drought or flooding, they do not vacate these locations. Therefore, they are uninterruptedly characterised by physical vulnerability.

- When people live in marginal areas, essential infrastructural facilities and services such as roads, electricity, hospitals, schools, sanitation and communication services are also not adequately available for them. As poor people are not an attractive target for the authorities in terms of revenue collection etc., investment on the infrastructural facilities in poverty-stricken locations is also minimal. Under these circumstances, poor people become the regular victims of the disasters unlike rich people who live in physically less vulnerable areas.

The poor content themselves in living in calamity-prone areas and cope with disasters as they occur

3.1.2. Economic vulnerability

Economic vulnerability refers to the exposure of the people to the disastrous shocks due to meagre income and deficient resources.

- Most of the livelihoods of poor communities are fragile, short-lived, and less productive, hence, these communities are not privileged enough to escape from the poverty cycle.

- Poor people who live in disaster-prone areas do not go for long-term investments as livelihood risk is relatively high. Temporary livelihoods exist basically below subsistence level and do not produce surplus for savings and investments for future. Consequently, communities of disaster-prone areas live under poverty condition for generations and generations with no improvement.

- When livelihoods produce surplus for future investment, recurrent disasters take away both the surplus and the original assets putting people again in a vicious circle of poverty.

Linking Poverty, Vulnerability, and Disaster Risk **171**

- Unlike other communities, disaster-prone communities need to replace their assets such as tools, livestock etc., repeatedly in the production cycle to maintain their meagre income for the survival.
- The economy of disaster-prone communities is weakened by inadequate and unfair economic services too. They have low access to institutional credit due to low creditworthiness and high premium for insurance, if available, due to high production risk. All these factors and issues collectively and negatively impact on the economy of poor increasing their economic vulnerability.

3.1.3. Social vulnerability

Social vulnerability, in its broadest sense, is the societal, cultural, institutional and political dimensions of vulnerability that exposes people to the adverse impacts of economic shocks and natural hazards (Weichselgartner, 2001).

- When people are socially vulnerable, they have a little say on the policy decisions made for them and have little access to influence

decision-makers. Social vulnerability is an inherent characteristic of poor irrespective to the locations which they live.

- The poor, who live in disaster-prone areas, are more vulnerable socially than the poor in other areas due to specific contextual issues. Due to recurrent disasters, often times, people of disaster-prone areas are displaced. Broken families, misplaced friends and neighbours, loss of employment and assets, collapsed institutions and disrupted societal links are regular features of disaster-prone regions and these factors collectively contribute to increase the social vulnerability of the communities.

4. Disaster risk

Disaster is defined as disruption of the functioning of a society due to widespread human, material, economic or environmental losses and impacts, which exceeds the ability of the affected society to cope using its own resources. Disasters are products of so many complicated socioeconomic factors other than natural factors. In other words, natural disasters impact differently on rich and poor as the disaster risk is an intricate function of three major factors; hazard, vulnerability and capacity of people.

- Hazard is a natural phenomenon which should impact people on the same magnitude. However, it does not happen that way because when hazard becomes a disaster, it is no longer a natural occurrence.
- Vulnerability is the nature and extent of exposure of people to probable damage.
- Capacity is the ability of the people to cope with the damage. When vulnerability is high, and disaster risk is high, then resilience is low. When capacity is high and disaster risk is low, then resilience is high. This logical explanation tells us that probable damages by disasters can be minimised by taking suitable action to handle different factors that contribute to vulnerability or capacity. Thus, impact of similar nature of disaster varies from country to country, and community to community, depending on the measures taken to reduce the vulnerabilities or to increase the capacity of people.

5. Conclusion

Poverty determines the economic and social destiny of people. In the context of risk management, vulnerability is nothing but characteristics of poverty that contribute to increase the disaster risk. Hence, in analysing vulnerability, the nature, depth, trend, intensity, and related issues of poverty need to be examined. No matter what approach is used to examine poverty (biological approach, inequality approach or relative deprivation approach), vulnerabilities remain the key features of poverty. The poor are always vulnerable and oftentimes, vulnerable people are poor. This is the inseparable connection between poverty and vulnerability. In general, when vulnerability is high, disaster risk is high and persistent poverty is inevitable. Thus, in order to reduce the disaster risk, different dimensions of vulnerabilities are to be looked into. In other words, vulnerability of communities is to be reduced to the lowest degree or converted to community resilience by taking appropriate actions to address poverty in a sustainable manner.

References

Anderson, M. B. and P. J. Woodrow (1998). *Rising from the ashes: Development strategies in times of disaster*. London: IT Publications.

Conway, T. and A. Norton (2002). "Poverty, risk and rights: New directions in social protection", *Development Policy Review* 20(5).

ITDG (2005). *Livelihood approach to disaster management: A policy framework for South Asia*. South Asia and Rural Development Policy Institute. Islamabad, Pakistan: Pictorial Printers.

Perera, I.R. (1998). "Open economy and absolute poverty", *Govikatayuthu Adyana* 03(02). Colombo, Sri Lanka: Hector Kobbekaduwa Agrarian Research and Training Institute.

Republic of Indonesia (1998). *Indonesia's poverty alleviation policy reform*. The State Coordinating Ministry for People's Welfare & Poverty Alleviation. Government Publication.

World Bank (1995). "Poverty problem in Sri Lanka", Adapted from the *Sri Lanka: Poverty Assessment*, in *Economic Review* 21(05). People's Bank Publication: Colombo: World Bank.

―――. (2008). *World Development Report 2008*. Agriculture for Development. Washington, D.C: Oxford University Press.

Weichselgartner, J. (2001). "Disaster mitigation: The concept of vulnerability revisited", *Disaster Prevention and Management* 10(2): 85-94.

12
The Impact of Climate Change on the Vulnerable

Climate change has a devastating impact on the vulnerability of the poor.

Source:

Adapted from: Part of a series examining the impact of climate change and poverty. Department for International Development (DFID) in the U.K.

1. Introduction

Climate change adds urgency to understanding and addressing the poor's vulnerability to current and future climate variability. Vulnerability is an indication of people's exposure to external risks, shocks and stresses and their ability to cope with, and recover from, the resulting impacts. Changing vulnerabilities can often explain how people move in and out of poverty suddenly and seasonally or gradually over time. Vulnerability may differ within people's lives. It also differs across groups within communities or individuals within a household, owing to their livelihood activities or social standing. People may use a number of strategies to move out of poverty, but without reducing their vulnerability, they can easily slip back into poverty at a later date.

2. Impact of climate change on the vulnerability of the poor

Climate change predictions include an increase in the number, frequency, and intensity of climatic shocks, along with gradual changes such as temperature increases, sea level rise and alterations in seasonal patterns. The following are the effects of climate change and their corresponding effects on vulnerability.

2.1 Increase in frequency of climate extremes

The increased frequency of climate extremes reduces the time for poor households to recover from one climatic shock to another. There is also the risk of unprecedented shocks, such as the flooding experienced in East Africa following an extended drought. Traditional coping strategies may not be sufficient in this context and will lead the poor to rely on ad hoc and unsustainable responses. This not only reduces resilience to the next climatic shock but also to the full range of shocks and stresses that the poor are exposed to.

2.2 Increase in climate variability

Rainfall is predicted to become more variable over a number of different regions. This has implications for managing water security, agricultural outputs, and food security and climate-related diseases.

2.3 Systematic changes in the climate in vulnerable locations

Systematic changes in mean temperatures and precipitation are projected to affect the most vulnerable locations. Areas already suffering from drought are expected to have an increased risk of drought due to climate change. Heavily populated coastal areas are most at risk from increased cyclone activity and sea level rise.

It is estimated that the number of people at risk from annual flooding will increase from 75 to 206 million, given a 40 cm rise in sea levels and further increases in coastal populations. Ninety per cent of those at risk will be within Africa and Asia.

An OECD report on the impacts on climate change in Nepal found that rising temperatures have increased glacial retreat and glacial lake outburst floods, which reduce the availability of water and energy (from hydropower), both of which are important factors in vulnerability reduction.

Climate change has a negative effect on the productivity of fisheries.

2.4 Degradation of ecosystem goods and services

The effects of climate change on natural resources will further reduce poor people's resilience, as they are often reliant on such resources for employment and food security. Desertification and salinisation are potential adverse impacts of climate change on agricultural land. For example, it has been estimated that a 2-3.5°C increase in temperature in India could result in a 9-25 per cent decline in farm revenues. Additionally, climate plays a role in the productivity of fisheries as seen during the dramatic decline of the Peru anchovy fishery during El Niño episodes and the reduction in fish biomass following Hurricane Lenny in St. Lucia.

FIGURE 12.1
Impact of climate variability and climate change on vulnerability

Changes in mean climate variability, extreme events and sea level rise	Effects on livelihoods	Impacts on vulnerability
Increased temperature and changes in precipitation reduces agricultural and natural resources.	Direct impacts of climatic shocks and stresses such as livelihood assets, health, food and water security.	Increased vulnerability due to: - Lower capacity to prepare; - Lower capacity to cope; and - Lower capacity to recover from climatic and non-climatic shocks and stresses.
Changes in precipitation, run-off and variability leads to greater water stress.	Increased pressure on coping strategies and social protection measures.	
Increased incidence or intensity of climate related extremes such as water stress.		
Temperature, water and vegetation changes resulting in increasing prevalence of disease.	Reduced ability to recover due to increased frequency of climatic shocks or increased intensity of climatic stresses.	

2.5 Climate change impacts on economic growth

Climate change is also likely to have a number of impacts on economic growth. This will affect the poor by reducing their ability to diversify livelihoods and their ability to afford sustainable and unrestricted coping strategies.

3. Coping with climate change

People develop coping strategies to deal with climate change as with other shocks or stresses. These include building social networks as forms of insurance, traditional forecasting in order to be prepared for climatic changes and ingenious means of protecting assets. In Asia, for example, people use floating seed beds in times of floods. However, the poor's range of coping strategies is naturally more restricted by their lack of assets and by the other stresses on their livelihoods.

As manifestations of climate change becomes more prevalent, current coping machanisms are likely to become overwhelmed. This includes both formal systems (e.g. social assistance), and informal systems (e.g. social networks).

4. The country-level response

The changing climate has consequences for the poor in developing countries. It is therefore necessary to ensure that policy and institutional frameworks and mechanisms are strengthened to support the poor's range of response options to climate change. The best way will be to support the reduction of vulnerability to the current climate, ensuring that this is not a stand-alone activity but integrated into sustainable development and poverty reduction.

5. Ways to decrease vulnerability to changing climate

5.1 Understand climate variability within the context of vulnerability

Vulnerability analysis, feeding into poverty reduction strategies and other macro-economic planning tools, needs to take the level and type of impacts of climate variability into account. Additionally, any action taken

180   Strengthening Resilience in Post-disaster Situations

Supporting the flow of climate information to the poor helps improve their response capability.

to reduce specific impacts of climate variability needs to be designed and undertaken with an understanding of the overall vulnerability context, not forgetting the special needs of vulnerable groups such as women, children and the elderly.

Vulnerability assessments should be combined with hazard information to assess the level of risk to communities. This hazard information can be obtained from seasonal weather forecasts and longer-term climate predictions.

5.2 Support the response-capability of the poor by strengthening their assets

- *Social capital*: supporting social networks that provide safety nets;

- *Natural capital*: protecting the resilience of natural systems to support livelihoods of the poor;
- *Physical capital*: assisting the poor to make their physical capital more climate-resilient;
- *Human capital*: supporting the flow of climate information to the poor; and
- *Financial capital*: supporting the poor to reduce and spread their financial risks.

There are a number of tools that can help to achieve these aims including disaster preparedness and social protection. It will also be necessary to incorporate vulnerability reduction into wider policies and programmes.

5.3 Support disaster preparedness and response

Of particular importance will be the empowerment of communities to take collective action to prepare for, cope with and recover from shocks and stresses and to improve information on impending shocks and stresses.

5.4 Support social protection measures

Their design requires a clear understanding of the vulnerability context for all groups, which should, by its nature, take climate variability into account. There may also be an increasing role for financial insurance mechanisms in assisting the poor to spread and reduce their risks.

5.5 Incorporate vulnerability reduction into wider policies by:

- Improving the flexibility and responsiveness of food security mechanisms to climate shocks, e.g. food intervention systems;
- Improving the resilience of the poor's health and the health care system;
- Ensuring macroeconomic policies reduce poor people's vulnerability; and
- Protecting natural resources in poor people's livelihood and coping strategies.

13
Understanding Adaptation and Mitigation

Adaptive capacity refers to the ability of systems to adjust to external disturbances such as climate change. In Vietnam, stilt houses were built to adapt to the changing environment

Source:

Adapted from: Institute for Social and Environmental Transition, 2008.

1. Basic concepts

It is important to build adaptive capacity and resilience in order to respond to the uncertainties and systemic changes anticipated as a consequence of climate shifts. But what do these terms mean and what are their key indicators in development situations?

1.1 Development indicators

In most contexts, development indicators focus on factors that relate to well being, economic productivity, and the environmental sustainability of the resource base that populations depend on for their livelihoods. Indicators typically include poverty levels, access to basic services (e.g., water, health care, etc.), economic productivity, income distribution, educational levels, and so on. These indicators often vary on the basis of gender, caste, community or other subgroups within regions.

1.2 Adaptive capacity

In development contexts, adaptive capacity refers to the ability of socioeconomic, institutional, and cultural systems to adjust to external disturbances such as climate change, natural disasters, or degradation in the water resource base. In general, it also refers to the ability of such systems to adjust to other forms of disturbance, such as economic fluctuations.

Socioeconomic, institutional and cultural systems with high levels of adaptive capacity are able to reconfigure themselves in response to disturbances without sustained declines in development indicators. Systems with lower levels of adaptive capacity experience long-term declines in some or all of these key development indicators when subject to disruption.

1.3 Adaptation

The process of adaptation involves changes (either fundamental or incremental) in livelihood and other systems that enable, at a minimum, maintenance of the pre-disruption status of development indicators.

Understanding Adaptation and Mitigation

> Adapting, coping and resilient systems
>
> Systems are well adapted to their context when they score high in relation to development indicators and are able to absorb 'normal' disturbances such as storms, floods, and droughts that fall within historical norms, with relatively minor disruptions and without sustained declines in such indicators.
>
> On the other hand, systems are merely coping if they survived but incurred substantial losses in development indicators either for the region as a whole or for sub-groups within it. Coping strategies focusses on techniques (such as reductions in food intake and sale of assets) that, while they may allow populations to survive, leave them impoverished after a disaster.
>
> Resilient systems are able to absorb fluctuations or disturbances without fundamental changes in their structure or in key development indicators.
>
> All 'well adapted' systems are resilient to 'normal' fluctuations in their environment. But not all resilient systems are well adapted. Those that score poorly in relation to development indicators can still be quite resilient; but because they do not meet basic development criteria, are maladapted rather than well adapted.

1.4 Coping

Coping (as opposed to adaptation) focusses more on survival regardless of changes in development indicators.

1.5 Resilience

Resilience is the ability to absorb disturbances, to be changed and then to reorganise and still have the same identity (retain the same basic structure and ways of functioning). It includes the ability to learn from the disturbance (Resilience Alliance, *http://www.resalliance.org/564.php*).

2. Two ways of adapting to climate change

2.1 Proactive Adaptation

Proactive adaptation is the ability of key actors to identify emerging hazards and take proactive steps to change policies, implement projects, establish warning systems or take other forms of action to reduce the anticipated impacts of climate change. The wide array of targeted projects for climate change such as early warning, drought mitigation, flood management and so on would all fall under the umbrella of proactive adaptation.

2.2 Autonomous adaptation

In contrast, autonomous adaptation is the adjustment or reshaping of livelihood and other systems by numerous dispersed entities in ways that, while they are not proactively planned by key actors, do serve to reduce risks or the impact of climate change. Economic diversification, for example, often occurs as a consequence of many factors operating within local societies (changes in trade, demographics, transport, communication, etc.). Diversification is a major factor reducing the risks associated with climatic variability and change but is generally not a directly planned response to specific climate related hazards. Instead, risk reduction is a property that emerges in an unplanned manner as individuals, companies, and other entities identify opportunities and respond to a wide variety of risks or constraints.

3. Risk reduction

3.1 Disaster risk reduction

Disaster risk reduction is any set of direct or indirect intervention that functionally reduce the vulnerability of individuals, households, communities, businesses, economic and livelihood systems, regions and other entities to severe disruption when extreme events occur. As with responses to climate change, disaster risk reduction can occur either through planned interventions or autonomously as an emergent characteristic of land use, infrastructure and livelihood systems.

3.2 Planned interventions to reduce disaster risk

This include a wide variety of activities: from the design of physical infrastructure to the development of 'hazard awareness' in vulnerable populations. Such interventions are typically targeted in relation to specific populations identified as vulnerable and toward specific hazards.

3.3 Autonomous and emergent forms of risk reduction

This refer to the characteristics of regions or systems that, while generally not specifically designed to reduce hazard risks, do increase the ability of vulnerable populations to deal well with such risks. The importance of such emergent features in relation to hazards is clearly evident in most disaster situations.

In the Nepal Terai, for example, research has documented that populations with access to a diverse array of organisations (government, non-government, business, etc.) are able to access diverse sources of support during floods; while those lacking access to diverse organisations are not. None of the organisations accessed were initially developed to support flood relief but their mere existence provided a nucleus for organisation

and response when needed. Institutional diversity can, as a result, be seen as an emergent characteristic of systems that may, in a generic manner, contribute to disaster risk reduction. Similar examples exist in relation to social networks, communication and transport systems, and infrastructure.

4. Reducing vulnerability to reduce risks

As with climate change, most strategies for disaster risk reduction currently focus on specific interventions designed to reduce the vulnerability of specific target groups. Changes in behaviour that emerge autonomously (i.e., in an unplanned manner as populations respond to risk) along with the emergent characteristics of systems, such as diversification may, however, play a much greater role in determining vulnerability than interventions targeted to reduce vulnerability.

4.1 Vulnerability

The term 'vulnerability' is intended to reflect both physical exposure to hazards and the complex web of interconnected economic, political, cultural and institutional factors that create social vulnerability. Definitions of vulnerability are often divided into terms of social vulnerability and physical vulnerability.

4.2 Physical vulnerability

This refers to the built environment (lifelines, such as water or electricity), the capability of structures to withstand the energy loads associated with extreme events while protecting the occupants, and the location of structures (CDRSS, 2006).

4.3 Social vulnerability

Weisner *et al.* (2004) defined social vulnerability as 'the characteristics of a person or group and their situation that influence their capacity to anticipate, cope with, resist and recover from the impact of a natural hazard.' Adger (1999) further describes social vulnerability as 'the exposure of groups or individuals to stress as a result of social and environmental

Understanding Adaptation and Mitigation

Factors affecting social vulnerability

Social vulnerability is dependent on a number of factors including usage and access rights to natural resources, access to political power and representation (Mustafa, 2002; CDRSS, 2006), cultural constructions surrounding gender, age, beliefs and norms, and physical vulnerability, among others.

Furthermore, certain characteristics such as age, gender, race and class, are strongly associated with social vulnerability. Disaster researchers have noted that women and children, in many situations, tend to have more of a reduced capability to cope with and respond to extreme events.

The gendered differences in vulnerability are often due to 'differences in socioeconomic status, domestic responsibilities and power, access to and control over resources such as land ownership and community organizations, and the intersection of gender with age, health, and safety' (Enarson and Morrow, 1998). In social structures where women are not allowed to own property, have little economic independence, or are illiterate, women are less likely to be able to cope with extreme events or have the means to recover after an event has occurred.

change, where stress refers to unexpected changes and disruption to livelihoods.'

5. Relationship between physical and social vulnerabilities

Physical and social vulnerability are inexorably linked. The degree to which the built environment can withstand disruptions caused by natural disasters is determined by human decisions to build in certain areas, the types of materials used, and by policies governing construction and placement of infrastructure. The factors giving rise to physical vulnerability depend on the ability of certain groups to anticipate disasters and cope with the potential outcomes of a hazard event.

All of these concepts of vulnerability recognise that it is a dynamic condition dependent on multiple factors and that the degree of vulnerability can change. As physical conditions or social conditions change, an individual or group or structure might become more or less vulnerable.

6. Determining vulnerability

Disaster researchers are beginning to recognise that labelling a population as being vulnerable without considering that population's capabilities, might lead to misidentification of needs and points of vulnerability reduction. Describing people as vulnerable without assessing their capacity to cope with extreme events is to label people as passive victims.

Non-governmental organisations (NGOs) and community based organisations have adopted capacity and vulnerability (VCA) frameworks in which a full assessment of disaster risk and capabilities are done. There are multiple VCA methodologies, all of which aim to identify vulnerable groups or individuals. Davis *et al.* (2004) describe the goal of VCA as:

> to identify specific vulnerable groups/individuals, based on key social characteristics such as gender, age, health status, disability, ethnicity and so forth. The process also includes an analysis of patterns of density, livelihood security and occupational activities that increase the vulnerability of certain households and communities. Capacity assessment aims at identifying a wide diversity of resources: community coping strategies, local leadership and institutions, existing social capital which may contribute to risk reduction efforts, skills, labour, community facilities, preparedness stocks, a local evacuation plan, etc.

Thus, VCA methodologies are emerging as more comprehensive approaches for assessing vulnerability and determining the capabilities of various groups and individuals to cope with, respond and recover from disasters.

References

Adger, W.N. (1999). "Social vulnerability to climate change and extremes in coastal Vietnam", *World Development* 27(2): 249-69.

Committee on Disaster Research in the Social Sciences (CDRSS) (2006). "Future challenges and opportunities", *Facing hazards and disasters: Understanding human dimensions*, Washington, D.C.: The National Academies Press.

Davis, I., B. Haghebaert and D. Peppiatt (2004). "Social vulnerability and capacity analysis: An overview", *Discussion Paper* prepared for the ProVention Consortium Workshop. Geneva: ProVention.

Enarson, E. and B.H. Morrow (eds.) (1998). *The gendered terrain of disaster*. Miami: Laboratory for Social and Behavioral Research, Florida International University.

Wisner, B., P. Blaikie, T. Cannon and I. Davis (2004). *At risk: Natural hazards, people's vulnerability and disasters* (2nd ed.), New York: Routledge.

14

Demistifying Terminologies and Definitions: Same Terms, Different Meanings?

The awareness of differences and identification of a common ground between DRR and CCA communities is essential.

Source:

Jorn Birkmann, *Gerd Tetzlaff and Karl-Otto Zentel (Eds.) Addressing the Challenge: Recommendations and Quality Criteria for Linking Disaster Risk Reduction and Adaptation to Climate Change.*

1. Introduction

There is a growing recognition that climate change adaptation (CCA) and disaster risk reduction (DRR) strategies need to be linked to address the challenges of sustainable development, resilience and human security. However, before the challenges and the synergies in CCA and DRR could be addressed, there is a need to harmonise meanings of terminologies used in these two different fields of expertise.

Terms play an important role for conceptualising problems and developing solutions but CCA and DRR discourses use competing terms and concepts. The DRR community has primarily focussed on major disasters and sudden-onset hazards—often from a social science perspective—while the CCA community has developed its terminologies based mainly on research with a natural science perspective. Even if both communities use similar terms such as hazard, vulnerability, mitigation, adaptation and resilience, they may attribute different meanings to the same term.

For example, both communities focus on vulnerability reduction but do they refer to the same thing? If not, then practical actions and coherent

To be able to achieve a common goal, the DDR and CCA must first agree on common definitions and terms.

strategies might be at risk since a different understanding of what vulnerability means implies consequently also different approaches for strategies and solutions.

Since both CCA and DRR communities aim to raise awareness, increase cooperation and dialogue between experts, policymakers and practitioners on climate change related risk and extreme events, the awareness of differences and identification of a common ground on terms is an essential task.

Table 14.1 summarises differences and similarities in key terms as used in CCA and DRR communities, and presents important recommendations on what needs to be done to develop congruent terminologies and thus, more integrated concepts.

2. Coping and adaptation

Aside from the need to harmonise meanings of the same terms used by CCA and DRR communities, there is also a compelling need to distinguish more precisely between the use of 'coping' on one hand and 'adaptation' on the other. Coping is used by DRR community mainly to describe response processes to actual or potential hazard impacts. The concept of adaptation was developed by CCA community as a second strategy towards the challenges of climate change in contrast to climate change mitigation or reduction of greenhouse emissions. However, these two terms have increasingly been used interchangeably, their different qualities and characteristics often disregarded.

Interestingly, the different definitions of coping and adaptation are often linked, such that adaptation should allow societies to better cope with stressors. An important difference is that 'adaptation' implies adjustments to changing conditions or a changing environment, while coping in DRR community involves reactions, decision-making, and dealing with the hazard impact. This does not necessarily imply an adjustment.

Furthermore, most definitions of coping and adaptation leave the time dimension relatively open. Adaptation, for example, could encompass a range of actions and measures over various time frames. In contrast, Vogel and O'Brien (2004) and Birkmann (2009) stress that coping and

TABLE 14.1
Climate change terminologies, similarities, and recommendations

Disaster risk reduction community (DRR community)	Climate change adaptation community (CCA community)	Similarities/differences	Recommendations
Adaptation In current documents, adaptation is linked to three activities in DRR such as: a) risk assessment, b) early warning systems, and c) sector-specific risk reduction plans (UN/ISDR; Submission to the UNFCCC; Status of Implementation of Article 4, Paragraph 8 of the Convention, Decision 5/CP.7 and Decision 1/CP.10. However, a more in-depth definition is not provided. In DRR research, adaptation can be understood as e.g., the change or adjustment of livelihoods to the altered conditions in order to maintain major activities during extreme events without losing assets and capital. In contrast to coping, adaptation is determined by medium- and long-term adjustments (Vogel/O'Brien, 2004) and corresponds with the notion of change (Birkmann, 2009).	Adaptation to climate change refers to adjustment in natural or human systems in response to actual or expected climatic stimuli or their effects, which moderates harm or exploits beneficial opportunities. (IPCC, 2007).	The DRR community has not sufficiently defined adaptation in terms of extreme events and disaster risk yet. The CCA definition would also be a good starting point for the DRR community.	Differences between adaptation and coping should be made clear. The areas where adaptation should be considered in DRR need to be extended, e.g. also disaster aid and reconstruction (water, sanitation, shelter) should consider aspects of climate change adaptation in the future.

contd...

Demistifying Terminologies and Definitions: Same Terms, Different Meanings? 197

Disaster risk reduction community (DRR community)	Climate change adaptation community (CCA community)	Similarities/differences	Recommendations
Coping/coping capacity The means by which people or organisations use available resources and abilities to face adverse consequences that could lead to a disaster. The strengthening of coping capacities usually builds resilience to withstand the effects of natural and human-induced hazards (UN/ISDR, 2004). Strategies and measures that act directly upon damage during the event by alleviating or containing the impact or by bringing about efficient relief (Thywissen, 2006). Coping is mainly impact related and rather short-term, compared to adaptation (Birkmann, 2009).	Coping is a function of; perception (of risk and potential avenues of action—the ability to cope is information contingent); possibilities (options ranging from avoidance and insurance, prevention, mitigation, coping); private action (degree to which special capital can be invoked); and public action (e.g., Webb and Harinarayan, 1999; Sharma et al., 2000 quoted in IPCC, 2001).	The DRR community links coping capacity to a hazard and its impacts whereas the term coping is used in the CCA community in a broader sense encompassing risk perception, options of individuals to act and public actions.	Coping should be used to describe short-term actions that are more spontaneous than strategic adaptation. Coping is hazard specific and hazard related. Adaptation is broader and should encompass a long-term perspective.
Extreme event DRR distinguishes mainly extreme events according to: a) sudden-onset hazards, and b) creeping changes. Furthermore, extreme events are also classified with regard to: 1) sudden-onset hazards such as floods, droughts, windstorms, and 2) extreme temperatures.	An extreme event is an event that is rare within its statistical reference distribution at a particular place. Definitions of 'rare' vary, but an extreme weather event would normally be as rare as or rarer than the 10th or 90th percentile. By definition, the characteristics of what is called 'extreme'.	The CCA community defines extreme event primarily based on its statistical occurence (rare events) while the DRR community mainly focusses on different hazard and disaster types and their chain of development.	The statistical focus on rare events might be misleading for the future, since extreme events become more frequent. Therefore, the definition should be broadened including aspects of the DRR community, such as the characteristics of an extreme event (extreme weather, magnitude, etc.).

contd...

...contd...

Disaster risk reduction community (DRR community)	Climate change adaptation community (CCA community)	Similarities/differences	Recommendations
Preparedness			
Activities and measures taken in advance to ensure effective response to the impact of hazards, including the issuance of timely and effective early warnings and the temporary evacuation of people and property from threatened locations (UN/ISDR, 2004).	Not a key term for the CCA community	The term is hardly defined for CCA. Adaptation is used fairly uncritically for preparedness.	The relation between preparedness and adaptation needs to be clarified.
Resilience			
The capacity of a system, community or society potentially exposed to hazards to adapt, by resisting or changing in order to reach and maintain an acceptable level of functioning and structure. This is determined by the degree to which the social system is capable of organising itself to increase its capacity to learn from past disasters for better future protection and to improve risk reduction measures (UN/ISDR, 2004). The capability of a system to maintain its basic functions and structures in a time of shocks and perturbations (Adger et al., 2005; Allenby and Fink, 2005).	The ability of social or ecological system to absorb disturbances while retaining the same basic structure and ways of functioning, the capacity for self-organisation, and the capacity to adapt to stress and change. (IPCC, 2007).	Definitions are similar but the DRR definition stresses the adaptation/learning process.	For concepts and strategies there is a need to specify what type of basic structures and functions need to be maintained during the time of shocks and stresses (extreme events). Remark: resilience should not be confused with robustness or stability. Change is also an important pre-requisite for more resilience. Resilience should be distinguished from, but linked to adaptation.

contd...

Disaster risk reduction community (DRR community)	Climate change adaptation community (CCA community)	Similarities/differences	Recommendations
Response			
The provision of emergency services and public assistance during or immediately after a disaster in order to save lives, reduce health impacts, ensure public safety, and meet the basic subsistence needs of the people affected. (UN/ISDR, 2009).	Not a key term for the CCA community	The term hazard is hardly defined for CCA	Response might be a term that needs to be added into the vocabulary of the IPCC and CCA when dealing with extreme events.
Risk			
The probability of harmful consequences or expected losses (deaths, injuries, property, livelihoods, economic activity disrupted or environment damaged, etc.) resulting from interactions between natural or human-induced hazards and vulnerable conditions. Conventionally risk is expressed by the notion: risk=hazards x vulnerability. Some disciplines also include the concept of exposure to refer particularly to the physical aspects of vulnerability.	The concept of risk combines the magnitude of the impact (a specific change in a system caused by its exposure to climate change) with the probability of its occurence. (IPCC, 2007).	The IPCC definition completely neglects the vulnerability ('internal') side of the risk which is included in the DRR definition. The term risk is defined very narrowly in the CCA community and encompasses mainly the magnitude, impact and frequency as key characteristics.	The harmonisation of both definitions is crucial. A concentration of emphasis on hazard aspects within the CCA community is not helpful, since most important also for adaptation strategies is the identification, measurement and assessment of vulnerability as an important component of risk. Impacts are different from vulnerabilities CCA and IPCC should put more emphasis on vulnerability in their risk definition.

Planting deepwater rice: Is this coping or adaptation to climate change? The DRR and CAA communities should know.

adaptation imply different timescales. While coping is often short term and linked to the ability to survive and cope with the impacts of a hazardous or extreme event, i.e., eating fewer meals during a drought, the aim of adaptation strategies is to maintain the 'standard of living.' It therefore requires planned action with a long-term perspective (Birkmann and Fernando, 2008; Birkmann, 2009; Vogel and O'Brien, 2004).

Figure 14.1, underlines the differences between coping and adaptation. It also stresses that adaptation and strategies need to have different qualities than just spontaneous coping actions. Coping may involve improvisation. Adaptation is part of local knowledge. Coping refers mainly to a feedback process that is directly linked to actual or potential hazard impacts, while adaptation is determined by medium- and long-term adjustments and reorganisation processes that correspond to the notion of change (Birkmann et al., 2009).

Demistifying Terminologies and Definitions: Same Terms, Different Meanings?   **201**

FIGURE 14.1
Coping and adaptation, impact and change

```
                            Impact
             Directly related to the hazard damage and
                             losses:
                    Immediate consequences
                              ▲
                    Corresponds
                        with
      Coping          ◄─────────┼─────────►    Adaptation
    Immediate                   │             Medium- and long-term
Related hazard impacts          │           reorganisation adjustment
during the disaster or crisis   │   Corresponds
                                ▼       with
                            Change
                        Turning point-
              different development path than before
```

Source: Birkmann, 2009 and Birkmann *et al.*, 2009.

To cope with the low level of production due to climate change, more and more countries are importing rice.

3. Recommendations

The DRR and CCA communities need to address terminology as one important challenge when linking the two fields. A potential way out of the dilemma of the incongruent use of terminology might be a process-oriented understanding of terms and defining them in relation to each other. For example, the definitions of preparedness and coping capacities in the DRR community and of adaptation in the CCA community seem to have much in common. Hence, a comparison of their respective scope and meaning could lead to a more precise understanding of e.g. the differences between coping and adaptation.

This process-oriented focus could function as a bridge between climate change modelling and impact assessments on the one hand and the risk perspective of the DRR community—including hazard, vulnerability, coping and adaptation aspects—on the other. Finally, it could facilitate understanding of vulnerability and adaptation as dynamic processes in contrast to static conditions.

References

Adger, W.N., N.W. Arnell, E.L. Tompkins (2005). "Successful adaptation to climate change across scales", *Global Environmental Change* 15 (2): 77-86.

Allenby, B. and J. Fink (2005). "Toward inherently secure and resilient societies", *Science* 309: 1034-36. August.

Birkmann, J. (2009). "Regulation and coupling of society and nature in the context of natural hazards—Different theoretical approaches and conceptual frameworks and their applicability to analyse social-ecological crises phenomena", in H.G. Brauch, U. Oswald Spring, P. Kameri-Mbote, C. Mesjasz, J. Grin, B. Chourou, P. Dunay and J. Birkmann (eds.), *Coping with global environmental change, disasters and security threats, challenges, vulnerabilities and risks*. Springer.

Birkmann, J. and N. Fernando (2008). "Measuring revealed and emergent vulnerabilities on coastal communities to tsunami in Sri Lanka", *Disasters* 32(1): 82-105.

Birkmann, J., P. Buckle, J. Jäger, M. Pelling, N. Setiadi, M. Garschagen, N. Fernando and J. Kropp (2009). "Extreme events and disasters: A window of opportunity for change? Analysis of Changes, Formal and Informal Responses After Mega Disasters", in *Natural Hazards*. In press.

Birkmann, J. (2006). *Measuring vulnerability to natural hazards*. New York: United Nations University Press.

Bohle, H.-G. (2001). "Vulnerability and criticality: Perspectives from social geography", in IHDP Update 2/2001, Newsletter of the International Human Dimensions Programme on *Global Environmental Change* 1-7. Heidelberg.

Cardona, O.D., J. E. Hurtado, A.C. Chardon, A. M. Moreno, S. D. Prieto, L. S. Velasquez and G. Duque (2005). "Indicators of disaster risk and risk management", Program for Latin America and the Caribbean, Summary Report for World Conference on *Disaster Reduction, IDB/IDEA Program of Indicators for Disaster Risk Management.* National University of Colombia /Inter-American Development Bank. Available at via *http://idea.manizales.unal.edu.co/ProyectosEspeciales/adminIDEA/ CentroDocumentacion/DocDigitales/documentos/IADBIDEA%20Indicators%20-%20Summary %20Report%20for%20WCDR.pdf*

Cutter, S. (2003). "The vulnerability of science and the science of vulnerability", *Annals of the Association of American Geographers* 93(1): 1-12.

Downing, T.E. (2004). "What have we learned regarding a vulnerability science?", *Science in Support of Adaption to Climate Change.*

Intergovernmental Panel on Climate Change (IPCC) (2009). "Managing the risks of extreme events and disasters to advance climate change adaptation", *Scoping Paper - IPCC Special Report.*

―――. (2007). "Climate change 2007: Impacts, adaptation and vulnerability", *Working Group II*, Cambridge.

―――. (2001). "Climate change 2001", *Third Assessment Report*, 3 Volumes. Cambridge.

Thywissen, K. (2006). "Components of risk". (Studies of the University: Research, Counsel, Education) No. 2/2006, UNU-EHS, Bonn.

UN/ISDR (2009). "Terminology: Basic terms of disaster risk reduction 2004". Accessible at *http://www.unisdr.org/eng/library/libterminology-eng-2004.htm, 2009-05-31*

―――. (2008). "Status of Implementation of Article 4, Paragraph 8 of the Convention, Decision 5/CP.7 and Decision 1/CP.10", Submission by the United Nations International Strategy for Disaster Reduction Secretariat on behalf of the International Strategy for Disaster Reduction System to the UNFCCC Subsidiary Body for Implementation. Accessible *http://www.unisdr.org/eng/risk-reduction/ climatechange/docs/ISDR_System_ Submission_SBI_Adaptation.pdf, 2009-05-27*

―――. (2004). *Living with Risk - A global review of disaster reduction initiatives.* Geneva.

Vogel, C. and K. O'Brien (2004). "Vulnerability and global environmental change: rhetoric and reality", *Aviso* 13.

15

A Framework on Addressing Climate Change Adaptation and Vulnerabilities

Varying views about climate change vulnerability entail different problem diagnosis and cures.

Source:

Philip K. Thornton *International Livestock Research Institute (ILRI)*.

1. Introduction

The impacts of climate change spares no one. But the marginalised sectors and regions are projected to be hit the most with their limited capacities and resources to respond to climate change impacts, hence, increasing their vulnerability. A basic and urgent concern therefore is to know varying views about vulnerability as various approaches entail different diagnoses of a problem and different cures.

In this sense, two interpretations of vulnerability come into play. The 'starting point approach' looks at vulnerability as a general characteristic of human and physical systems, formed by multiple factors and processes. It maintains that vulnerability determines the adaptive capacity. While the 'end point approach' views vulnerability as a residual of climate change impacts minus adaptation. It considers climate change as the main problem and that adaptations and adaptive capacity determine vulnerability.

The basis for much activity lies in identifying and treating two types of vulnerability: biophysical vulnerability, or the sensitivity of the natural environment to an exposure to a hazard; and social vulnerability, or the sensitivity of the human environment to the exposure. An impact is then seen as being a function of hazard exposure and both types of vulnerability. The key point is the need to combine notions of biophysical and social vulnerability with understanding of the risks or hazards faced.

A more integrative approach to assessing climate change and the complex system of mitigation, adaptation and vulnerabilities is provided by Fussel and Klein (2006).

2. Adaptation policy assessment framework

Füssel and Klein (2006) present an adaptation policy assessment framework (Figure 15.1). It started off by explaining that emissions of greenhouse gases (GHG) occur, changing their levels of concentration in the earth's atmosphere leading to changes in climate. The basic assumption is the higher the level of emissions, the greater the changes in climate and climate variability. Currently, the amount of atmospheric GHG emissions is undetermined. Nevertheless, literatures tell of long-term

A Framework on Addressing Climate Change Adaptation and Vulnerabilities

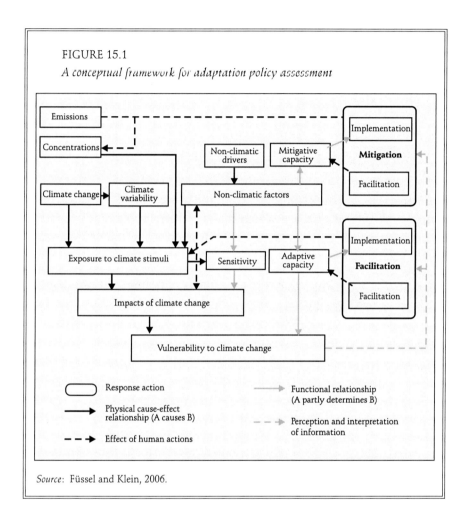

FIGURE 15.1
A conceptual framework for adaptation policy assessment

Source: Füssel and Klein, 2006.

climate impacts (e.g., gradual rises in temperature over decades, extreme climate events) and of small-scale impacts (e.g., year-to-year rainfall variability and the frequency, intensity, and location of extreme events). All these things, together with various non-climatic factors, expose places or systems to hazards or risks. Anthropogenic climate change has considerable spatial heterogeneity, so exposure to climate change depends to a large extent on location. Exposure is also determined by a wide range of non-climatic factors, which may be environmental, economic, social, demographic, technological and political.

Non-climatic factors can also affect the sensitivity of a system to climatic stimuli as well as its exposure. The sensitivity of a system refers to the usually complex 'dose-response relationship' that exists between its exposure to climatic stimuli and the resulting impacts. For example, a household that moves from one location to another purely for economic reasons may change its exposure to climatic stimuli. The distinction between changes in sensitivity and changes in exposure can become blurred.

The impacts of climate change are, thus, the consequences that arise from natural and human systems, which may be either direct (e.g. increasing heat stress on a plant) or indirect (shifts in the distribution of a livestock disease vector such as ticks).

The arrow from the impacts box to the vulnerability box indicates that the potential impacts of climate change on a particular system (in connection with its adaptive capacity) determine the vulnerability of that system to climate change—it does not imply that impacts cause vulnerability.

Füssel and Klein (2006) also suggest simplifications in the framework in as far as vulnerability should be viewed: one is that the original includes notions of scale in it (which have been omitted here for clarity) and the other is little sense of time.

But Füssel (2007) suggests that notions of external shocks and inherent coping ability can be folded into the framework (i.e., shorter term ideas of coping on a day-to-day basis, as well as longer term adapting to change), and sees coping ability as a co-factor of vulnerability rather than as an alternative definition. The concept of coping can then be integrated into the framework and lies over parts of the adaptation box, as well as the adaptive capacity box, in effect.

Once specific systems have been recognised as being vulnerable, various responses (Box 15.1) may be triggered at different levels in the hierarchy, be they personal, communal, or policy-related, for example.

Systems have adaptive capacity, which describes their ability to modify behaviour so as to cope better with changes in external conditions when they arise. Adaptive capacity and vulnerability are thus negatively correlated (the former improves as the latter declines, and vice-versa).

Many things contribute to adaptive capacity, such as economic resources, technology, information and skills, infrastructure, institutions, and equity, for example. The ability of systems to cope with current climate variability is an important indicator of their capacity to adapt to future climate change.

BOX 15.1
Two basic human responses

1. Mitigation refers to actions that limit the level and rate of climate change. Mitigation options include reducing greenhouse gas emissions directly (using bike instead of car) and reducing their concentration in the atmosphere (planting more trees to sequester carbon).

2. Adaptation refers to adjustment in natural or human systems in response to actual or expected climatic stimuli or their effects, which moderates harm or exploits beneficial opportunities. Adaptation to climate change comprises a broad range of actions (e.g. planting a variety of heat-tolerant beans).

In agriculture, mitigative and adaptive measures include agroforestry practices such as improved fallows which can sequester more carbon than can be done in croplands alone, and which are also able to maintain crop yields in poor seasons, thus increasing the resilience of the system (Verchot et al., 2007).

Two types of adaptation activities are distinguished in the framework. 'Facilitation' refers to activities that enhance adaptive capacity, such as scientific research, data collection, awareness raising, capacity building, and the establishment of institutions, information networks, and legal frameworks for action. 'Implementation' refers to activities that actually side-step adverse climate impacts by reducing exposure or sensitivity to climatic hazards (such as moving your house from a flood plain to halfway up a hillside), or by moderating relevant non-climatic factors. There are two-way arrows between adaptive capacity and adaptation: adaptive capacity determines the feasibility of particular adaptation options, and is itself influenced by actions that can facilitate adaptation.

The mitigation piece of the framework has the same structure as the adaptation piece. Mitigative capacity can be enhanced by facilitation measures, such as the establishment of a carbon trading scheme. Mitigative capacity is affected by various non-climatic factors. An example would be the effectiveness of a carbon trading scheme in reducing GHG emissions, which is to a large extent determined by the effectiveness of appropriate institutional arrangements in both the places where the carbon is being traded and where it is being sequestered.

3. Conclusion

Despite the volume of literature on vulnerability and vulnerability assessment that already exists, a considerable amount of work remains to be done on the frameworks. There is a real need for a clear nomenclature to make assessments consistent and coherent. There are also issues that need to be resolved concerning cross-scale assessment and the role of specific actors at different scales. A lot of work has been done, but there is a lot still to do.

References

Füssel, H.M. and R.J.T. Klein (2006). "Climate change vulnerability assessments: An evolution of conceptual thinking" *Climatic Change* 75: 301-29.

Füssel, H.M. (2007). "Vulnerability: A generally applicable conceptual framework for climate change research", *Global Environmental Change* 17: 155-67.

Nakicenovic, N., J. Alcamo, G. Davis, B. de Vries, J.Fenhann, S. Gaffin, K. Gregory, A. Grübler, T.Y. Jung, T. Kram, E. Lebre la Rovere, L. Michaelis, S. Mori, T. Morita,

A Framework on Addressing Climate Change Adaptation and Vulnerabilities 211

W. Pepper, H. Pitcher, L. Price, K. Riahi, A. Roehrl, H.H. Rogner, A. Sankovski, M. Schlesinger, P. Shukla, S. Smith, R. Swart, S. van Rooijen, N. Victor and Z. Dadi (2000). Emissions scenarios: A special report of the intergovernmental panel on climate change (IPCC). Cambridge: Cambridge University Press, p.509. Available online at *http://www.grida.no/climate/ipcc/ emission/index.html*

O'Brien, K., S. Eriksen, A. Schjolden and L. Nygaard (2004). "What's in a word? Conflicting interpretations of vulnerability in climate change research", *CICERO Working Paper* 2004:04. Norway, Oslo: Centre for International Climate and Environmental Research.

Thornton, P.K., P.G. Jones, T. Owiyo, R.L. Kruska, M. Herrero, P. Kristjanson, A. Notenbaert, N. Bekele, A. Omolo, V. Orindi, A. Ochieng, B. Otiende, S. Bhadwal, K. Anantram, S. Nair, V. Kumar and U. Kelkar (2006). "Mapping climate vulnerability and poverty in Africa", Report to the Department for International Development, ILRI, Nairobi, Kenya, May, p.200. Available online at *http://www.dfid.gov.uk/research/ mapping-climate.pdf*

Thornton, P.K. (2009). "Visualizing vulnerability and impacts of climate change", *Rural Poverty and Environment Working Paper* 23. Ottawa, Canada: IDRC. Available online at *http://www.idrc.ca/en/ev-144541-201-1-DO_TOPIC.html*

Verchot, L.V., M. Van Noordwijk, S. Kandji, T. Tomich, C. Ong, A. Albrecht, J. Mackensen, C. Bantilan, K.V. Anupama, C. Palm (2007). "Climate change: Linking adaptation and mitigation through agroforestry", *Mitigation and Adaptation Strategies for Global Change* 12: 901-18.

Vincent, K. (2004). "Creating an index of social vulnerability to climate change for Africa", *Working Paper* 56. Tyndall Centre for Climate Change Research.

16
Framework for Disaster-Resistant Sustainable Livelihoods

Source:

Dilhani Thiruchelvarajah *Practical Action, Sri Lanka 2010.*

This article presents a framework for policy and practice for the application of a livelihood-centred approach to disaster risk management in South Asia. The framework is aimed at disaster management and development policymakers and practitioners (government and NGO) in the region, and at international donors. It takes Department for International Development's (DFID) 'Sustainable Livelihoods Framework' as a starting point, and moves a step further to become a framework of application specific to the geophysical, socioeconomic and political considerations of the subcontinent.

1. Disaster-resistant sustainable livelihoods (DRSL)

This framework (Figure 16.1), captures the major issues for consideration in achieving sustainable disaster risk reduction and poverty reduction.

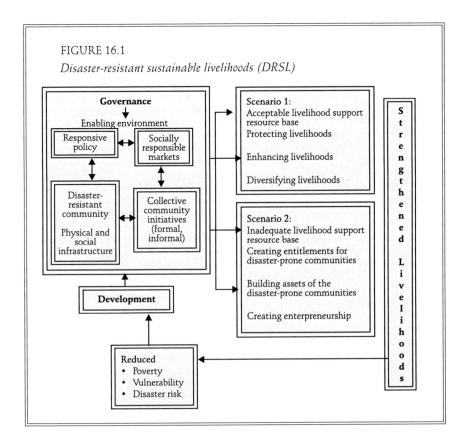

FIGURE 16.1
Disaster-resistant sustainable livelihoods (DRSL)

Framework for Disaster-Resistant Sustainable Livelihoods 215

Individuals who have adequate livelihood resource base have easy access to assets such as skills and knowledge.

The framework recognises that assets (natural, physical, financial, social and human) are the foundations of livelihood strategies and outcomes. It outlines that in South Asian countries there are two major scenarios observed in relation to livelihood assets:

1.1 First scenario

Among communities, households and individuals who have access to assets such as land, water, skills and whose assets are functional in terms of carrying out various livelihood activities, and have the potential to generate livelihood outcomes come under threat during disasters.

During disasters, the assets and livelihood outcomes come under threat. If remedial measures are not taken in time, the asset base can collapse. In such a scenario there is a need to:

- protect livelihood assets;
- strengthen livelihood assets; and
- diversify livelihood options.

1.2 Second scenario

Among communities, households and individuals who possess minimum assets, who are deprived of an adequate livelihood base and where livelihood options are too marginal to support subsistence become extremely vulnerable to disasters. In this scenario there is need to:

- create entitlements;
- build assets; and
- encourage diversified livelihood options.

Most importantly the DRSL framework notes that assets do not turn into livelihoods automatically. An enabling environment is essential for this. For instance, the availability of land and the skills possessed by an individual do not ensure household livelihood security unless the land is arable and brought under productive use by employing the requisite skills.

There are four prerequisites for creating the desired 'enabling environment' in the context of South Asia:

1.2.1 Disaster-resistant physical and social infrastructure

Physical infrastructure includes culverts, bridges, water structures, drainage channels, and roads. Social infrastructure includes knowledge/information, life-saving services, access to productive resources, marketing, and social networks.

1.2.2 Collective interest community institutions

These are formal and informal groups and networks aimed at articulating the community's interests and demanding governments' accountability. They include kinship, family, faith groups, ethnic groups, political organisations, welfare organisations, local government bodies, NGOs, and CBOs (community-based organisations).

1.2.3 Responsive governance

It should be emphasised that a mere mobilised community or a self-help group (SHG) may not be able to win back its basic rights (entitlements to assets, land rights, health, education, and other services) unless governance structures are sensitive to its needs and responsive to its demands. Governance principles, polices, and practices are the most fundamental element in ensuring an enabling environment that turns assets into livelihoods.

1.2.4 Socially responsible markets

Monopolistic and discriminatory market mechanisms negatively affect agricultural economies. The livelihoods of rural communities who are connected to the global economy are vulnerable to fluctuations in world commodity prices. When low commodity prices coincide with natural hazards, rural livelihoods come under high stress. Fluctuations can be felt directly by those who extract a livelihood from the sale of primary resources (farmers, fishermen and foresters), but also by the rural landless

Formal groups help articulate community interests and demand government accountability that will help in strengthening of livelihoods.

who are reliant on selling their labour and may be the first to suffer in an economic downturn. Therefore, market regulations in favour of agricultural economies are required to stabilise rural livelihoods.

2. Application of the DRSL framework

Application of the framework within the context of South Asia can be realised by:

a. approaching disasters through livelihoods;

b. approaching public policy through communities.

2.1 Approaching disasters through livelihoods

The Livelihood Options for Disaster Risk Reduction (LODRR) programme experience suggests a fundamental change in the way disasters are analysed and addressed. It advocates a process-based analysis, where the problem of disasters is approached with the analysis of hazards, risk, resources and asset bases, vulnerabilities and capacities within the socioeconomic context at the community level, with the engagement of the communities throughout the process. The analysis must be location and hazard-specific, providing the basis for identifying the most suitable livelihood options for given locations, hazards, resources and communities. It has the potential to reduce levels of risk and poverty by being specific and appropriate (in terms of both physical and social infrastructure), opening avenues for the expansion of communities' asset bases and diversified livelihood options, thereby expanding resilience and reducing poverty.

The LODRR programme has applied this approach in designing and implementing pilot demonstrations in 11 locations in South Asia. The lessons from the micro level are sufficiently valid and powerful to apply at a larger scale, and to call for changes at the policy level. Current policy and governance changes taking place in the region (decentralisation of governance structures, community-based poverty reduction strategies, and PRSPs) open windows of opportunity for the application of this approach.

Framework for Disaster-Resistant Sustainable Livelihoods

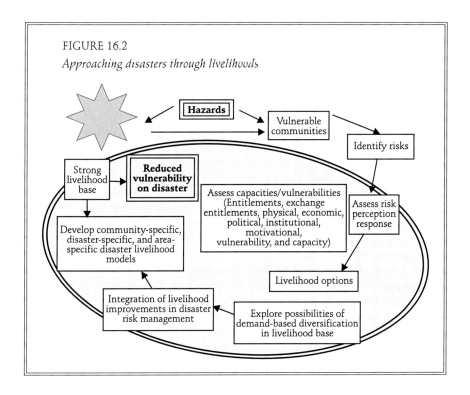

FIGURE 16.2
Approaching disasters through livelihoods

2.2 Approaching public policy through communities

The experience of the LODRR programme calls for a change in the policies and processes of disaster management. This changed scenario places livelihood issues at the centre of disaster management programmes. Given the political realities of governance patterns in South Asia (dominated by top-down administration and officials who are comfortable following familiar systems) these changes can be prompted effectively by undertaking community-specific, area-specific and hazard-specific demonstrations of livelihood-centred disaster risk reduction programmes. These demonstrations, when developed as a learning model in coordination with local governments, have the potential to generate policy response at higher levels.

The LODRR programme has also found that government institutions in South Asia are not receptive to new ideas or different approaches unless they are backed by tangible dividends. To adjust to the perceptual

Appropriate livelihoods reduce the level of risk, poverty and enhance disaster resiliency.

preferences of decision-makers, it is appropriate to start by prompting communities to identify needs, get into mobilised groups, enhance their negotiating power and build local leadership to influence public policy at the local level. The programme experience is that policy changes from above without the participation of first respondents (the community) will not bear the desired fruits and will face complications in implementation. Therefore the suggestion is that to make any substantial change in the way disasters are managed, for the 'paradigm shift' from emergency management to disaster risk management to take place in practice, policy reviews and revisions should be approached through pressure/demands from the community level.

This requires disaster management interventions to take at-risk communities as the starting point, enabling the preparation of plans which reflect the dynamics of local disaster risk, local resources, asset bases, and the livelihood scenario and then link it up with the procedural regimes of governance. Communities should be empowered and local leadership should be strengthened to take up disaster risk- poverty-livelihood issues with local, district, state/provincial, and national governments to negotiate for needed support.

Community mobilisation in disaster risk reduction, when translated into public action, makes political and government institutions more responsive and accountable and also helps to challenge and change stereotyped disaster responses from the government.

Here, the starting point is local government institutions and officials, with whom the mobilised communities can forge interactions to achieve a win-win situation. The new form of community-government liaison is mutually beneficial for both parties: for communities by getting disaster risk and poverty issues addressed, and for the local government officials as an opportunity to affirm their position through the dynamics of interactions and the positive outcomes of the investments. The rest of the upward convincing of district, province and national level institutions and policy can then be a combined effort of communities and local governments through the demonstrated evidence.

References

UNDP (2004). *Reducing disaster risk: A challenge for development*. Geneva: Bureau of Crisis Prevention and Recovery. February. p. 66.

17

Adaptation: The Context for Change

Source:

L.N. Stern *Adapted from: "The Context of Change," in Closing the Gaps: Disaster Risk Reduction and Adaptation to Climate Change in Developing Countries. Report of the Commission on Climate Change and Development, Ministry of Foreign Affairs, Sweden, 1-4 pp.*

1. Introduction

Adaptation is about capacity building and adaptive capacity increases with human development. As such, it is highly context-specific. In countries and communities where human development indicators are low, priority must be given to strengthening the adaptive capacity of both people and institutions. If capacity is already present, then actions should emphasise climate-specific adaptation measures.

The United Nations Framework Convention on Climate Change (UNFCCC) defines adaptation[1] as 'a process through which societies make themselves better able to cope with an uncertain future. Adapting to climate change entails taking the right measures to reduce the negative effects of climate change (or exploit the positive ones) by making the appropriate adjustments and changes.'

Adaptive measures seek to address climate change impacts by, for example, a new seawall, crop insurance schemes, research on heat- and

Some drivers of human development include:

- Wealth or access to assets provides buffer and serves as a back-up that enable people to recover during crises. Assets may be financial or material, directly accessible or through insurance, and may come from family and kin or through government social protection schemes.

- Health safeguards the individual's productive capacity and the integrity of families. Clean water, effective sanitation, safe childbirth, and the right kind and amount of food intake so that children can grow to their full potential, are some variables under health.

- Education gives people access to information, knowledge of their options, and the ability to make informed choices.

- Governance or the fullness of an institutional environment provides the means through which people can access resources, articulate needs, and exercise their rights.

---

1. UNFCCC (2007): 12.

Adaptation: The Context for Change

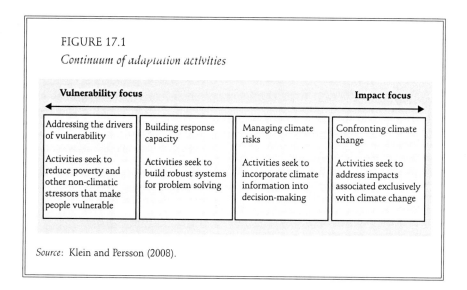

FIGURE 17.1
Continuum of adaptation activities

Vulnerability focus			Impact focus
Addressing the drivers of vulnerability	Building response capacity	Managing climate risks	Confronting climate change
Activities seek to reduce poverty and other non-climatic stressors that make people vulnerable	Activities seek to build robust systems for problem solving	Activities seek to incorporate climate information into decision-making	Activities seek to address impacts associated exclusively with climate change

Source: Klein and Persson (2008).

drought-tolerant crop varieties, agricultural diversification, vaccines, upgraded drainage systems, enhanced water use efficiency, enlarged reservoirs, or revised building codes. Building adaptive capacity aims to deal with the multiple drivers of vulnerability including poverty. Between building adaptive capacity and instituting adaptive measures, there exists a continuum of adaptation activities (Figure 17.1).[2]

2. Adaptation *vs.* mitigation

Currently, policymakers and scientists have their eyes focussed more on mitigation over climate change adaptation. Many environmentalists fear that any focus on adaptation would take attention away from the urgent need to mitigate. This trend has changed.

In late 2007, the Swedish government launched the Commission on Climate Change and Development (CCCD) to examine climate change adaptation and its links to development and disaster risk reduction. It was also tasked to issue policy recommendations on how the resilience of vulnerable communities and countries can be strengthened through official development assistance (ODA).

---

2. Ibid.

Initially, the Commission said that a rapid move toward a low-carbon global economy needs to be realised urgently, efficiently, and with great determination. Such recommendation falls under mitigation. However, mitigation—specifically, efforts to decrease the amounts of greenhouse gases (GHGs) released into the atmosphere—has yet to begin at the required scale and pace. GHG emissions have increased steadily since the UNFCCC was adopted in 1992. As such, anthropogenic emissions have already changed the climate and the long residence of GHGs in the upper atmosphere will continue to alter the climate in the next decades, if not centuries, to come. Civilisations need to adapt to these changes.

On the other hand, the Fourth Assessment Report of the Intergovernmental Panel on Climate Change (IPCC) noted that even if the global society succeeds in reducing GHG emissions, some climate change impacts are now unavoidable and solutions must be identified in order for people to adjust.

As a result, the 2007 Bali Action Plan (BAP) called for a strong and extremely ambitious step on adaptation, prioritising the urgent and immediate needs of vulnerable developing countries including the full range of risk reduction measures.

3. A sense of urgency

Since the IPCC's Fourth Assessment Report, a number of scientists and scientific organisations have published papers suggesting that climate change is happening much faster than the panel suggested. The 2009 meeting of the American Association for the Advancement of Science (AAAS) discussed how carbon emissions have grown from 0.9 per cent per year in the 1990s to 3.5 per cent per year since 2000.

The European Union has set a goal of not allowing the temperature increase to exceed 2° C. Many scientists argue that an increase of 2° C is too high because staying under that target means cutting at least 80 per cent of GHG emissions by 2050. Also, positive feedback mechanisms such as reduced albedo and melting of permafrost are likely to accelerate climate change. At this rate, the planet Earth seems to be moving toward dangerous tipping points.

> Projected impacts of climate change in various continents
>
> Africa: Analysts project that by 2020, between 75 and 250 million people will be exposed to increased water stress due to climate change. In some countries, yields from rainfed agriculture could reduce by up to 50 per cent.
>
> Asia: Freshwater availability, particularly in large river basins, is projected to decrease by the 2050s. Diarrheal diseases associated with floods and droughts are expected to increase in East, South and Southeast Asia.
>
> Latin America: Crop and livestock productivity are both expected to decrease with adverse consequences on food security. Changes in precipitation patterns and the disappearance of glaciers in the Andes are projected to significantly affect water availability for human consumption, agriculture, and energy generation.
>
> Small island states: Sea level rise threatens vital infrastructure, settlements, and facilities that support the livelihood of island communities. By mid-century, climate change is expected to reduce water resources in many small islands of the Caribbean and Pacific to the point where they become insufficient to meet the demand during low rainfall periods.
>
> *Source*: IPCC (2007).

Evidence is also growing that the absorptive capacity of carbon sinks such as oceans and terrestrial ecosystems is diminishing. Deforestation, soil erosion, overfishing, and bad management of freshwater resources have further reduced the capacity of Earth's systems to respond to future shocks.[3]

Last March 2009 in Copenhagen, 2,500 researchers from some 80 countries presented their most recent findings. Recent observations confirm that, given high rates of observed emissions, the worst-case IPCC scenario trajectories (or even worse) are being realised. For many key parameters, the climate system is already moving beyond the patterns of natural

---

3. Klein and Persson (2008).

variability within which our society and economy have developed and thrived. These parameters include global mean surface temperature, sea level rise, ocean and ice sheet dynamics, ocean acidification, and extreme climatic events. There is a significant risk that many of the trends will accelerate, leading to an increasing risk of abrupt or irreversible climatic shifts.[4]

4. A complex web of crisis

The impacts of climate change mix and overlap with the impacts of other contemporary global issues such as rising food prices, financial crisis, energy shortages, ecosystem degradation, and demographic changes.

---

4. Millennium Ecosystem Assessment (2005).

Climate change is also catalysed by the use of carbon based energy and the conversion of forests to farmland.

The persistence of these scarcities and extent of interconnectedness are still unclear. What is known is that these strong connections mean that the scarcities must be managed in a highly coordinated manner. The next-generation technologies, protective investments, and shifts in policies need to correct economic and financial imbalances that can also serve the collective purpose of protecting the environment from overexploitation. Failure on one will be failure on both.

5. A pro-poor adaptation framework

Every nation and individual must adapt to climate change. This truth is manifested by various disasters such as the 2003 heat waves in Europe and North America and Hurricane Katrina in 2005 which affected the United States. Although single events can rarely be attributed to climate change, it is evident that the poor, even in wealthy countries, are the most affected by extreme weather events. A pro-poor climate change adaptation framework must be in place because:

- climate change threatens poverty reduction and the achievement of the Millennium Development Goals (MDGs);
- poor people depend directly on endangered ecosystems and their services for their well-being;
- poor people lack the resources to adequately defend themselves or to adapt rapidly to changing circumstances; and
- poor people's voices are not sufficiently heard in international discussions, particularly in climate change negotiations.

6. A new development path

The development path under climate change will involve rethinking and reformulation on four levels: speed, scale, focus and integration.

6.1 Speed

No time should be wasted—climate change is happening faster than science predicted. The IPCC prediction that between 75 million and 250 million people in Africa will be exposed to increased water stress by 2020 allows for little more than a decade of development.

6.2 Scale

Development must decrease the vulnerability of the world's poorest, especially the 'bottom billion'. The number of people in danger of the negative impacts of climate change is growing; responses must match the scale.

6.3 Focus

Development should be centred on managing risks, building the resilience of the poorest, and enhancing the ecosystem functions. This requires focus and emphasis in all sectors, including but not limited to:

- food security, with special attention to poor people's access to markets;
- water, with special attention to equity of access and the efficient use of rainwater;
- natural resources management, with special attention to land tenure and user rights of the poor;
- energy, with special attention to a decentralised mix of renewable and low-carbon options for the poor;
- migration, with special attention to policies that benefit the most vulnerable regions and minimise transaction costs on remittances; and
- disaster risk reduction, with special attention to the most vulnerable groups and settlements and to rehabilitated infrastructure that contributes to reduced risk.

6.4 Integration

Adaptation, mitigation and human development goals are closely interrelated. Mitigation measures such as afforestation, reforestation, and avoided deforestation are also effective adaptation measures as they improve economic and ecological resilience. The concept of sustainable

development united the two concerns of environment and development; likewise, the new development path unites environment, development and climate change (adaptation and mitigation).

7. Conclusion

Climate change adaptation is about development wherein the capacity to manage risk determines progress. Adaptation is more than climate-proofing development efforts and ODA. The Commission explains that it requires action, additional funding, as well as deep cooperation between the rich and poor nations and between rich and poor people within nations. It needs sustainable development—a development which addresses the needs of the present in ways that do not compromise the ability of future generations to meet their needs.

References

Institute of Physics (IoP) (2009). "Climate change: Global risks, challenges & decisions", Synthesis Report. Key Messages from the International Scientific Congress. University of Copenhagen, 10-12 March.

Intergovernmental Panel on Climate Change (2007). "Climate change 2007", Synthesis Report. Contribution of Working Groups I, II and III to the Fourth Assessment. Geneva.

Klein, R.J.T. and Å. Persson (2008). "Financing adaptation to climate change: Issues and priorities", in H. McGray, A. Hammill, R. Bradley, E.L. Schipper and J. Parry (eds.), *Weathering the storm: Options for framing adaptation and development*. Washington, DC: World Resources Institute; Brussels: Centre for European Policy Studies.

Millennium Ecosystem Assessment (2005). "Ecosystems and human well-being", Synthesis Report. Washington, DC: Island Press. p.160.

UN Framework Convention on Climate Change (UNFCCC) (2007). *Climate change: Impacts, vulnerabilities and adaptation in developing countries*. Bonn. p.12.

18
Looking at Climate Change Adaptation from Various Realms

Varying views about climate change vulnerability entail different problem diagnosis and cures.

Source:

L.N. Stern Adapted from: "The Context of Change," in Closing the gaps: Disaster risk reduction and adaptation to climate change in developing countries. Report of the Commission on Climate Change and Development, 2009, Ministry of Foreign Affairs, Sweden.

The 2007 Commission on Climate Change and Development (CCCD) recognises that climate risk management requires appropriate adaptive capacity. As such, the adaptive capacity of businesses and ecosystems should interact and complement individual's adaptive capacity to maximise gains in the long run. This section discusses various adaptation mechanisms that may be employed in the light of climate change.

1. Adaptation mechanisms

1.1 Adaptation through market-building

Much adaptation occurs in the private sector—as evidenced by the choices, investments, and actions of individuals, households, and businesses. To adapt effectively and efficiently, businesses need sufficient assets, health, knowledge, and policy support to invest in adaptive behaviour and increase adaptive capacity. Businesses also need to be in markets that allow easy planning and execution of adaptive behaviour. The private sectors also serve as facilitators of adaptation by offering a wide range of goods and services (e.g., banking systems, agricultural risk insurance etc.) and expertise on adaptation options and resources can be marketed to improve the way people adapt.

Businesses must adapt to and try to exert some control over customers, suppliers, lenders and regulators, as well as to the various climate change effects. They must adapt to and exert some control over markets, while the markets themselves operate within the sphere of higher scales of influences and regulations and depend on these for their adaptive capacity. To the extent that businesses trade relevant goods and services to individuals and households, there is an adaptation market. Access is determined largely by price mechanisms, but state interventions can help break down problems of market segmentation and inaccessibility for the poor.

Access to information about climate variability and change is one adaptation-relevant commodity that people and businesses will invest on to help reduce the unpredictability of climate events and trends. Governments can and do provide such information. Similarly, expertise on adaptation options and resources can be marketed to improve the

Adopting farmers' best practices: The case of fondo de mitigación del riesgo agrícola (FMRA), Bolivia

Fundación PROFIN has developed an innovative, index based insurance scheme being piloted in four provinces in Bolivia. It combines incentives for risk reduction and a flexible, people-centred index mechanism. In this scheme, the trigger is based on the production levels of reference farming plots in areas that are geographically similar in terms of temperature, precipitation, humidity, and soil type. Farmers identified as good practitioners by their peers cultivate the reference plots. The scheme is based on the fact that these farmers have established reputations within their communities for their skills and knowledge. It is also assumed that the yields on their plots can serve as reliable indicators of whether production levels have been adversely affected by environmental factors (thus triggering an insurance payout) or by other factors within a farmer's control. This reduces the moral hazard in the scheme, and the reference farmers also serve as technical assistance agents to promote ideas for increasing yields and reducing disaster risks and impacts. The system encourages other farmers to match the reference farmers in implementing efforts to reduce the effects of drought, excess rains, hailstorms, and frost because those farmers run the risk that their own plots will be significantly affected while the reference farmers' plots will be less affected.

Source: Fondo de Mitigación del Riesgo Agrícola, at *www.fundacion-profin.org/fmra.html*

>
> A win-win relationship for the ecosystems and the poor: The case of Tigray province, northern Ethiopia
>
> Collaboration between local farmers and experts on a farming-with-nature project led to higher yields and groundwater levels while increasing household income and livelihood opportunities for women.
>
> The project involved using compost to increase yields, selecting a diversity of wild plant species to decrease the need for fertilisers, making trench bunds to hold water and reduce soil erosion.
>
> *Source*: Lundberg and Moberg (2008).

way people adapt. However, the products most useful for adaptation by the poor are often not provided in forms they can afford or obtain. This differential access to insurance, credit and technologies accentuates the 'adaptation gap' between the poor and the better-off. In some cases, products are widely available but need to be tailored to the specific needs and limited means of the poor. This could mean a major redesign of both the products and their markets. In other cases, new products to support the poorest may need to be developed.

1.2 Adaptation through ecosystem protection

Ecological systems are highly resilient. However, tipping points occur when the cumulative effects of both slow and fast environmental changes and disturbances reach the ecosystem's maximum thresholds. This then results in dramatic and often rapid negative changes in ecological systems.

Small events such as droughts, floods, or pest outbreaks can trigger ecological changes that are difficult or even impossible to reverse. With the onset of climate change, ecosystems that have co-evolved with the previous climate are likely to change. If these changes bring rapid losses of ecosystem services, human adaptation options are reduced. However, the people can help protect and improve local ecosystems in ways that also support livelihoods and adaptive capacities.

Increasing the flow of ecosystems services could help disadvantaged groups deal with future impacts of climate change. These strategies can lead to risk reduction and can also contribute to attempts to promote a transition to sustainable poverty alleviation in rural communities. However, the 'win-win' relationship between investments in ecosystem services and improved livelihoods of the poor is not cheap, quick, easy, or inevitable. The rich can capture the benefits of these efforts, and in some cases the outcomes have not lived up to expectations.

Coping strategies can also create 'lose-lose' situations. The poorest of the poor, who are unable to grow traditional crops, are often forced to extract too much resource from the ecosystems for food, fiber, and fuel wood. This accelerates ecosystem degradation and in the long run, the qualitative changes in the ecosystem leave them worse off.

The indirect impacts of climate change can lead to acts that rapidly undermine the resource base of local communities. These include extraction of non-renewable fossil groundwater, rapid land use change as a result of investments in biofuels, or increased fishing pressure as a response to changing market prices.

Efforts to conserve ecosystems while adapting to climate change often also involve trade-offs, as highlighted by research at the World Resources Institute (WRI). Managing the trade-offs and effectively using adaptation strategies to realise both livelihood and ecosystem objectives require an understanding of factors that undermine resilience such as biodiversity loss, habitat fragmentation, and a range of management and governance issues. Social sources of resilience including diverse and effective institutions, social capital, and ecological understanding are also critical assets under consideration.

1.3 Adaptation through social protection

Social protection refers to a broad set of policies and programmes to reduce poverty and vulnerability by diminishing people's exposure to risks, encouraging effective labour markets, and building people's capacity to protect themselves against hazards and loss of income. It entails direct and predictable transfers of resources to the poor. At present, social transfer policies tend to ignore the long-term risks associated with climate change. Studies in three countries namely Mexico, Brazil and Ethiopia revealed that state-managed social transfer programmes improve the adaptive capacity of the poor. Social transfers are used to scale up access to and demand for equitable health and education services. The problem now is that social transfers need significant amounts of additional revenue; the transfers must also be long term and predictable.

Governments and donors need to assure policy coherence between social transfer programmes, wider service sector initiatives, and climate adaptation plans within and among sectors (health, education, social welfare and agriculture). Managing social transfers to improve climate adaptive capacity will require institutional strengthening that may call for a parallel technical assistance programme.

2. From adaptive capacity to capacity development

Institutions at local, national and international levels shape adaptation. Adaptive capacity at the local scale depends on the capacity for planned

adaptation at wider scales. To improve people's adaptive capacity, three main institutional ingredients are necessary: capacity development, inclusive governance, and ownership.

2.1 Capacity development

Capacity development goes beyond the technical cooperation and training approaches associated with 'capacity building.' Public, private, and civil society organisations will never become sustainable or responsive if they are created solely to implement a project. Generally, stand-alone training projects have had little discernible or sustainable impact on capacity development.

Top-down 'capacity building' efforts which merely tries to transfer the 'right answers' to the people were the common trend used in addressing a certain issue. The Paris Declaration on Aid Effectiveness (2005) attempts to move away from this paradigm by stressing that a nationally and locally-owned capacity development process should be at the core of any sustainable change effort. Sadly, this has not been reflected in most of recommendations for climate change adaptation.

2.2 Inclusive governance

As an adaptation tool, inclusive governance can help reduce vulnerability but it is highly dependent on many factors, which include capacities to do the following:

a. Providing leadership

In some cases, the responsibilities of the local governments increase without corresponding human or financial resources. Efforts to decentralise responsibilities for adaptation need to reflect inevitable human resource gaps, lags in organisational and institutional reform, and unfulfilled promises of financial devolution. The principle of subsidiarity is essential, in that local actors and local governance are the most important basis for action as guided by existing conditions in the area. Adaptation efforts must be based on the opportunities and constraints facing local governments and their relations with line ministries. There is a need to understand how climate change adaptation could be better managed through decentralised structures by suggesting new priorities for research, evaluation, and public administration reform.

b. Actively engaging civil societies

Civil society organisations (CSOs) act as watchdogs to see that rules concerning a certain issue (e.g. climate change adaptation) are enforced. They can also mobilise populations to act accordingly and some even take on significant roles in service provision. They help fill the gaps created by a weak local government. Local-national-global networks of civil society organisations can help address the gaps between national and local governments by creating multi-stakeholder platforms for debate about how to manage climate risks.

c. Providing access to important services

Access to information is very crucial considering that climate change poses a lot of uncertainty. People need the right information to make their own informed decisions. Often, national policies and commitments to reduce vulnerability are not transferred to the local level. Aid can help overcome some of these obstacles and has been effective in producing risk maps, initiating participatory planning, and building basic human resources. The greater challenge is in establishing effective links between national climate change efforts and the municipalities, districts and provinces that must move from words to actions.

d. Mobilising a dynamic business environment

Businesses serve as engines for development by providing jobs, tax revenues, and technology. It is the biggest investor in areas such as technology transfer. Adaptation efforts of businesses are mostly directed on climate-proofing factories, stores, and field operations. The business community must be present in forums for planning and decision-making. It must be noted, however, that the risks they address are not merely profit-oriented but also contribute to risk reduction at the individual level.

3. Ownership

Good governance for adaptation is ultimately manifested in the ownership of policy objectives. Ownership relies on coherence among formal regulatory frameworks and central investment strategies (among international and donor agencies) *vis-à-vis* local plans and processes. It is a precondition for long-term sustainability but not a guarantee. The design of National Adaptation Programmes of Action (NAPAs) by least-developed countries best illustrates this. It shows that there is a missing link between short political cycles and the need for investments in long-term risk reduction.

However, another concern that needs to be addressed is the task of bridging the local-national divide which continues to be a big challenge and the prevalence of the top-down effort in some climate change adaptation agenda.

> *Bringing together ownership and sustainability: The NAPA experience*
>
> National Adaptation Programmes of Action (NAPAs) are nationally owned but they are rarely integrated into national planning and budgetary processes. They have not received significant levels of financing because they are seen as exemplifying the kind of 'projectisation' that the Paris Declaration seeks to eliminate and replace with programmatic approaches. Furthermore, the little funding they have received has largely been outside of the mainstream aid process, which also contributes to limited integration.
>
> The NAPAs and many other small adaptation and disaster risk reduction initiatives, particularly those promoted by civil society, are a first step towards local learning about how to manage climate change. They can create concrete, on-the-ground examples of actions to address what is widely seen as a rather abstract set of hazards. Ultimately, the NAPAs are more about capacity development than they are about furthering adaptation.
>
> Without this first step, there will be great difficulties in taking a far larger second step, since the modest levels of ownership that they represent will be extinguished. Support for initiatives such as the NAPAs must then be situated within a much more comprehensive dialogue on adaptation. This means that although these 'pilot projects' may not actually be scaled up, they should be used as capacity development exercises and as a platform for establishing a broad dialogue about what climate change adaptation implies.

In this scenario, local actors will have no power to design their own strategies in climate change adaptation. Opportunities for local people to take ownership of their own climate change agenda will not be integrated into their own development strategies.

Development communities tend to perceive that more emphasis is given to macro-level technical models compared with on-the-ground realities in terms of climate change adaptation. Thus, there is a need to devise and use monitoring and evaluation (M&E) techniques to determine whether people are managing the environment in prescribed ways. Adaptation requires a learning approach combined with systems by which affected populations can begin to hold their political leadership (and aid donors) to account for whether vulnerability has been reduced. As such, three principles should guide in monitoring and evaluation of adaptation efforts: impartiality, independence, credibility and usefulness.

References

Lundberg, J. and F. Moberg (2008). *Ecological in Ethiopia - Farming with nature increases profitability and reduces vulnerability*. Stockholm: Swedish Society for Nature Conservation.

Geoghegan, T., J. Ayers and S. Anderson (forthcoming). "An assessment of channels to support climate adaptation by the poorest", *Working Paper*. London: International Institute for Environment and Development.

19

Assessing Resilience and Vulnerability: Principles, Strategies and Actions

Successful management of the hazards, risks, impacts and consequences is not possible without community commitment and involvement.

Source:

Philip Buckle, Graham Marsh and Sydney Smale Adapted from: *Assessing Resilience and Vulnerability: Principles, Strategies and Actions. Guidelines prepared for Emergency Management, Australia.*

Understanding resilience and vulnerability is a key element of effective disaster management. Resilience and vulnerability are interactive with each other across social levels and across space and time (Figue 19.1). In particular, a resilience and vulnerability assessment should be able to:

1. Identify the strengths of particular areas, communities or groups, in terms of resources, skills, networks and community agencies. These strengths and local capabilities may be used and further developed to minimise the negative consequences of an emergency. Prevention and preparedness activities, as well as recovery activities can be supported.

2. Identify vulnerabilities of particular areas, communities or groups, so that these can be managed in terms of prevention and preparedness activities, response activities and recovery programmes. By identifying risk and vulnerabilities prior to an event, local managers will have the opportunity to plan to avoid or to minimise the negative consequences of emergencies and disasters.

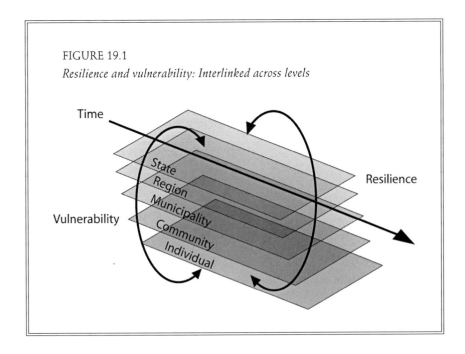

FIGURE 19.1
Resilience and vulnerability: Interlinked across levels

Figure 19.2 shows the elements of resilience and vulnerability.

1. Resilience and vulnerability: Principles and issues

When considering individual, groups, and community, there is a set of principles that support resilience and reduce vulnerability. Successful management of the hazards, risks, impacts and consequences is not possible without community commitment and involvement.

The affected community will expect to contribute to their own preparedness and response capability, and especially their own recovery. If denied this opportunity, they may establish their own structures and processes to achieve that end.

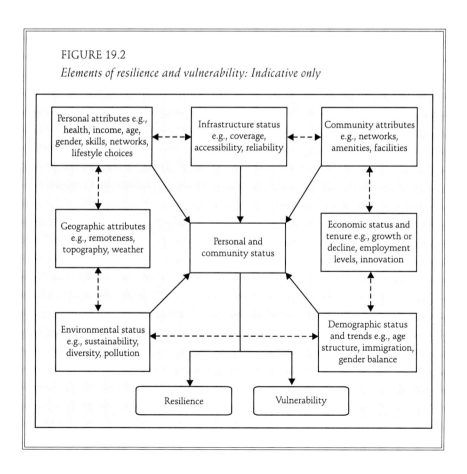

FIGURE 19.2
Elements of resilience and vulnerability: Indicative only

1.2 Information and knowledge

Access to the following information is needed before proceeding to a vulnerable community:

- risks and appropriate protective action;
- support measures, eligibility conditions, and application procedures;
- bio-psychosocial reactions that can be expected and coping mechanisms to these reactions in themselves; and
- problem's cause and its implications to the lives of the people.

1.3 Resources

- resources, advice, expertise, personnel, goods, and funds available to support mitigation and safety measures;
- financial assistance for eligible parties to help restore losses such as grants, loans, and insurance, where appropriate; and
- physical goods, such as temporary accommodation, essential household items, temporary public transport, tools and other items.

1.4 Management capacity

- time and opportunity to undertake recovery activities;
- physical capacity, which may include the support of other people, machinery or support where there is a particular need;
- access to services through establishing transport systems, locating service centers close to affected areas or access to translators, interpreters or other language and media services; and
- expertise access to specialist services, such as tradesmen, financial counsellors, and other professional services.

1.5 Support

- personal support such as outreach services, personal advisers and counsellors, specialist support services, advocates and gatekeepers; and

- community support like the appointment of community development officers.

1.6 Participation

- consultation in developing and implementing assistance and recovery programmes;
- encouragement in making a contribution to policy and programme development; and
- engagement in monitoring and auditing the progress of recovery.

2. Resilience and vulnerability assessment basics

- Resilience and vulnerability assessment is a necessary component of effective emergency management planning. However, it is unlikely that any assessment, or community audit, will capture every potential need or identify every person who, in some circumstance, may be exposed to a risk or to the possibility of some loss.
- After an emergency, it will be necessary to scan the affected area, through information campaigns, outreach programmes, letter box drops and other methods, to identify people who require assistance.
- Resilience and vulnerability analysis need to be conducted with sensitivity and proper regard to people's privacy and their right not to disclose information. There are also legal and other requirements to maintain proper standards of confidentiality when dealing with information from the public.
- Each event is unique and will generate its own set of vulnerabilities. Each individual, family and community is different and may be vulnerable, or have resilience in different ways.
- Conducting a resilience and vulnerability analysis is not an end in itself. The purpose is to highlight issues, needs and concerns and to work to effect change—to improve resilience and/or to reduce vulnerability.

- As part of this emergency management activity, one should look at the renewal and development possibilities in an area. This applies particularly after an event, when it is important to move forward, rather than to simply try to repair the damage. Recovery may offer developmental opportunities that would otherwise not occur.

3. Setting a framework for assessment

When beginning a process to assess resilience and vulnerability, it is important to establish a framework for the assessment. A first step in this will be to decide upon the elements to be examined. Typically, these will include at least the following:

3.1 Impact of hazard

Although many hazard impacts result in similar consequences, such as injury and the loss of residence, there may be differences between them. Bushfires, for example, usually offer less warning time than floods and may totally destroy houses. Floods, however, cause damage but are unlikely to result in the total loss of a residence.

3.2 Locality

Each locality has unique characteristics and may have vulnerabilities and strengths in a different combination to others. It is therefore important to clearly define the area being assessed.

3.3 Scale

The size of the area chosen is also important. A large area may combine smaller areas that have little in common and will therefore skew results if aggregated data is used. Scale also applies to the level of aggregation area—that is, the extent to which it may be useful to generalise about an area.

3.4 Demographics

An analysis may be undertaken for a particular group within a given area. The results of this analysis will be useful only for that group of people.

4. Resilience and vulnerability assessment approaches

4.1 Functional or management focus

This identifies the functional characteristics of an entity that can be dealt with in a practical way and manages both resilience and vulnerability so that the former does not diminish and the latter does not increase (Figure 19.3).

4.2 Demographic analysis

The method seeks to identify classes of people at risk. Broad, but useful for developing generic policies and programmes. More specific assessment and development applications are still required.

4.3 Hazard analysis

This assesses the impacts of particular types of hazards. Useful in assessing risk at a location but it usually focusses on the hazard rather than its impacts on people and its social and economic consequences.

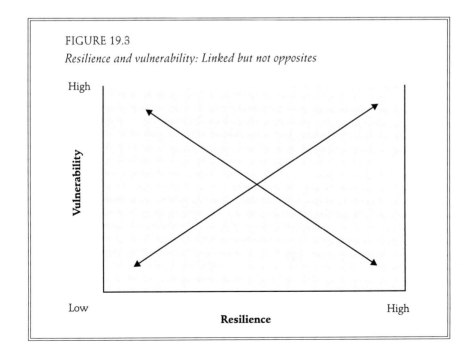

FIGURE 19.3
Resilience and vulnerability: Linked but not opposites

4.4 Area analysis

This approach, not often used, looks at the whole range of risks and hazards for an area. This sometimes combines disparate elements and can be complex and hard to translate into effective policy and programmes but is gives a idea of the risks faced in a jurisdiction.

4.5 Time scale analysis

This approach, using some of the methods mentioned above, analyses and describes risks, resilience and vulnerability as they unfold and develop over time. A variant of this is to look at the risks that attach to particular recurrent periods, such as climatic seasons or drought.

4.6 Needs analysis

This approach can provide useful information linking risks, consequences and significant needs such as medical care, food, water, shelter, income security, etc.

4.7 Political economy

This approach looks at the broad trends (not at particular issues or actions) in the political and economic structure, life and values of a society. This is useful for identifying some contextual issues.

4.8 Capability analysis

This approach, similar to our functional approach, looks at capacity (usually of communities) to provide support and services to members of the community.

Elements from all these approaches can be used to clarify issues and strategies to reduce vulnerability and to increase resilience.

5. Elements of a good resilience and vulnerability approach

The basis of any approach or method should include:

- Clarification of objectives and targeted outcomes.
- Development of a framework of analysis and understanding.

- Indication of data and information requirements.
- Identification of personal and community needs.
- Identification of developing policy options.
- Indication of appropriate strategies and options, and
- Assistance in developing service and programme standards and audit criteria.

> This material is part of a document developed from an earlier set of guidelines prepared by the Department of Human Services, Victoria. It was prepared following reviews of some major emergencies including the 1997 Dandenong Ranges Bushfires, the 1998 East Gippsland Floods and the 1998 Victorian Gas Crisis. Members of the department, Victoria State Emergency Service, local government, specialist practitioners and others were involved in their preparation. It came from an Emergency Management Australia (EMA) funded project which examined resilience and vulnerability, particularly for individuals, groups and communities. A wide range of people, representing numerous distinct communities as well as a range of local, government and emergency management agencies were involved in an extended interview, discussion and consultation process for this project.

20

Analytical and Assessment Tools and Methods

Focus group discussion helps explore particular issues and problems and reveals attitudes of people towards them.

Source:

Philip Buckle, Graham Marsh and Sydney Smale Adapted from: Assessing Resilience and Vulnerability: Principles, Strategies and Actions - Guidelines Emergency Management Australia. May 2001.

1. Introduction

There is a range of techniques and methods that may be used to identify problems and solutions pertinent to resilience and vulnerabilty. These methods are not exclusive and in many cases complement each other. All of them have their uses. All of them have limitations. A key limitations to assessment is the nature of change. Change, whether as desired outcomes or undesired results, is rarely linear and is often driven by complex forces. Prediction is a hazardous task. Rather than prediction, it is safer and often more productive to aim for a general situational assessment of what may occur in the future. For example, it is clear that the proportion of older people in the population is increasing. What this means, however, for health care, recreation, family relationships, household size, and social welfare support is much less clear.

2. Brainstorming

This method is applied most often as a group exercise where group members think imaginatively and without restrictions on issues, problems, and resources. The results will usually need to be ordered and assessed and evaluated for their applicability. This approach is useful for developing creative and unexpected solutions, for identifying otherwise unseen issues.

3. Cost benefit analysis

This method typically relies on the measuring of financial and economic cost and benefits of a particular activity. Attempts are made to identify costs and benefits and these are then weighed against each other.

This method's strengths are in its systematic approach and in its efforts to attribute a common unit of measurement to items being considered. Its limitations lie on things that cannot be assessed in terms of economic cost.

4. Delphi techniques

This approach uses the skills, knowledge and experience of people with acknowledged expertise to assess issues and solutions. Cross- checking occurs through a review process. This technique may be applied as a group process, commentary from various sources or on iterative process of exchange and comment. This approach is useful for drawing upon detailed knowledge and skilled capacity.

5. Environmental impact assessment

An assessment using various techniques and methods of enquiry to estimate the impacts, both positive and negative, of development or other activity on environmental systems and species.

6. Focus groups discussions and workshops

Focus groups discussions are meetings of selected people (or people selected from specific sections of the community) who, under the management of a facilitator, explore a particular issue in a structured or semistructured way to reveal values, attitudes or behaviour.

7. Mapping

Data of virtually any sort can be mapped, especially using modern geographic information systems. Most entities can be mapped, from demographic data to values and attitudes. Mapping is useful as a visualisation tool. It allows the reader to easily see differences between areas and allows trends to be identified.

8. Profiling

Profiles may be generated of demographic, sociocultural, economic or environmental issues. A profile gives a snapshot for a particular area, community or group of people at a particular point in time.

9. Scanning and scoping

These are alternative terms for broad brush viewing of the subject in which one is interested—view the context rather than the detail to identify broad structures or trends. It relies to a degree on intuition expertise and sensitivity to contextual cues.

10. Social impact assessment

An asessment, using various techniques and methods of enquiry to estimate the impacts, both positive and negative, of development or other activity on social structure and life.

11. Surveys, questionnaires and interviews

Surveys and questionnaires are tools used to ascertain perceived facts, opinions and values or types and levels of activity. They may collect useful data that needs to be collated and interpreted. Response rate may be low. Interviews may be structured or semi-structured which allows the researcher to explore particular issues in detail.

12. Trends extrapolation

Trends extrapolation requires the identification of existing trends or processes or directions of change and then extending these into the future. Date sources will usually be quantitative (say demographic data) or systematically listed qualitative data say responses to standard questionaires or surveys. This method is useful for projecting into the future and estimating, on local or broad scales, what changes or develoments might be expected.

13. Cross-impact analysis

The outcome of processes such as Delphi techniques may be a list of possible futures or possible consequences. Cross-impact analysis is a technique for assessing the consequences for each identified outcome on each other. This may be conducted through a cross-impact analysis matrix. This may be done either using probability ratings where these are available or by using qualitative estimates.

14. Scenario analysis and planning

Much planning, hazards analysis and risk assessment, even much social and demographic analysis, has been undertaken on the basis of a 'snapshot' approach. Basically this entails assessing a situation as it 'appears' at a given point in time. This approach is relatively fast, easy, cheap, and can easily be conducted. For these reasons, it is a useful audit tool at critical times. It is deficient for longer term planning and for more detailed and deeper understanding of these things that create risk, generate vulnerability and work for or against achieving increased resilience. This method also relies heavily on 'expert intution'; local or expert knowledge of existing conditions. Both forms of knowledge is maintained and is accesible in such circumstances however, such knowledge may be diminished though personnel leaving the area or through the death or illness of those with such knowledge. Mechanisms should be done in place to ensure such knowledge is maintained and is accessible in such circumstances.

However, there are many features of social life, environmental processes, organisational behaviour, human interaction, and social processes and development that are counter-intuitive or not easily revealed.

Scenario analysis is a technique for delving more deeply and more intricately into complex and changing situations. It requires establishing first an issue to be explored, whether it is the impacts of a particular hazard or the range of impacts of a particular hazard or the range of impacts on an area.

Secondly, establishing of the exploratory framework is for the area to be investigated, prevailing social and economic trends, the time frame to be considered and other broad scale issues. Then a group of participants must be gathered. The skills, interests and responsibilities of the participants will be determined in part by the purpose and desired outcomes of the scenario exercise. A facilitator or exercise director should be selected and briefed and chosen for their standing and expertise.

The scenario is then run according to the framework previously agreed. It requires the participants to (be encouraged) to think imaginatively and boldly to uncover all relevant, but often unexpected, issues and factors. The role of the exercise director is critical in being able to encourage imaginative approaches to analysis, review and assessment. The scenario provides an opportunity for people to consider an issue imaginatively and critically and to feed from the ideas and propositions and insights generated by other participants.

15. Conclusion

Perhaps all we can say with confidence is that change will occur, while acknowledging that planning has to occur as rigorously and comprehensively as possible. All assessment, being more or less estimates, need to be monitored and updated regularly. Finally, assessment, predictions and projections are not deterministic. Within limits, we have the capacity to change or modify expected outcomes to more closely meet government and social objectives.

21
Participatory Community Risk Assessment

In PRA, data collection and analysis are done by the local people, with outsiders facilitating rather than controlling.

Source:

Adapted from: *Community as First Responder: Building CBDM Capacity at a District Level.* SEEDS, New Delhi, India.

1. Introduction

Community based disaster management (CBDM) is a process where a community at risk are actively engaged in the identification, analysis, treatment, monitoring and evaluation of disaster risks. Tools are essential to aid this process. This article focusses on tools for assessment and related issues.

2. Tools for risk assessment

Participatory community risk assessment (PCRA) is the core of the disaster risk reduction (DRR) process. Risk assessment needs an understanding between the community, the implementing organisation, and the field practitioners. For carrying out the risk assessment process, certain tools can be used. These are termed as the participatory rural appraisal or PRA tools.

PRA is a methodology which has gained prominence due to the paradigm shift from top-bottom approach to bottom-up approach. The vulnerable communities and groups are the main actors in disaster management. They have knowledge about their locality, history of disaster in their place and how disasters have affected them over time. For collecting this information PRA methods are used. It is a means of collecting different kinds of data, mobilising intended communities and evoking their participation, and opening ways in which these communities can participate in decision-making. The innovative nature of PRA has helped people to express themselves and expose various dimensions of their lives. In PRA local people are not seen as beneficiaries, but as partners in the research and development (R&D) process. In PRA, data collection and analysis are undertaken by local people, with outsiders facilitating rather than controlling.

PRA has been applied in every domain of development and community action, both urban and rural. PRA techniques are equally applicable in urban settings but it has been felt that urban areas need a different approach because of their complex nature. Though being utilised in a variety of settings, their significance is highlighted more in rural areas.

The purpose of PRA is to enable development practitioners, government officials, and local people to work together and plan appropriately according to the needs of the local people. The convenient entry points for conducting PRA exercise are the local community leaders like, the *sarpanch* or the youth. However, one must be sensitive enough to include people from every section of the community including the women and the aged. Also, different set of people might have different outlook towards an issue, like, procuring water might be the foremost problem for women but for the male livelihood tops the agenda. Therefore, diverse viewpoints must be incorporated while conducting PRA.

PRA techniques can be combined in a variety of ways, depending on the topic under study. While some of the tools are generic which can be used at the initial stage of an intervention, certain tools are more specific which are used during the implementation of CBDM. Before implementing the direct PRA methods in the field certain methods, like reviewing the secondary data, may be used to support the learning process.

2.1 Review of secondary data

- It includes collection and review of existing data and information relevant to the area.
- Review of secondary data is useful to get an initial picture of the situation of the target community both in socioeconomic and institutional context.
- Secondary data can be processed in two stages:
  i. Identification and compilation of the material;
  ii. Analysis of the collected information.

2.2 Direct observation

- It concerns obtaining first hand information by systematically and directly observing the events. For example, relationship between the people of a community.
- Observations are analysed afterwards for patterns and trends.
- Hypothesis arising out of verbal information should be cross-checked.

- This tool can be used strategically in any phase of the implementation of CBDM.

Given are the few direct PRA methods which can be modified for developed into new tools.

2.3 Semi-structured interviews

- Semi-structured interviews are carried out in an informal and conversational way.
- The individual interviews and group interviews form the part of semi-structured interviews. It is advisable to take not more than an hour for an individual interview and not more than two hours for a group interview.
- The interviews should be structured in advance.
- It is important to include women as respondents and take their perspective on issues.
- Semi-structured interviews can be conducted at individual level, houshold level or community level.
  - *Individual or key informant interviews*: These are interviews with selected individuals who have long experience with community or specialised knowledge.
  - *Household interviews*: Households may be selected from each socio-economic category in the population and interviews can be conducted in these specially selected houses.
  - *Focus group interviews*: Focus group interviews are conducted with a special focus on a group of people (6 to 10 in number) who have certain factors in common. This helps to get specific views and perspectives on certain issues. Focus group discussions are conducted based on age group, owners of specific resources, people involved in a specific role or activity.

2.4 Mapping

To start with the process, one should obtain a base map of the region from the local authorities. Also try to get satellite images and maps of the particular area. To enrich the maps with detailed attributes Geographical Information System (GIS) mapping can be used. Along with these maps, the community must also be involved in drafting a map of their own area. Mapping helps in visually representing the physical attributes and various resources in the community. The maps are of different kind like hazard maps, social maps, resource maps, vulnerabilty maps, etc. Community and resource maps can be combined with a hazard map to come up with a local vulnerability map where one can easily identify settlements, resources and infrastructure threatened by a certain hazard (Figure 21.1).

2.5 Activity calendars

An activity calendar is useful in getting insights into the type of activities implemented by a community or household in a day, month, season or a year. It helps to understand the problems related to the activities

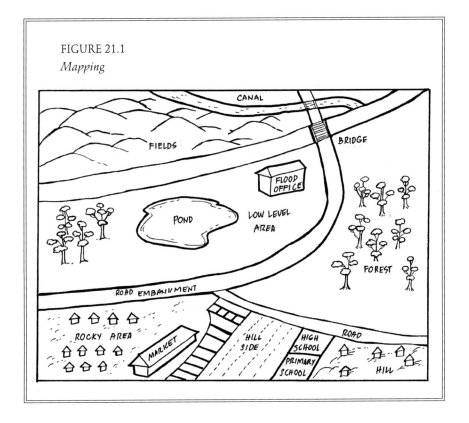

FIGURE 21.1
Mapping

performed and a comparison can also give leads on changing patterns and trends. Activity calendars can be made for any period of time. The most useful are the daily activity pattern and the seasonal activity calendar.

2.5.1 Daily activity profile

The male and female members of the community are separately asked in chronological order about their usual activities during a day, the duration of these activities and the location of where the activity is being implemented. Activities should be inclusive of productive, reproductive and sociocultural aspects. The purpose is to understand the working pattern, peak work loads and other problems related to the activities. Also, the daily activity profile gives a clear idea of the time which may be utilised for the proposed programme activities.

2.5.2 Seasonal calendars

The seasonal calendars help to visualise the timing and duration of hazards across the year. Seasonal calendars depict various physical and socioeconomic phenomena occurring in various months. It shows the main activities, problems and opportunities throughout the annual cycle and help identify the months of greatest vulnerability, difficulty and variations.

2.6 Matrix ranking/scoring

Matrix ranking is used to find out the degree to which different hazards affect people, property, community resources, infrastructure, and other element of the community. A rank or score is given to disasters to find out which disaster has the most or least effect on the community.

2.7 Transect walk

Transect walk refers to walking systematically with the people of the community through the area and discussing various aspects of specific locations. The discussion can cover the area's physical environment, land uses, amenities and the needs and problems related to them. It also helps to identify danger zones, safe areas, natural resources and land used patterns. Transect walk is usually done during the initial phase of an invention.

Participatory Community Risk Assessment

FIGURE 21.2
Venn diagram

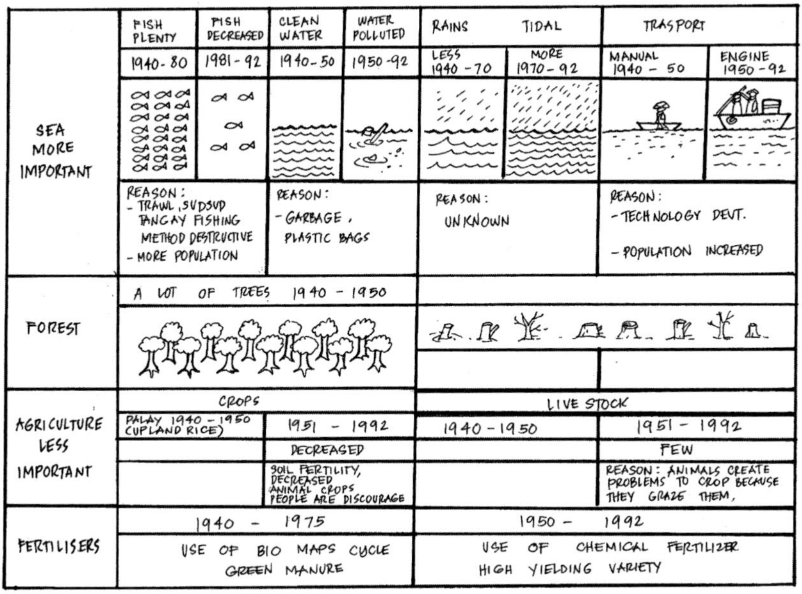

	FISH PLENTY	FISH DECREASED	CLEAN WATER	WATER POLLUTED	RAINS	TIDAL	TRASPORT	
	1940-80	1981-92	1940-50	1950-92	LESS 1940-70	MORE 1970-92	MANUAL 1940-50	ENGINE 1950-92
SEA MORE IMPORTANT	⋈⋈⋈ ⋈⋈⋈ ⋈⋈⋈ ⋈⋈⋈ ⋈⋈⋈	⋈ ⋈ ⋈ ⋈ ⋈	~~~~ ~~~~	~~~~ ~~~~	,,, ,,,	//// ////	🧍	🚤
	REASON: - TRAWL, SUDSUD TANGAY FISHING METHOD DESTRUCTIVE - MORE POPULATION		REASON: - GARBAGE, PLASTIC BAGS		REASON: UNKNOWN		REASON: - TECHNOLOGY DEVT. - POPULATION INCREASED	
FOREST	A LOT OF TREES 1940 - 1950 🌳🌳🌳🌳🌳🌳				🌲 🌲 🌳 🌲 🌳 🌲 🌳 🌲			
AGRICULTURE LESS IMPORTANT	CROPS				LIVE STOCK			
	PALAY 1940-1950 (UPLAND RICE)	1951 - 1992			1940 - 1950		1951 - 1992	
		DECREASED					FEW	
		SOIL FERTILITY, DECREASED ANIMAL CROPS PEOPLE ARE DISCOURAGE					REASON: ANIMALS CREATE PROBLEMS TO CROP BECAUSE THEY GRAZE THEM.	
FERTILISERS	1940 - 1975				1950 - 1992			
	USE OF BIO MAPS CYCLE GREEN MANURE				USE OF CHEMICAL FERTILIZER HIGH YIELDING VARIETY			

2.8 Venn diagram

Venn diagrams (Figure 21.2) or chapati diagrams are used to generate insights on the relative importance that local and external institutions have in the community. The participants use their own criteria to determine effective and ineffective services. Circles of different sizes are drawn which depict the importance of the issues. The diagram also indicates the relationship of the community with different institutions, denoted by distance between the circles.

2.9 Historical profile

In a historical profile, the villagers provide an account of how different aspects of a particular area have changed over the years. It helps a great deal in revealing the changes and trends that have come about over a time, especially in terms of the intensity and number of disasters. Historical profile may be measured through the methods of historical transect and time line.

2.10 Livelihood analysis

This helps in learning about people's lives and the intricacies of the economic structure prevailing in the community. It depicts the different livelihood pattern of a particular area, how they are affected by disasters and what are the coping strategies of the people (Figure 21.3).

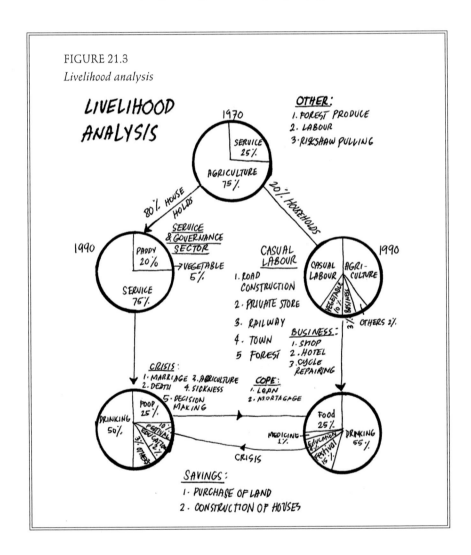

FIGURE 21.3
Livelihood analysis

With the use of these tools, the process of risk assessment is carried out. Risk assessment integrates the results of scientific knowledge, secondary data and community peceptions. The assessment process is carried out by first, assessing the capacities and vulnerabilities; second, stakeholder analysis; and third, by identifying and coordinating with community based organisations.

3. Methodology for risk assessment

Community risk assessment can be carried out as a prelude to all programmes and activities. It uses the various participatory tools to get an understanding of people's exposure to natural hazards at the grassroot level. It is also called vulnerability and capacity assessment (VCA).

VCA is concerned with collecting, analysing and systematising information in a structured way on a given community's vulnerability to hazard. VCA analyses the factors that generate the vulnerability of the community members and looks into the root cause of these. These vulnerabilities can be categorised into physical, social and economic. Capacity assessment is used to identify resources and strengths of the community used to deal with and respond to crisis. Here, capacities refer more to people's abilities to recover after the impact of disasters and during periods of stress.

Main objectives of VCA:
- To reduce the impact of hazards through mitigation, prediction and warning, and preparedness.
- To build capacities that would help reduce country's vulnerability.
- To reduce the ways in which people are affected by poor governance, discrimination, inequality and inappropriate access to resources and livelihood; and
- To tackle the root causes that lead to vulnerabilty.

VCA can be successful only with the active participation of all community members. It is comprehensive and covers all important variables in a community. It gives equal consideration to the physical/material, social/ organisational and attitudinal/motivational aspects of a community according to class, gender, age, etc. VCA is carried out in the following manner:

3.1 Collection and analysis of data

Collecting data on the demographic, socioeconomic and infrastructural details of a given area is a pre-requisite for the commencement of an intervention. These details would build the capacity profile of the community and also help in realising the gaps within the capacities of the community. The gaps, once realised, should be strengthened for reducing vulnerabilities.

The demographic details include the size of population, geographic boundaries, male-female ratio, number of children, single parent, physically challenged, etc. The socioeconomic details include division of the community on the basis of social classes, number of people living

below poverty line, rate of literacy, education facilities, occupational patterns, etc. Critical infrastructure facilities include a number of key facilities available in the community, which encompasses health centres, education centres, public buildings and other public facilities. The assessment of these facilities should be done qualitatively in terms of its structural strength. The total number of health centres, schools, etc. should also be mentioned in order to deduce whether these facilities are adequate or inadequate.

4. Hazard, risk and impact assessment

Based on the hazard profile (Figure 21.4) of the community, the risks are to be detailed out and the impacts required to be assessed and treated. Hazard, risk and impact assessment is facilitated through community workshops and focus group discussions. Assessing the hazards, risks and impacts involves the following process:

FIGURE 21.4
Community hazard profile

Hazards	Date of event	Impact of hazards
Earthquake	2006	Minor shaking. No major loss
Forest fire	Frequent during hot summers	Loss of vegetation and grass resulted in a lack of fodder for livestock and increase of wild animal terror in agricultural land and danger to houses located near forest.
Landslide	Frequent during rainy season and major destruction in 2008	Damage of roads and agricultural land, natural spring, water pollution. Blockage of main roads and link roads for remote location villages.
Human animal conflict	Frequent, three cases in year 2008	Losses of agricultural crops, domestic animal loss, threat to human being and their livestock. Increase in wasteland and pastureland. Terror to human being mainly children and women due to wild animals.

4.1 Hazard profile

Based on the data on hazard zonation, the hazard profile of the intervention area must be defined. The risk profile built for the primary area of intervention may also be utilised for outlining the hazard profile.

4.2 Impact of past hazards

To know the frequency of disasters, hazards are listed down along with the dates of occurrence. Along with this, the impact of each hazard should also be noted.

4.3 Elements at risk

Against each hazard, it is required to identify the elements which are at highest risk, that is, who is at risk? It can be community, infrastructure, houses, children, women or livelihood. The level of risk for each of this elements should then be classified and marked into high, low or moderate (Figure 21.5).

4.4 Impact assessment and risk treatment options

It is important that the identified risks and impacts are treated to reduce the vulnerability of the community. After identifying, the probable risks the treatment options are listed below. The risks require to be treated in a way that it enhances the coping mechanism of the community. The risk treatments may be divided on the basis of different phases of disaster; that is, before, during, after and non-disaster situation (Figure 21.6).

5. Identify problems, solutions and responsibilities

After having identified the hazards and resources, the community should be consulted to identify the problems, solutions and the entity that would be responsible to solve it. For adequate participation of all stakeholders, a workshop can be conducted putting the community at centre stage. The workshop must involve the local organisations, government level officials and community members. The participatory joint workshop with the community, local government authorities and NGOs gives the opportunity to create a consensus on the identification of solutions and

FIGURE 21.5
Impact of past hazards

Sr. no.	Hazards	Who are at risk?	Risk rating		
			High	Moderate	Low
1	Earthquakes	Entire village *panchayat* community especially women, children, people with disabilities and old age, property, community infrastructure including community resources.	Yes	-	-
2	Forest fire	Cash crops, grazing land and agriculture land near forest areas, wild animals and forest vegetation, livestock fodder, natural water resources.	Yes	-	-
3	Landslide	Community, infrastructure, water bodies, livestock, communication, mobility and agricultural land, cash crops, etc.	Yes	-	-
4	Human-animal conflict	Farmers and their livelihoods as well as livestock and agriculture land and villages conflict.	Yes	-	-

also helps define roles and responsibilities of various stakeholders in the community.

i. The workshop should be well represented by the community representatives like community leaders, women, and individuals from various occupational background, elderly members, youth clubs, and government officials working in the intervention area.

ii. The facilitator should group the members such that the groups are well represented by all the stakeholders.

iii. The group members should individually identify the problems. Depending on the number of people in the workshop, the facilitator can ask to identify one or two or maximum five problems faced by the community.

iv. Once the problems are identified, they can then be broadly categorised into different heads such as health, education,

FIGURE 21.6
Impact assessment and risk treatment options

Hazards	Risk	Risk treatment
Earthquake	Damage to life and property	- To create awareness among people regarding earthquake resistant structures, both public and private - Awareness camps on disaster management for the entire *gram panchayat* - Training of task forces and disaster management - Community members on disaster management - Construction of alternate safe routes and safe shelters to deal with emergency situations
Forest fire	Grassing land, forest vegetation, wild animals and crisis of domestic fuel wood and community	- To construct firewalls in forest fire prone-areas - Maintenance of forest by cleaning of pine trees and waste dry leaves with joint initiative of *panchayat* and forest department - Roping of trees other than pine - Awareness in villages regarding forest fire - Training of Task Force to prevent forest fires
Landslide	Infrastructure, resources, roads, agricultural land and community transportation and mobility blocked in remote locations	- Construction of retaining walls in landslides zones - Roping of trees nearby roads and risk management during construction of link roads to avoid landslides - Proper rain water drainage management to avoid landslide during rainy season; awareness camps on landslides
Human-animal conflict	Villagers and their livestock, cash crops and livelihood	- Construction of safety walls to avoid wild animals in agriculture land - Growing of fruit trees instead of pine trees in forest - Sterilisation and export of monkeys should be started to reduce human-animal conflict from this area - Joint initiative of village *panchayat* and forest department to reduce human-animal conflict.

agriculture, transport and communication, and water and sanitation.

v. Now each group must be alloted one problem and asked to detail out the sub-problems and their solutions.

vi. Lastly, they should identify who would be solving these problems, an NGO, community or local government authority.

(For a complete discussion of the process developed by SEEDS, please refer to the website *www.seedsindia.org or write to info@seedsindia.org*)

22

Integrating Ecosystems Management with DRR

Rapid environmental assessments (REA) are useful in assessing environmental situation post disaster.

Source:

Sudmeier-Rieux, K. and N. Ash Adapted from: (2009). *Environmental Guidance Note for Disaster Risk Reduction: Healthy Ecosystems for Human Security.* IUCN.

Ecosystems help reduce the risk of disasters, offering local communities a certain level of protection. Efforts can be made to enhance their protective benefits. Similarly, in a post-disaster situation, environmental concerns must be addressed to reduce possible negative impacts.

1. Pre-disaster

- Prevention, mitigation and preparedness stages should ensure that proper environmental practices are followed; that value and restore ecosystems, especially wetlands, coastal ecosystems and forests on steep slopes as natural buffers. Specific projects may include wetland restoration, tree planting, and restoring coastal open spaces.
- Disaster risk reduction planning should include coordination with environmental ministries, in addition to disaster management and land use planning authorities.
- It should be ensured that existing legislation is being followed and enforced, especially related to zoning and land use planning; for example, respecting coastal buffer zones and proper road building in mountainous areas to avoid landslides; and ensuring that land use planning is not damaging to ecosystems and human well-being.
- Education and training about the role of ecosystems and their multiple benefits for protection and human well-being must be conducted.

2. Post-disaster

- Response, recovery and rebuilding stages progress from quick relief to save lives to short and medium-term planning of lodging and livelihood solutions. Basic environmental concerns must be integrated into each of these stages, following the goal of 'reducing the underlying risk factors'.

    Basic environmental considerations can be included in contingency plans and standard disaster response procedures in order to avoid potential damage that can be incurred and impede long-term recovery.

- Pollution should be minimised and waste management be made effective; ensuring that waste does not contaminate waterways or wetlands areas and hazardous waste materials are secured;

Watershed management

Watershed management is necessary for agricultural, environmental and socioeconomic development. The physical and biological resources of watersheds provide goods and services to people, including water protection, attenuation of disasters by regulating runoff, protection of coastal resources and fisheries, protection of the environment and protection of productive lowlands. Watershed management programmes need to build on existing environmental initiatives.

- When located in flood plains, structures should be built to withstand flood damage, to prevent floodwater contamination, and to avoid disruption to river courses, river banks and vegetation;
- Intensive agricultural activity should not to be permitted on slopes greater than specified percentage reflecting land stability;
- Clear cutting of forests should be limited with forest conservation and sustainable forest management prioritised;
- Institutional bodies, such as river basin organisations should be formally established to address land use conflicts, and staff trained in conflict resolution;
- Public participation of both men and women should be increased in management decisions;
- Effective management plans and enforcement of environmental and zoning regulation are critical; and
- Regional environmental impact assessments are needed to ensure that cumulative impacts of economic activities are sustainable.

Forest management

Forest management is required to balance demand for forest products with the ecological requirements of forests, while ensuring other key benefits for livelihoods, notably by stabilising steep slopes and reducing soil erosion. Although listed separately here, forest management is often integrated into watershed management.

- Protect and improve the forest environment through increased vegetation;
- Help alleviate poverty by generating income through increased tree cover and related activities;
- Increase forest resources;
- Establish community-driven economic activities based on forest plantation;
- Increase multiple uses for land; and
- Create popular awareness about sustainable forest management.

Coastal zone management

Ecosystems such as coral reefs and coastal mangrove forests can adapt to change and recover from storms and floods and still provide services of protecting the coast and absorbing pollution. But once these ecosystems are put under pressure by coastal development, they may lose their resilience.

Coastal zone management strategies being considered in the Asia-Pacific region after the 2004 tsunami highlighted the continuum of inland areas, coasts and oceans. Below are some key entry points:

- Replant coastal forests and restoration of mangroves, which have been taken up as a part of the environmental recovery process;
- Restore and maintain the health of the coral reefs and seagrass beds;
- Maintain and/or develop mangrove belts as buffer zones for coasts and coral reefs; and
- Protect wetlands and watersheds to minimise sedimentation.

(Modified from DEWGA, 2008)

- Transitional shelters and settlements[1] must be located away from sensitive ecosystems and from areas that may put people in harm's way (such as flood plains, wetlands and animal habitats) while providing adequate sanitation facilities;
- It should be made sure that building materials are sustainably sourced (e.g., no mining of coastal sand dunes, mangroves, or coral reefs to rebuild houses);
- Damaged ecosystems with native species must be rehabilitated and the spread of invasive alien species (non-native species that can invade habitats and agricultural land) prevented.
- Special provisions should be made for women, children and other vulnerable populations, according to Sphere Handbook charter.[2]
- REAs[3] are useful in assessing the environmental situation post-disaster in a quick and low cost manner for more effective immediate and long-term recovery planning (Miththapala, 2008).

3. Key actions for ecosystem based DRR

Watersheds, forests and coastal zones are naturally linked. For example, without adequate upstream forest cover, sedimentation can create severe downstream pollution and damage to coastal vegetation and coral reefs.

References

Miththapala, S. (2008). *Integrating environmental safeguards into disaster management.* Volume 1 and 2. Ecosystems and Livelihoods Group. Colombo, Sri Lanka: IUCN.

DEWGA (2008). "Linking disaster risk reduction, envionrmental management and development practices and practitioners in Asia Pacific region: A review of opportunities for integration", *Working Paper.* Available at: *www.dewga.net/Data/Publication/Stocktaking%20Paper_Version% 206%20080825.pdf*

---

1. For more information see: (*www.sheltercentre.org*)
2. *www.sphereproject.org*
3. *www.abuhrc.org*

23
Ecosystem-based Disaster Risk Reduction

Ecosystem-based DRR recognises that ecosystems are connected through biodiversity, water, land, air and people.

Source:

Sudmeier-Rieux, K. and N. Ash Adapted from: (2009). *Environmental Guidance Note for Disaster Risk Reduction: Healthy Ecosystems for Human Security.* IUCN.

Ecosystem management is central to building resilience of communities and nations under the Hyogo Framework for Action (HFA), especially HFA priority 4.[1] Therefore, ecosystem-based disaster management policies, practices and guidelines need to be an integral part of national disaster risk reduction (DRR). Ecosystem-based disaster management refers to decision-making activities that take into consideration current and future human livelihood needs and bio-physical requirements of ecosystems, and recognises the role of ecosystems in supporting communities to prepare for and cope with disaster situations.

This is of particular relevance to the field of disaster risk management as it is a meeting point for enhanced livelihood security for the poor and long-term management of ecosystems. It is a strategy consistent with the Ecosystem Approach of the Convention on Biological Diversity (EA-CBD), for the integrated management of land, water and living resources for human benefits as well as conservation goals. Ecosystem-based DRR recognises that ecosystems are not isolated but connected through the biodiversity, water, land, air and people that they constitute and support (Shepherd, 2008). Sustainable ecosystem management is based on equitable stakeholder involvement in land management decisions, land use trade-offs and long-term goal setting. These are central elements to reducing underlying risk factors for disasters and climate change impacts. How can we integrate ecosystem management and disaster risk management?

Although disaster risk management, ecosystem management, development planning (and climate change adaptation) institutions each have their

Four previously separate institutional spheres need to converge to form new procedures for integrated disaster risk management. Ecosystem management becomes central to disaster risk reduction and climate change adaptation without which goals of human security and sustainable development cannot be achieved.

---

1. WCDR (2005).

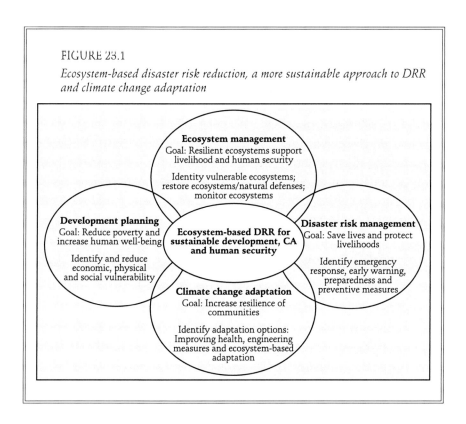

FIGURE 23.1
Ecosystem-based disaster risk reduction, a more sustainable approach to DRR and climate change adaptation

own specific set of stakeholders, goals and actions, a number of these are interrelated (see Figure 23.1). They each seek the overarching goal of sustainable development, human well-being and human security. Improved dialogue and specific coordinating mechanisms are being created between these spheres, although more effort is needed to achieve greater convergence. Likewise, conservation programmes can benefit by including risk and climate change considerations into project planning and monitoring. Below are examples of specific actions that can be taken towards bridging the gap between ecosystem-based management and disaster risk management.

1. Giving explicit consideration to ecosystem-based DRR

Many countries have already recognised the need for legislation and zoning regulations that support sustainable development and environmental principles. However, where legislation often fails is in the implementation and enforcement, leading the way for unsustainable and risk-building practices, such as locating housing in dangerous places. Policies and financial incentives can be offered for investing in ecosystem protection, such as 'payments for ecosystem services', or through new carbon market and other schemes such as REDD[2], which aim to reduce environmental degradation. Incorporating environmental concerns into contingency plans for disaster response is intended to follow the principles of 'do no harm' to long-term recovery (i.e., improper waste management practices that pollute waterways, or locating transitional shelters and settlements in floodplains or elephant pathways) and aim to rebuild back better.

2. Appropriate national and local governance and policies

- Recognise the value of ecosystems as necessary for disaster risk reduction;

- Grant relevant legal authority to environmental, planning and disaster management agencies to coordinate and enforce sustainable environmental DRR policies and procedures;

- Seek to integrate national adaptation processes, such as NAPAs with DRR and environmental national strategies; and

---

Payments for ecosystem services

These financial mechanisms are increasingly being used successfully to finance ecosystem conservation and restoration. Examples include payments to a community to maintain forest cover in sensitive water recharge areas, or on steep slopes to reduce the occurrence of landslides or downstream flooding.

The beneficiary community or other third party would pay for the benefits generated.

(IUCN-UNEP, 2007)

---

2. REDD: Reducing Emissions from Deforestation in Developing Countries.

- Encourage new financial incentives for investments in sustainable ecosystem management that emphasise ecosystems as part of disaster risk planning, possibly financed through payments for ecosystem services.

3. Implementing environmental monitoring and enforcing sustainable land use planning

Progress can be made by integrating land use planning and environmental monitoring into disaster management such as maintaining wetlands for flood risk reduction. Environmental monitoring implies maintaining baseline data on ecosystem health and tracking trends in environmental degradation, such as deforestation and drought, and restoration. Integrated risk assessments can be designed as a useful tool to couple physical risk, vulnerability and environmental assessments. They go beyond the environmental impact assessment (EIA) and strategic environmental assessment (SEA) processes, which are conducted for new development projects.

4. Integrated mechanisms and procedures

Integrated mechanisms and procedures are useful to:

- Promote and enforce integrated land use (spatial and temporal) planning and zoning that include protection of ecosystems (e.g., Integrated Coastal Zone Management (ICZM), Integrated Water Resources Management (IWRM), and forest management plans) and risk assessments;
- Conduct environmental monitoring and assessments (ecosystem baseline data, EIAs, SEAs for new development projects and programmes);
- Conduct integrated risk assessments (coupling physical risk, vulnerability, and environmental assessments);
- Implement ecosystem restoration and rehabilitation that follow clear technical guidance and match local needs and priorities; and
- Incorporate environmental safeguards into disaster emergency response plans, such as rapid environmental assessments (REA).

5. Engaging with stakeholders

Ecosystem management practices are the most successful when they involve communities as stakeholders and land stewards, such as community-managed marine protected areas, or community forest user groups. These environmental mechanisms can become especially relevant and effective for DRR if they incorporate disaster risk assessments. To achieve this, there is a need to put into place mechanisms for consultations between environmental, planning and disaster management authorities. It is important to:

- Build dialogues and mechanisms for collaboration between environmental, planning and disaster risk management authorities and people affected by the decisions; and
- Include communities, especially women, minorities, and people with disabilities in designing and implementing the above procedures.

6. Knowledge creation and exchanges

Capacity-building through awareness-raising, education and training are critical to changing attitudes and behaviours towards more sustainable environmental practices.

As an example, ecosystem rehabilitation and restoration can be options in the aftermath of a disaster or to safeguard against new ones. However, successful ecosystem rehabilitation requires time, knowledge, resources and should be conducted in consultation with communities, appropriate technical advice, and based on local needs and priorities especially when natural restoration may be the most effective option.

Awareness raising, education, training and knowledge exchange can help to:

- Promote new knowledge creation and sharing among scientists, practitioners and communities;
- Recognise the value of local practices and knowledge; and
- Recognise the special role that women play as agents of change and stewards of natural resources and as being highly affected by extreme events.

References

Shepherd, G. (2008). *The ecosystem approach: Learning from experience*. Gland, Switzerland: IUCN. Available at: *www.iucn.org/about/union/commissions/cem/cem_resources/?1652/The-ecosystem-approach-learning-fromexperience*

IUCN-UNEP (2007). *Developing international payments for ecosystem services: Towards a greener world economy*. IPES Brochure, available online at: *http://cmsdata.iucn.org/downloads/ipes_brochure_0607_1.pdf*

WCDR (2005). "HFA Priority 4: Reduceing underlying risk factors", *Progress Report on the Implementation of the Hyogo Framework for Action, 2005-2015: Building the Resilence of Nations and Communities to Disasters*. World Conference on Disaster Reduction, 18-22 January, Kobe, Hyogo, Japan. *www.unisole.org*

24
Indicators for Use in Ecosystem-based Disaster Risk Reduction

Source:

Sudmeier-Rieux, K. and N. Ash Adapted from: (2009). Environmental Guidance Note for Disaster Risk Reduction: Healthy Ecosystems for Human Security. IUCN.

1. What are indicators?

The term 'indicators' refers to data of a quantitative or qualitative nature which can provide and communicate scientifically robust measures of the status or change in condition. They indicate the current status and any changes in a process or a system with respect to a given aspect of interest. An indicator is a pointer. It can be a measurement, a number, a fact, an opinion or a perception that points to a specific condition or situation, and measures changes in that condition or situation over time. Indicators facilitate a close observation about the results of initiatives or actions, and help to simplify the presentation of complex situations. They are very important tools to evaluate and follow up disaster risk reduction (DRR) processes, and are valuable tools to help achieve better results in projects or initiatives. A good indicator is considered SMART (specific, measureable, achievable, relevant and timely).

2. Several types of indicators

Qualitative indicators are measures that refer to qualities. They deal with aspects that are not directly quantifiable, opinions, perceptions or judgments from people about something, such as people's reliance on their boats as an instrument of economic independence. On the other hand, quantitative indicators are measures that directly refer to numbers or amounts, such as the number of women who own boats in a community. Each type of indicator—qualitative and quantitative— expresses different, complementary, needed dimensions about the situation of interest.

Progress or results indicators convey whether tangible results are being achieved, and process indicators indicate about the state of a process, such as stakeholder dialogue. The difference between the two may be time dependent. For example, a training workshop on environmental legislation and DRR in the short term may lead participants to attitude changes among participants and a process towards new legislation may be undertaken. Real progress resulting in new legislation and implementation mechanisms may take much longer and is dependent on other factors although the impetus may have come from the initial workshop.

3. Purpose and caveats of the suggested 'Indicators for Ecosystem-based DRR'

We have developed these indicators to offer guidance on example areas to focus policy and resources in order to make progress on achieving HFA Priority 4, 'reducing underlying risk factors' and in particular, 'sustainable ecosystems and environmental management'.

The indicators are both qualitative and quantitative, and mainly process-oriented. Caveats of the proposed (and any) indicators are multiple. They need to be configured to the local context in order to become SMART; they are not universal; they will not always apply to all countries, at all scales; they may not adequately reflect cultural considerations and specific contexts. However, the following list of indicators is intended to provide guidance for integrating ecosystem management into DRR policies and practices, a dimension that has not received adequate attention and practical guidance to date.

The suggested indicators can be used for further defining and refining nationally and locally relevant indicators. They have also been classified according to disaster risk management, vulnerability-related policies, operational mechanisms, knowledge and education, human well-being, ecosystem services, drivers of threats to ecosystem services and characteristics of disaster-resilient communities.

Important work has already been conducted in developing and testing relevant indicators for sustainable development and human well-being, ecosystem health, ecosystem services, disaster management. We have drawn upon many of these sources to develop this list of indicators relevant to ecosystem-based disaster risk reduction.

Increasing numbers of extreme events causing casualties and affecting populations are weather and climate related. However, climate change, although often cited as the culprit of rising numbers of disasters, is one of several factors increasing disaster vulnerability and environmental degradation. The risk of suffering from any particular disaster depends on the size and frequency of the hazard event but even more on the vulnerability of people, often linked to environmental degradation and governance issues. Disasters are not caused by extreme events themselves,

Examples of indicators for use in ecosystem-based DRR

1. Risk identification indicators

1.1 Systematic inventory of disasters and losses, including small events

1.2 Hazard monitoring and mapping

1.3 Vulnerability and risk assessments take into account monitoring of ecosystem conditions, ecosystem services and threats to ecosystems

2. Policy indicators linking ecosystem-based management to DRR

2.1 National platforms for DRR, HFA focal point and other national disaster risk institutions include environmental and planning ministries in decision-making and implementation

2.2 Legislative mechanisms effectively incorporate sustainable land use planning into DRR legislation

2.3 Zoning regulations take into account specific ecosystem considerations and enforcement

2.4 Cross-sectoral mechanisms effectively incorporate sustainable land use planning into DRR legislation

2.5 NAPAs and National Adaptation plans include DRR and sustainable environmental management actions

2.6 National biodiversity strategies and action plans include DRR considerations

2.7 National resources related policies and environmental legislation (forestry plans, ICZM plans, etc.) include and implement risk assessments

2.8 National sustainable development strategies include and implement risk assessments

2.9 Public and private infrastructure investments that include enforceable EIAs and risk assessments

2.10 Financial incentives in the form of tax rebates, subsidies, and other monetary and non-monetary rewards for investments in ecosystem restoration and sustainable environmental management that emphasise ecosystems as part of disaster risk planning

3. Ecosystem-based management and DRR

Risk assessments are integrated into:

3.1 IWRM programmes

3.2 ICZM programmes

3.3 Protected areas management

3.4 Community conservation areas—these include local communities in ownership of conservation projects

3.5 Community-managed marine protected areas

3.6 Forest management plans

3.7 Integrated Forest Fire Management (IFFM)

3.8 Forest landscape restoration areas

3.9 River basin organisations for improved river management through stakeholder involvement

3.10 Livestock management—establishment of grazing practices

contd...

...contd...

3.11 Fisheries management—establishment of quotas and regulations
3.12 Water management—equitable pricing and distribution schemes

4. Knowledge, participation and education
4.1 Public information and community participation are part of risk assessments
4.2 Non-state actors are involved in dialogue and implementation of DRR at the national and local levels, including civic groups, environmental, humanitarian and development agencies
4.3 Disaster practitioners and environmental managers are trained in integrated risk assessment, which include ecosystem management
4.4 Primary school children are educated in disaster preparedness and environmental stewardship

5. Human well-being and human security: Reducing exposure to disasters and vulnerability
Many excellent human well-being and human security indicators have already been developed, including from the following sources: UN Commission on Sustainable Development (CSD) indicators; human development index (HDI); human poverty index (HPI); gender related development index (GDI); governance index (Kaufmann); prevalent vulnerability index (PVI); Inter-American Development Bank (IDB)

6. Ecosystem health indicators by ecosystem type
6.1 General:
  6.1.1 Changes in native species richness
  6.1.2 Abundance of selected key species
  6.1.3 Change in threat status of species
  6.1.4 Number and area of protected areas
  6.1.5 Invasive alien species
6.2 Agro-ecosystems/forests:
  6.2.1 Land use changes
  6.2.2 Vegetation cover
  6.2.3 Per cent of land degradation
  6.2.4 Arable and permanent cropland area
  6.2.5 Reduced dependency on fertiliser and pesticide use
  6.2.6 Proportion of land area covered by forest
  6.2.7 Area under sustainable forest management
6.3 Wetlands/rivers:
  6.3.1 Per cent of area maintained as wetlands
  6.3.2 Riverbank vegetation maintained
  6.3.3 Water quality and turbidity
  6.4.4 River fragmentation
6.4 Water:
  6.4.1 Drinking water quality
  6.4.2 Bathing water quality

contd...

...contd...

6.4.3 Proportion of total water resources used
6.4.4 Water use intensity by economic activity
6.4.5 Wastewater treatment
6.5 Coastal/Marine:
    6.5.1 Area of healthy seagrass beds and marine algae
    6.5.2 Proportion of marine area protected
    6.5.3 Health of marine ecosystems, as measured by marine trophic index (MTI)
    6.5.4 Coverage of live coral reef ecosystems
    6.5.5 Area of healthy mangroves as buffer zones as measured by area, density and width

7. Monitoring of threats to ecosystems

7.1 Climate change impacts
7.2 Conversion of ecosystems for urbanisation and agriculture
7.3 Fragmentation of habitats
7.4 Slash and burn agriculture
7.5 Over harvesting of forest products
7.6 Desertification
7.7 Industrial logging/illegal logging
7.8 Over grazing/cattle ranching
7.9 Invasive alien species
7.10 Soil erosion
7.11 Eutrophication: Over-use of fertilisers

Sources: UN Commission on Sustainable Development (2007) Cardona; Inter-American Development Bank (2005); Millennium Ecosystem Assessment (2005); Convention on Biological Diversity-Environmental Vulnerability Index (2004).

but occur when a society's capacity to cope with an extreme event is overwhelmed or mismanaged. For these reasons, the terms 'natural disaster' and 'natural hazard' have increasingly become misnomers (Wisner et al., 2004; Abramovitz et al., 2002).

Unfortunately, available economic statistics on disasters do not reflect lost agricultural land and livelihoods in developing countries. The more common and chronic disasters—shallow landslides, recurring flooding, rising seawaters, drought, and impacts of invasive species— impose the greatest costs on poor populations, and yet are not mirrored in official statistics on disasters. These small, cumulative disasters are most often those grounded in land use and a pressure on natural resources, and are therefore often the most avoidable through appropriate ecosystem management.

Even if the number and frequency of extreme events increases, the magnitude of disasters can be reduced through adopting integrated approaches that combine development processes, DRR measures, and ecosystem management. Combining ecosystem restoration in degraded areas with long-term views of settlement design and planning includes investing in ecosystems as cost-effective, successful alternatives and complements to physical engineering structures. We consider this guidance note to be one contribution of practical ideas and indicators for how to shape an integrated approach to DRR. The 'Environmental Guidance Note for Disaster Risk Reduction' is in progress that will evolve with new experiences, success stories, lessons learned and good practices. However we are convinced that rather than controlling nature, which has all too often been the approach in the past, we have learned that we must work with nature if we are to keep ourselves safe while facing increasingly hazardous times.

References

Abramovitz, J., T. Banuri, P. Girot, B. Orlando, N. Schneider, E. Spanger-Siegfried, J. Switzer and A. Hammill (2002). "Adapting to climate change: Natural resource management and vulnerability reduction". *Background Paper. Task force on climate change: Adaptation and vulnerable communities*. Gland, Switzerland: IUCN; Washington, DC, USA: Worldwatch Institute; Winnipeg, Canada: IISD; and Somerville, USA: SEI-B.

Aguilar, L. (in press). IUCN/UNISDR "Gender indicators", Gland, Switzerland: IUCN.

Hewitt, K. (ed.) (1983). *Interpretations of calamity: From the viewpoint of human ecology*. Boston, USA: Allen and Unwin.

Wisner, B., P. Blaikie, T. Cannon and I. Davis (2004). *At Risk, natural hazards, people's vulnerability and disasters*. Second Edition. London, UK and New York, USA: Routledge.

25
Healthy Ecosystems and their Role in Disaster Risk Reduction

Source:

Sudmeier-Rieux, K. and N. Ash *Adapted from: (2009). Environmental Guidance Note for Disaster Risk Reduction: Healthy Ecosystems for Human Security. IUCN.*

1. Introduction

Ecosystems contribute to reducing the risk of disasters in multiple and varied ways. Well-managed ecosystems can reduce the impact of many natural hazards, such as landslides, flooding, avalanches and storm surges. The extent to which an ecosystem will buffer against extreme events will depend on an ecosystem's health and the intensity of the event. Degraded ecosystems can sometimes still play a buffering role, although to a much lesser extent than fully functioning ecosystems.

Ecosystems are defined as dynamic complexes of plants, animals and other living communities and their non-living environment interacting as functional units (Millennium Ecosystem Assessment, 2005). They are the basis of all life and livelihoods, and are systems upon which major industries are based, such as agriculture, fisheries, timber and other extractive industries. The range of goods and other benefits that people derive from ecosystems contributes to the ability of people and their communities to withstand and recover from disasters. The term 'sustainable ecosystems' or healthy ecosystems, implies that ecosystems are largely intact and functioning, and that resource use, or demand for ecosystem services do not exceed supply in consideration of future generations.

Healthy ecosystems comprise interacting, and often diverse plant, animal and other species, and along with this species and underlying genetic diversity, constitute the broader array of biodiversity.

'Biodiversity' is the combination of life forms and their interactions with one another, and with the physical environment, which has made Earth habitable for people. Ecosystems provide the basic necessities of life, offer protection from natural disasters and disease, and are the foundation for human culture (Millennium Ecosystem Assessment, 2005).

2. Benefits of ecosystem

The benefits that people derive from ecosystems, or 'ecosystem services' are often categorised into four types:

Supporting services: these are overarching services necessary for the production of all other ecosystem services such as production of biomass, nutrient cycling, water cycling and carbon sequestration.

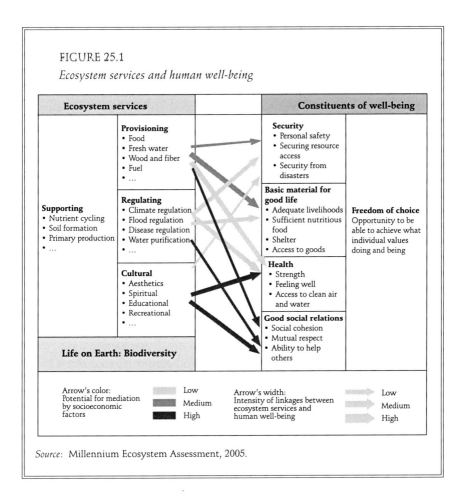

FIGURE 25.1
Ecosystem services and human well-being

Source: Millennium Ecosystem Assessment, 2005.

Provisioning services: these are the services we often consider as 'ecosystem goods' and products obtained from ecosystems to support livelihoods such as food, fibre, genetic resources, medicines, fresh water, etc.

Regulating services: these are the services that offer protection and otherwise regulate the environment in which people live, such as flood regulation, water filtration, pollination, erosion control, and disease regulation.

Cultural services: these are services supporting spiritual values, aesthetic, educational and recreational needs (Millennium Ecosystem Assessment, 2005).

The Millennium Ecosystem Assessment (MA), a five-year international assessment initiative, clearly demonstrated the strong and varied links between human well-being, human security, livelihoods, health and intangible benefits such as equality and freedom of choice, with ecosystem services. The MA also highlighted that ecosystem degradation is undermining this link due to a number of human activities, mainly:

- over-exploitation of resources or higher demand for ecosystem goods than can be sustained, such as overfishing;
- land use and land cover changes, or changes to habitats due to conversion to croplands and urbanisation;
- climate change impacts are affecting ecosystems and exacerbating environmental degradation;
- invasive alien species are introduced species that compete and encroach vigorously upon native species, with the potential to degrade ecosystem services and cause severe economic damage; and
- pollution, from chemical waste and agricultural inputs has severely degraded many ecosystem services, and continues to act as a major driver of change (Miththapala, 2008).

Degraded ecosystems reduce community resilience for sustainable development as well as disaster preparedness and recovery.

Ecosystem degradation and loss have led to serious impacts on human well-being: these include reduced availability of goods and services to local communities, increased spread of diseases and reduced economic opportunities. This, in turn, is leading to loss of livelihoods, and reduced food security (Miththapala, 2008).

Healthy ecosystems both reduce vulnerability to hazards by supporting livelihoods, while acting as physical buffers to reduce the impact of hazard events. As such, this 'natural infrastructure' is in many cases equally effective in reducing the impact of hazard events, and are often less expensive than human-built infrastructure. Disasters also hamper development goals, and yet few governments, donors and development organisations adopt a precautionary approach in the design and management of projects, and fewer still recognise the role and value of ecosystem management for reducing disaster risk (UNEP, 2007).

3. Ecosystems and Disaster Risk Reduction (DRR)

Five reasons why ecosystems matter to DRR:

- Human well-being depends on ecosystems that enable people to withstand, cope with, and recover from disasters. Disaster-resilient communities, especially in rural areas, are based on healthy ecosystems and diverse livelihoods.

- Ecosystems, such as wetlands, forests and coastal systems can provide cost-effective natural buffers against hazard events and the

---

Broadly defined, the total economic value of ecosystems includes:

Use values—

- Direct values: benefits derived from the use of environmental goods either for direct consumption or production of other commodities.

- Indirect values: benefits provided by ecosystem functions and services that maintain and protect natural and human systems such as maintenance of water quality and flow, flood control and storm protection; and

- Option values: the premium placed on maintaining an ecosystem service (i.e., a pool of species, genetic resources and landscapes) for future uses.

Non-use values—

- Bequest value: the willingness to pay to ensure that future generations inherit a particular environmental asset; and

- Instrinsic value—i.e., the value of biodiversity in its own right, independent of value placed on it by people.

(Emerton and Bos, 2004)

impacts of climate change. According to the World Bank (2004), investments in preventive measures including in maintaining healthy ecosystems is seven-fold more cost-effective than the costs incurred by disasters.

- There are clear links between resource degradation and disaster risk. Degraded ecosystems are unable to provide the benefits that help communities to reduce their vulnerability to disasters. In addition, many disasters are caused by reoccurring conflicts, which are based on competition for scarce natural resources and once a conflict has started it can also lead to additional environmental degradation.

- Healthy and diverse ecosystems are more robust to extreme weather events. Disasters can affect biodiversity through the spread of invasive species, mass species mortality, loss of habitat and poorly designed post disaster clean-up efforts. This may have a negative impact on progress toward achieving the objectives of the Convention on Biological Diversity (CBD) and Millennium Development Goals (MDGs); and

- Ecosystem degradation reduces the ability of natural systems to sequester carbon—exacerbating climate change impacted disasters.

The negative impacts of climate change and disaster events are more severe on vulnerable people, and at the same time, they are creating greater population vulnerability for those living in the conditions of socioeconomic exclusion, including women and children. This is due to increasing environmental degradation, populations living in more exposed areas, more extreme weather events and the social and governance factors that affect livelihoods.

Disasters are mainly social constructs; they are largely determined by how a society manages its environment, how prepared it is to face adversity and what resources are available for recovery. As pointed out by Moser and Satterthwaite (2008), the more assets people have, the less vulnerable they are and the greater the erosion of people's assets the greater is their insecurity. Such assets also include the access to healthy ecosystems.

Vulnerable populations are more at risk to disasters—those in rural areas are also heavily dependent on ecosystem services for their livelihoods and for physical protection. Therefore, investing in ecosystems and mainstreaming disaster risk and ecosystem management in development planning is likely to make a major contribution to the goal of achieving sustainable livelihoods for the poor.

4. Examples and values of protective ecosystem services

4.1 Regulating flood waters

Wetlands and peatlands provide storage space for flood waters, and there is growing evidence that maintaining vegetation and associated soil structure in local watersheds regulates the flow of rain water into streams and rivers, although this service can be overwhelmed with large-scale rainfall and flooding events. Sri Lanka's Muthurajawela marsh is a coastal peat bog covering over 3,100 hectares and an important part of local flood control as the marsh buffers and regulates flood water discharge into the sea. The annual value of this service was estimated at more than US $5 million, or US $1,750 per hectare (Emerton and Bos, 2004). Riparian and coastal vegetation also stabilises shorelines and riverbanks. The costs of losing vegetation along riverbanks has been estimated at up to US $425 per metre of bank (Ramsar Convention on Wetlands, 2005).

4.2 Reducing landslides, avalanches and rockfalls

In addition to providing improved aesthetics over engineering structures, forests are estimated to save between US $2-3.5 billion per year in disaster damage (UNISDR, 2004). Switzerland, for example, long ago recognised the value of 'protection forests' in reducing damage from avalanches, landslides and rock falls, and forests are a key part of the country's disaster prevention plan (Stolten et al., 2008). Healthy forests are less likely to be invaded by pests, invasive alien species and destroyed by natural hazards, and provide numerous additional benefits such as the storage of carbon, and the opportunity for recreation, timber production and non-timber products.

4.3 Improving coastal management and flood risk reduction

Intact coastal ecosystems—in particular mature, stabilised sand dunes, coral reefs, lagoons, salt marshes, and mangroves—play an important role in reducing flood damage during coastal storms (UNEP-WCMC, 2006). Coastal ecosystems are particularly effective in reducing flooding from small and medium-scale events. In addition to reducing coastal flooding, mangroves provide many other services, such as nursery habitat for fish and other marine species, fire wood, building materials and medicine which support the needs of communities for both DRR and development (ProAct Network, 2008).

An analysis of 141 countries in the period of 1981 to 2002 found that disasters (and their subsequent impacts) on an average killed more women than men, or they killed women at a younger age than men in societies where women's economic and social rights are not protected. (Neumayer and Plumper, 2007).

4.4 Drought, sand storm and fire regulation

Wildfires, wind erosion, severe drought are expected to increase worldwide as a result of climate change-induced weather changes. Wind erosion causes severe loss of topsoil, estimated at 161 tonnes of lost soil annually in Canada alone, and causing significant economic losses (ProAct Network, 2008). Ecosystems can act to buffer the processes of drought and desertification through shelterbelts, greenbelts, hedges and other 'living fences'. These buffers help break the force of winds, provide shade, stabilise dunes, maintain soil structures, trap water and restore organic material, rendering soil more favourable to agricultural practices. Fire is a natural part of many ecosystems, and can enhance vegetation by controlling invasive plants and enhancing regenerative processes, especially in grazing lands. Where a reoccurring feature, fire is best managed as a part of agro-ecosystems, creating firebreaks, and controlling understory vegetation (Goldammer, 1988; ProAct Network, 2008; Stolten et al., 2008).

References

Emerton, L. and E. Bos. (2004). "Value: Counting ecosystems and water infrastructure", *Water and Nature Initiative*. Gland, Switzerland: IUCN, The World Conservation Union.

Goldammer, J.G. (1988). "Rural land-use and fires in the tropics". *Agroforestry Systems* 6: 235-52. Available at: *http://www.fire.uni-freiburg.de*

Millennium Ecosystem Assessment (2005). *Ecosystems and human well-being: Synthesis*. Washington, D.C., USA: Island Press.

Miththapala, S. (2008). *Integrating environmental safeguards into disaster management*. Volume 1 and 2. Ecosystems and Livelihoods Group, Colombo, Sri Lanka: IUCN.

Moser, C. and D. Satterthwaite (2008). "Towards pro-poor adaptation to climate change in the urban centres of low- and middle-income countries", *Human Settlements Discussion Paper (Climate Change and Cities: 3)*. London, UK: IIED.

Neumayer, E. and T. Plumper (2007). "The gendered nature of natural disasters: The impact of catastrophic events of the gender gap in life expectancy, 1981-2002", *White paper*. London, UK: London School of Economics and Political Science.

ProAct Network (2008). *Environmental management, multiple disaster risk reduction and climate change adaptation benefits for vulnerable communities*. Tannay, Switzerland: ProAct Network. Available at: *http://www.proactnetwork.org*

Ramsar Convention on Wetlands (2005). An intergovernmental treaty designed to conserve and promote wise use of wetlands. *Resolution IX.4: The Ramsar convention and conservation, production and sustainable use of fisheries resources* (Resolution IX.9 (COP 9, Kampala, Uganda, 2005): The role of the Ramsar convention in the prevention and mitigation of impacts associated with natural phenomena, including those induced or exacerbated by human activities).

Stolten, S., N. Dudley and J. Randall (2008). *Natural security, protected areas and hazard mitigation*. Gland, Switzerland: WWF and Equilibrium.

UNEP (2007). "Environment and vulnerability: Emerging perspectives", UN/ISDR Working Group on Environment and Disaster Reduction. Geneva, Switzerland.

UNEP-WCMC (2006). *In the front line: Shoreline protection and other ecosystems services from mangroves and coral reefs*. Cambridge, UK: UNEP.

UN/International Strategy Disaster Reduction. (2004). *Living with risk: A global review of disaster reduction initiatives*. Geneva, Switzerland: UN/ISDR.

World Bank (2004). "Natural disasters: Counting the cost". Press release, March 2. Available at: *http://www.worldbank.orga po*

26
Adaptation to Climate Change

Source:

Adapted from: Issue # 29 April 2007 southasiadisaters.net prepared originally for Climate Risk Management: Asia Regional Workshop, April 23-26, 2007 held in Kathmandu, Nepal.

1. Introduction

Climate change is a global concern in the 21st century, foremost in South Asia, where the world's poor remain most susceptible to the potential damages and uncertainties inherent in a changing climate.

In the region, 'repeat droughts' or 'repeat floods' are becoming common in 2 out of 10 communities. There is a loss of landmass in coastal areas, repeated droughts, and chronic flooding, displacing individuals and communities. This, in turn, leads to economic and political costs, causing a projected 15 per cent to up to 30 per cent migration in South Asia settlements by 2015.

Aside from displacement, climate change adversely affects the different sectors of the economy such as agriculture and food, water and coastal resources, and human health. For instance, it may increase heat related human and animal diseases that lead to death, reduce people's livelihood and incomes, and alter regional food security because of decreased agricultural production. Hence, all losses associated with climate change may jeopardise the developing countries' efforts in achieving both their national and the Millennium Development Goals (MDGs).

2. Potential impacts of climate change on the MDGs

Climate change-induced losses may pose a serious threat in meeting the MDGs among the poor in Asia (Figure 26.1). Hence, the World Bank suggests that stakeholders should assess these risks and link options for climate change to development processes from the beginning.

3. Adaptation to climate change effects

Global warming can exacerbate the effects of natural hazards (Figure 26.2). The IPCC suggests these options for adapting to climate change-induced losses in different regions of Asia:

3.1 Agriculture

- *Boreal Asia*: Adopt suitable crops and cultivars; make optimum use of fertilisers and adaptation of agro-technologies.

Severity of droughts can be hightened by climate change

- *Arid and Semi-Arid Asia*: Shift from conventional crops to intensive greenhouse agriculture/aquaculture; protect against soil degradation.
- *Temperate Asia*: Adopt heat-resistant crops and water-efficient cultivars with resistance to pests and diseases; conserve soil.
- *Tropical Asia*: Adjust cropping calendar and crop rotation; develop and promote the use of high-yielding varieties and sustainable technological applications.

3.2 Water resources

- *Boreal Asia*: Develop flood protection systems in north Asia (because of permafrost melting and increased stream flow volume/ surface runoff) enhance management of international rivers.
- *Arid and semi-arid Asia*: Enhance conservation of freshwater supply.

FIGURE 26.1

Direct and indirect impacts of climate change on MDGs

MDG Goal 1: Eradicate extreme poverty and hunger

Direct Impacts:

- Climate change may reduce poor people's livelihood assets.
- Climate change may alter the path and rate of economic growth due to changes in natural systems and resources, infrastructure and labour productivity. A reduction in economic growth directly impacts poverty through reduced income opportunities.
- Climate change may worsen regional food security.

MDG Goal 2: Achieve universal primary education

Indirect Impacts:

- Links to climate change are less direct but loss of livelihood assets (natural, health, financial and physical capital) may reduce opportunities for full time education in numerous ways. Natural disasters and drought reduce children's available time (which may be diverted to household tasks) while displacement and migration can reduce access to education opportunities.

MDG Goal 3: Promote gender equality and empower women

Indirect Impacts:

- Climate change may exacerbate current gender inequalities. Depletion of natural resource and decreasing agricultural productivity may place additional burdens on women's health and reduce time available to participate in decision-making processes and income generation activities.
- Climate related disasters have been found to impact more severely female-headed households particularly where they have fewer assets to start with.

MDG Goals 4, 5 and 6: Health related goals:
- **Combat major diseases**
- **Reduce infant mortality**
- **Improve maternal health**

Direct Impacts:

- Direct effects of climate change may include increases in heat related mortality and illness associated with heat waves (which may be balanced by less winter cold related deaths in some regions).
- Climate change may increase the prevalence of some vector-borne disease (e.g., malaria and dengue fever), and vulnerability to water, food or person-to-person borne diseases (e.g., cholera and dysentery).

MDG Goal 7: Ensure environmental sustainability

Direct Impacts:

- Climate change may alter the quality and productivity of natural resources and ecosystems, some of which may be irreversibly damaged, and these changes may also decrease biological diversity and compund environmental degradation.

MDG Goal 8: Global partnership

Direct Impacts:

- Climate change is a global issue and responses require global cooperation, especially to help developing countries adapt to the adverse impact of climate change.

Adaptation to Climate Change

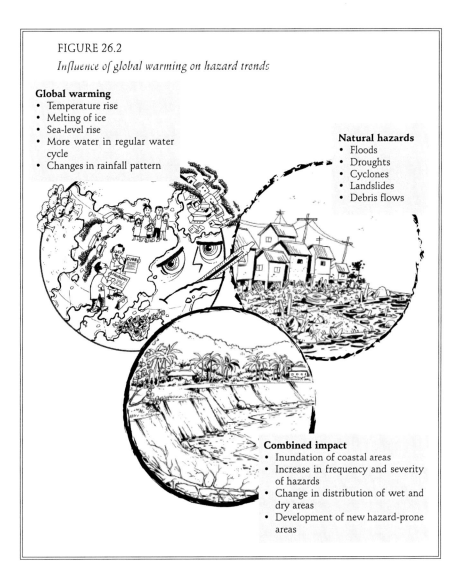

FIGURE 26.2
Influence of global warming on hazard trends

Global warming
- Temperature rise
- Melting of ice
- Sea-level rise
- More water in regular water cycle
- Changes in rainfall pattern

Natural hazards
- Floods
- Droughts
- Cyclones
- Landslides
- Debris flows

Combined impact
- Inundation of coastal areas
- Increase in frequency and severity of hazards
- Change in distribution of wet and dry areas
- Development of new hazard-prone areas

- *Temperate Asia*: Use flood and drought control measures; improve flood warning and forecasting systems.
- *Tropical Asia*: Develop flood and drought-control management systems; reduce future developments in floodplains; use appropriate measures for protection against soil erosion; conserve groundwater supply, water impoundments, and efficient water resource systems.

3.3 Ecosystems and biodiversity

- Assess risks to endemic species and ecosystems.
- Introduce integrated ecosystem planning and management.
- Reduce habitat fragmentation and promote development of migration corridors and buffer zones.
- Encourage mixed-use strategies.
- Prevent deforestation and conserve natural habitats in climatic transition zones inhabited by genetic biodiversity with potential for ecosystem restoration.

3.4 Coastal resources

- *Boreal Asia*: Modify infrastructure to accommodate sea-level rise.
- *Arid and semi-arid Asia*: Protect lakes and water reservoirs; develop aquaculture farming techniques.
- *Temperate Asia*: Follow setback examples for new coastal development; evaluate coastal subsidence rates in sensitive coastal regions; prepare contingency plans for migration in response

to sea-level rise; improve emergency preparedness for weather extremes (e.g., typhoons and storm surges).

- Tropical Asia: Protect wetlands and allow for migration; prepare contingency plans for migration in response to sea-level rise; improve emergency preparedness for weather extremes; evaluate coastal subsidence rates in sensitive coastal regions.

3.5 Human health

- Build heat-resistant urban infrastructure and take additional measures to reduce air and water pollution.
- Adapt technological/engineering solutions to prevent vector-borne diseases/epidemics.
- Improve health care system, including surveillance, monitoring and information dissemination.
- Increase infrastructure for waste disposal.
- Improve sanitation facilities.

3.6 Cross-cutting issues

- Continue monitoring and analysis of variability and trends in key climatic elements.
- Improve weather forecasting systems in the region.

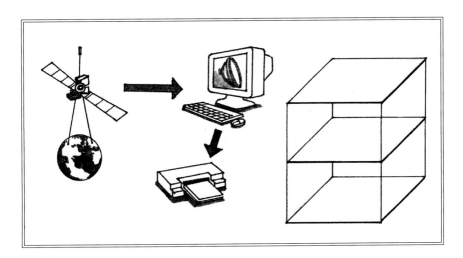

- Apply new techniques for confident projection of regional climate change and its variability, including extreme events.
- Improve coordination of climate change adaptation activities among countries in the region.
- Improve and implement reforms on land use planning.
- Keep the NGO community and the public aware of developments on risks of climate change and involve them in planning, adaptation and mitigation strategies.
- Take advantage of traditional knowledge in planning for the future.

4. Conclusion

Asia, especially South Asia, is home to the world's largest number of poor who are most at risk to climate change hazards. This is because most of them depend on natural resources for their food and livelihood, yet they have limited capacity to reduce and adapt to the risks. Hence, all stakeholders—including communities, local and international organisations, local governments, and civil societies—must engage in partnerships to provide holistic solutions in managing and adapting to climate change hazards. These should include human, institutional and financial adaptations.

While agro-climatic adaptations (e.g., cropping systems, reforestation, waste management, coastal system improvement, etc.) and management are important, the role of the local communities is critical. Experiences have shown the need to empower communities to recover from climate change hazards. This involves a 'bottom-up public dialogue in favour of the poor affected or vulnerable to climate change risk' where they can better articulate their needs.

Empowering affected communities also includes identifying more livelihood opportunities for affected communities such as ecotourism activities. Such efforts can be greatly supported by microfinance products and services. There are evidences showing that microcredit groups of the poor have done better in coping with natural disasters such as those from climate change.

While governments aim for policies related to climate change, multi-partnerships at the 'ground' level is vital. In fact, the World Conservation Union recommends a shift from mandate-centred aid as practiced by many agencies to need-based aid as exemplified by multi-agencies working with NGOs and civil societies that have more flexibility and grassroots reach.

Lastly, process documentation and local action research must be strengthened to help us better understand the impacts of climate change and adaptation at the grassroots level. Such information would be valuable inputs into national and global policies.

27
Capitalising on Similarities: Bridging Differences between CCA and DRR

Source:

Paul Venton (lead author) and **Sarah La Trobe** Linking climate change adaptation and disaster risk reduction, July 2008. Tearfund in collaboration with the Institute of Development Studies.

Climate change adaptation (CCA) and disaster risk reduction (DRR) have similar aims and mutual benefits. However, to date, the climate change and disaster risk management communities[1] have operated largely in isolation from each other. This situation must change as a matter of urgency. Adaptation and DRR policymakers, experts and practitioners must communicate and collaborate with each other to ensure a comprehensive risk management approach to development at the local, national and international levels of government.

Closer collaboration could reduce climate related losses through more widespread implementation of DRR measures linked with adaptation. It could result to more efficient use of financial, human and natural resources, and increase the effectiveness and sustainability of both adaptation and DRR approaches. Raising awareness of the similarities and differences between climate change adaptation and DRR will help highlight the benefits of an integrated approach to the two.

1. Similarities

1.1 Similar aims

Both adaptation and DRR seek to build resilience in the context of sustainable development. CCA requires the reshaping and redesigning of development, social and economic practices to respond effectively to new or anticipated environmental changes. Likewise DRR seeks to influence development decision-making and protect development aspirations from environment related risk. The effectiveness of both adaptation and DRR are limited if they are not viewed within the broader context of sustainable development.

McGray et al. (2007) at The World Resources Institute (WRI) present a model of adaptation which helps to illustrate how closely DRR is linked with adaptation. The WRI frames adaptation as a 'continuum of responses to climate change,' divided into four types of adaptation efforts, ranging from 'pure' development activities at one end of the continuum to very explicit adaptation measures at the other. The four types are:

---

1. Policymakers, experts, academics and practitioners.

In seeking to increase poor people's resilience to climate change and disasters, local communities must participate in CCA and DRR decision-making.

a. addressing the drivers of vulnerability, e.g. factors making people vulnerable to harm;

b. building response capacity (laying the foundation for more targeted actions);

c. managing climate risk (reducing the effects of climate change on resources/livelihoods); and

d. confronting climate change (highly specialised activities, such as relocating communities in response to sea level rise).

While DRR measures typically fall under the middle two categories of building response capacity and managing climate risk, they can fit into every category of the continuum, addressing drivers of vulnerability (e.g. diversifying livelihood strategies in flood-prone areas) as well as confronting climate change (e.g. reducing the risk of glacial lake outburst floods).[2]

1.2 Mutual benefits

To reduce vulnerability to hazards, the DRR community implements a variety of measures which support adaptation in two ways: (1) reducing climate related disaster risk, and (2) offsetting the long-term implications of climate change. For example, reforestation (a key 'DRR' measure) will lessen the impact of a flood, but will also offset long-term soil degradation and help control local temperature and rainfall; improvements to the health sector in developing countries will help safeguard health in times of flood, and where there is lack of clean, safe drinking water.

In the same way that DRR supports adaptation, measures more typically associated with adaptation to climate change such as addressing the impact of glacial retreat or salt water intrusion onto agricultural land, will support DRR through reducing long-term vulnerability and influencing development potential. With similar aims and mutual benefits, the relevance of DRR to the design and implementation of adaptation policies and measures cannot be over-emphasised. As Sperling and Szekely (2005) state, 'To be effective, efforts to respond to the exceptional challenges posed by a changing climate must build on and expand the existing capability of disaster risk reduction, and should not be undertaken in isolation from this wider agenda.

The disaster risk management community not only has transferable, practical experience in addressing hazards, it also has strong and well-established local and regional institutions which are currently lacking in the field of adaptation.

1.3 Non-structural measures

Non-structural measures refer to policies, knowledge development/ awareness and methods and operating practices, including participatory

---

2. See Figure 7 in McGray *et al.* (2007): "A Continuum of Adaptation Activities: From Development to Climate Change".

mechanisms, which can reduce risk and related impacts.[3] These non-structural measures are well placed to serve both a DRR and a CCA agenda. The dynamism associated with training and awareness-raising means that people and institutions can apply skills and knowledge in different circumstances as they emerge. For example, awareness-raising as a component of an early warning system to cope with current flood risks is well placed to form an effective basis under a different future flood scenario.[4]

1.4 Poverty reduction and underlying risk

For both adaptation and DRR, poverty reduction and sustainable natural resource management are essential components of reducing vulnerability to hazards and climate change.[5] Addressing underlying risk factors is critical for effective poverty and vulnerability reduction.[6] Underlying risk relates to the interaction of a range of factors including globalisation processes, demographic trends, economic development, and trade patterns, which have an impact on exposure and vulnerability to hazards.

In principle, both adaptation and DRR aim to address such macro-level influences. However, in practice, perspectives on underlying risk do not yet go deep enough into the social, economic and political realms where risk is generated for the poor and most vulnerable. As such, a shared challenge for the climate change and disaster risk management communities is ensuring that adaptation and DRR commonly address the root causes of risk, not merely the symptoms.

1.5 Mainstreaming

It is increasingly recognised that adaptation and DRR must be integral components of development planning and implementation, to increase sustainability. In other words, these issues need to be 'mainstreamed' into

---

3. See *www.unisdr.org/eng/library/lib-terminology-eng%20home.htm*
4. This approach would be similar to building livelihood assets in each of the five areas of the sustainable livelihoods framework—human, natural, financial, social and physical capital.
5. Thomalla *et al.* (2006).
6. Addressing underlying risk was adopted as one of the five 'Priorities for Action' of the Hyogo Framework for Action 2005-2015.

Traditional knowledge is an important starting point for developing DRR strategies.

national development plans, poverty reduction strategies, sectoral policies and other development tools and techniques. At the World Conference on Disaster Reduction (WCDR) in 2005, governments agreed to adopt a mainstreamed approach to DRR.[7] To date there has been no such formal international-level agreement on mainstreaming adaptation. However, in 2005 the Commission for Africa made a significant recommendation that 'donors make climate variability and climate change risk factors an integral part of their project planning and assessment by 2008.'

1.6 Converging political agendas

Tanner (2007) states, 'The basis for adapting to the future climate lies in improving the ability to cope with existing climate variations. By improving the capacity of communities, governments, or regions to deal with current climate vulnerabilities, for instance through existing DRR,

---

7. Hyogo Framework (2005).

activities, their capacity to deal with future climatic changes is likely to improve.

In the policy debate on climate change there has been growing recognition of the importance of adaptation, and within this, the need to improve the capacity of governments and communities to address existing vulnerabilities to current climate variability and climatic extremes. This development has taken place in parallel to the shift from disaster management to disaster risk management, which is more anticipatory and forward-looking.[8] CCA and DRR, therefore, have merging remits and highly significant converging political agendas.

2. Differences

2.1 Hazard types

Climate-related, or 'hydro-meteorological' hazards only represent one type of hazard dealt with by the disaster management community. The full range of hazards that DRR can encompass includes natural (e.g., geological, hydro-meteorological and biological) or those induced by human processes (e.g., environmental degradation and technological hazards). Therefore, DRR expands beyond the remit of climate change adaptation (Figure 27.1).

Similarly, CCA moves outside the realm of most DRR experience, to address longer term impacts of climatic change such as loss of biodiversity, changes in ecosystem services and spread of climate-sensitive disease. These issues are typically positioned at the far end of the WRI's adaptation continuum, and are less likely to be addressed by the DRR community.

Both the climate change and disaster risk management communities must recognise that adaptation and DRR have these more exclusive elements, to avoid perpetuating the erroneous view that all adaptation and DRR is the same. However, recognition of exclusive elements should not detract from efforts to develop a more integrated approach, as the majority of adaptation and DRR measures have mutual benefits, offsetting both climate and disaster related risks.

---

8. Thomalla *et al.* (2006).

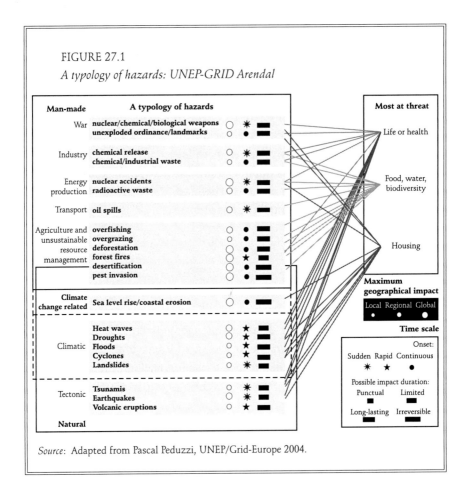

FIGURE 27.1
A typology of hazards: UNEP-GRID Arendal

Source: Adapted from Pascal Peduzzi, UNEP/Grid-Europe 2004.

2.2 Time scale

Thomalla *et al.* observes that much of the difference between adaptation and DRR relates to a different perception of the nature and timescale of the threat:

> Disasters caused by extreme environmental conditions tend to be fairly distinct in time and space (except for slow-onset or creeping disasters like desertification) and present a situation where the immediate impacts tend to overwhelm the capabilities of the affected population and rapid responses are required. For many hazards there exists considerable knowledge and certainty (based on historical experiences) about the event characteristics ... as well as exposure characteristics. Most impacts of climate

> Rationale for closer collaboration:
>
> - The institutional frameworks, political processes, funding mechanisms, information exchange fora and practitioner communities have developed independently and remained largely separate to date.19
> - There is no evidence of a systematic integration of disaster risk management and climate change adaptation in terms of concrete project activities.
> - Climate change is often housed in environmental or meteorology departments of governments. Government departments responsible for poverty and DRR are in some cases aware of vulnerability to extreme climate events, but have no means of coordination. This leads to the development of parallel effort in all three areas.

change, meanwhile, are much more difficult to perceive and measure, since the changes in average climatic conditions and climatic variability occur over a longer period …[9]

DRR focusses on reducing foreseeable risks based on previous experience, whereas adaptation originates with environmental science, predicting how climate change will be manifested in a particular region over a longer time period. Consequently, DRR is more likely to struggle to integrate risks that have yet to be experienced,[10] whereas this is a core component of an adaptation strategy with its focus on shifting environmental conditions.[11] However, according to Sperling and Szekely (2005), DRR is increasingly incorporating scientific advances and consequently is gaining a longer-term perspective. Indeed it must, if DRR measures are to be sustainable in the face of climate change.

---

9. Ibid.
10. For example, it is not uncommon for a housing damaged in a flood to be rebuilt disregarding risks associated with its location in a seismically active zone, unless an earthquake has been experienced in recent history.
11. Few *et al.* (2006).

TABLE 27.1
The key differences between DRR and CCA measures and approaches

Differences		Signs of convergence
DRR	Climate change adaptation	
Relevant to all hazard types	Relevant to climate related hazards	n/a
Origin and culture in humanitarian assistance following a disaster event	Origin and culture in scientific theory	Climate change adaptation specialists now being recruited from engineering, watsan, agriculture, health and DRR sectors
Most concerned with the present- i.e., addressing existing risks	Most concerned with the future i.e., addressing uncertainty/new risks	DRR increasingly forward-looking Existing climate variability is an entry point for climate change adaptation
Historical perspective	Future perspective	As above
Traditional/indigenous knowledge at community level is a basis for resilience	Traditional/indigenous knowledge at community level may be insufficient for resilience against types and scales of risk yet to be experienced	Examples where integration of scientific knowledge and traditional knowledge for DRR provides learning opportunities[1]
Structural measures designed for safety levels modelled on current and historical evidence[2]	Structural measures designed for safety levels modelled on current and historical evidence and predicted changes	DRR increasingly forward-looking
Traditional focus on vulnerability reduction	Traditional focus on physical exposure	n/a
Community-based process stemming from experience	Community-based process stemming from policy agenda	n/a
Practical application at local level	Theoretical application at local level	Climate change adaptation gaining experience through practical local application
Full range of established and developing tools[3]	Limited range of tools under development	None, except that climate related disaster recognition that more adaptation tools are needed
Incremental development	New and emerging agenda	n/a

contd...

...contd...

Differences		Signs of convergence
DRR	Climate change adaptation	
Political and widespread recognition often quite weak	Political and widespread recognition increasingly strong	None, except that climate related disaster events are now more likely to be analysed and debated with reference to climate change[4]
Funding streams ad hoc and insufficient	Funding streams sizeable and increasing	DRR community engaging in climate change adaptation funding mechanisms

Note:
1. Cronin et al. (2004).
2. Plus a determination on the level of acceptable risk: For the Netherlands the impact of flooding is enormous, and therefore flood defences are engineered to withstand very unlikely conditions, whereas for another region, the cost of such measures may be considered out of proportion with the additional safety level achieved.
3. For example: early warning systems; seasonal climate forecasts and outlooks; insurance and related financial risk management; building design codes and standards; land use planning and management; water management including regional flood management, draining facilities, flood prevention and flood-resistant agricultural practices; and environmental management, such as beach nourishment, mangrove and wetland protection, and forest management. (See climate and disaster risk reduction UN/ISDR 2003).
4. For example: Hurricane Katrina in the USA in 2005 or flooding/heat wave in Europe in 2002 and 2003 respectively.

2.3 Level of significance placed on existing capacities

Building resilience is a basis for both DRR and CCA. However, for DRR the emphasis is on determining existing capacity so as to anticipate, resist, cope with and recover from the impact of hazards. 'Traditional knowledge' on such matters is therefore an important starting point for developing DRR strategies. However, 'traditional knowledge' may be limited in its effectiveness at dealing with an exacerbation of existing problems, or with 'non-traditional problems', such as those to be experienced for the first time through climate change. A blend between traditional knowledge and an understanding of the projected impacts of climate must be sought.

2.4 Design limits for structural measures

Structural measures refer to any physical construction to reduce or avoid negative impacts of hazards, including engineering measures and construction of hazard-resistant and protective structures and infrastructure.[12] Under a DRR initiative based upon present and historical experiences, there is a greater likelihood that design limits for structural measures, such as flood embankments, will not be adequate in the face of climate change. Similar issues could be faced when considering changes in the frequency and severity of storms, drought, and other climate related phenomena, including sea level rise. Initiatives focussed on climate change adaptation are more likely to design structural measures with consideration for new, predicted impacts.

2.5 Comprehensiveness of measures to reduce vulnerability

The environmental science basis, from which climate change adaptation is emerging, means that adaptation largely focusses on shifting environmental conditions. Without such a strong environmental perspective and background, DRR is more likely to also consider and address social, physical and economic factors. Furthermore, through its inter-disciplinary analysis of conditions across all these categories, the disaster risk management community is more capable of recognising the wider constraints that determine vulnerability. This may account for why the adaptation community tends to place strong emphasis on developing hazard forecasting and early warning systems,[13] whereas DRR, by its nature, extends beyond disaster preparedness measures alone. Table 27.1 highlights the key differences between DRR and CCA measures and approaches, and indicates signs of convergence between the two disciplines.

3. Recommendations

Communication and collaboration between the climate change adaptation and disaster risk management communities must be improved.

3.1 Climate change adaptation community

- Use the guidance of the Hyogo Framework for Action (HFA) 2005-2012 agreed by 168 governments in Kobe, Hyogo, Japan in 2005, to

---

12. See www.unisdr.org/eng/library/lib-terminology-eng%20home.htm
13. Thomalla *et al.* (2006).

facilitate a comprehensive, system-wide risk-reducing approach to climate change adaptation.[14]

- Ensure a strong focus on DRR within the adaptation pillar of the post-2012 climate change framework.
- Use existing DRR tools that are proven effective in dealing with weather related events that will be exacerbated by climate change.
- Ensure adequate focus on the socio-economic and political dimensions of managing climate risks, in consultation with the disaster risk management community.
- Ensure that adaptation is informed by successful community-based experiences in vulnerability reduction.

3.2 Disaster risk management community

- Demonstrate and promote the role of DRR in CCA policies, strategies and programmes. Make DRR information and tools more accessible for CCA negotiators and managers.
- Ensure that all DRR policies, measures and tools account for new risks and the aggravation of existing risks posed by climate change.
- Actively engage in and seek to influence climate change policy at international, national, and local levels.

3.3 Both communities

- Increase awareness and understanding of adaptation and DRR synergies and differences at all levels.
- Encourage systematic dialogue, information exchange and joint working between climate change and disaster reduction bodies, focal points and experts, in collaboration with development policy-makers and practitioners.

---

14. Few *et al.* (2006).

References

Commission for Africa (2005). "Our common interest", Report of the Commission for Africa. Available online: *http: nepad.org*

Cronin, S.J., D.R. Gaylord, D. Charley, S. Wallez, B. Alloway and J. Esau (2004). "Participatory methods of incorporating scientific with traditional knowledge for volcanic hazard management on Ambae Island, Vanuatu", *Bulletin of Volcanology* 66: 652-668.

Few, R., H. Osbahr, L.M. Bouner, D. Viner and F. Sperling (2009). *Linking climate change adaptation and disaster risk management for sustainable poverty reduction: Synthesis report*. A study carried out for the Vulnerability and Adaptation Resource Group, VARG, Washington DC.

Hyogo Framework (2005). *The Hyogo Framework for Action 2005-2015: Building the resilience of nations and communities to disasters*. Ninth Plenary Meeting, WCDR. 22 January. Hyogo, Japan: UN.

McGray, H., A. Hammill and R. Bradley (2007). *Weathering the storm: Options for framing adaptation and development*. Washington, DC: World Resources Institute.

Sperling, F. and F. Szekely (2005). *Disaster risk management in a changing climate*. Washington, DC: VARG.

Tanner, T. (2007). "Screening climate risks to development cooperation", in Focus Issue 02.5. Brighton: IDS. November.

Thomalla, F., T. Downing, E. Spanger-Siegfried, H. Guoyi and J. Rockström (2006). "Reducing hazard vulnerability: Towards a common approach between disaster risk reduction and climate adaptation", *Disasters* 30 (1): 39-48.

UN/ISDR (2003). *Climate and disaster risk reduction*.

28

Disaster Risk Reduction and Climate Change Framework: The Cordaid Lens

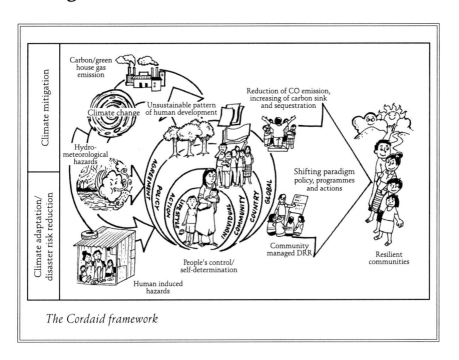

The Cordaid framework

Source:

Rustico Biñas, *Cordaid Resilient Communities: Navigating in the Fast Changing Climate*

1. Introduction

Community Managed Disaster Risk Reduction (CMDRR) is an emerging approach in response to climate change where Cordaid's development partners facilitate the empowering process to build resilient communities. The CMDRR process allows communities to actively identify, analyse and provide solutions to the disaster risks they are vulnerable to. Communities also identify, implement and evaluate disaster risk reduction (DRR) measures.

DRR was identified as one of the 10 programme areas in the Cordaid Strategic Plan 2007-2010. This reflects Cordaid's acknowledgement of the need for proactive rather than reactive response to the increasing hazard events in different parts of the world. Cordaid envisions DRR as an emerging framework and a tool for development. Furthermore, Cordaid recognises its indispensable role in addressing impacts of various hazards, with climate change as a compounding factor that has adversely affected the entire world, especially its vulnerable communities.

Cordaid is an international development organisation that focusses on emergency aid and structural poverty eradication. Ninety years in this work has enabled it to build a network of almost 1,000 partners in 36 countries in Africa, Asia, Latin America and Central and Eastern Europe. (*http://www.cordaid.nl/*)

2. The Cordaid framework

The framework of Cordaid on climate change mitigation and adaptation and DRR considers the different factors that affect society at various levels. Responses that will be developed will be geared towards building the resilience of communities to hazard events.

Cordaid believes that climate change and disasters are both social constructs (man-made). Therefore, the people themselves are the focal point of processes to address the impact of climate change. This is the basis for the shift of Cordaid's paradigm as reflected in its policy, programmes and actions.

The shift in paradigm entails recognition that current practices and lifestyle contribute to climate change and expects a transformation that should result from it. The culture of consumerism that contributes to the increase in emission of greenhouse gases into the atmosphere must be addressed because these harmful gases are causing change in climate. The effects of climate change are translated into various hydro-meteorological hazards like drought, hurricane and floods that further aggravate human-induced hazards like conflicts and civil unrest and societal hazards.

The climate mitigation portion of the framework generally means reducing carbon emission and increasing carbon sink and sequestration. In order to reinforce measures to mitigate the impacts of climate change, activities and concerns must be accomplished at different levels of engagement. At the global level, Cordaid will support policy negotiations through

A treadle pump can supply water in place of diesel- and electric-powered water pumps, a simple way to reduce carbon footprint.

Solar panels are expensive but they can efficiently provide electricity needs in the long term.

reduction of carbon and chlorofluorocarbons (CFC) emissions, and carbon sequestration. At the national level, country partners should contribute in terms of developing policy agenda and legislation to cut carbon and CFC emissions. Community action should be focussed on increasing carbon sink and sequestration, and lastly, at the individual level, Cordaid advocates for and educates people towards change in lifestyle to reduce carbon footprint.

Climate adaptation and DRR are means to strengthen the resilience of communities and to build a safe society. Cordaid approaches these by responding to the hydro-meteorological hazards caused by climate change and human-induced hazards. In this context, Cordaid will help partners to link their work to the National Adaptation Plan for Action (NAPA) and Strategic National Action Plan (SNAP-DRR) for Disaster Risk Reduction

In Bangladesh, environment-friendly brick making technology has been introduced as an alternative to fixed chimney kilns to reduce CO_2 emissions.

platform. Furthermore, Cordaid will push for strengthening community readiness for hazard management, prevention and mitigation at the community level. At the individual level, Cordaid will promote human rights as a foundation for safety, hazard specific knowledge, and enhanced attitude and skills for survivability.

3. Conclusion

Cordaid believes that all action points have to shift paradigms and mindsets from cost benefit analysis to a strong sense of moral obligation to cut carbon emissions. Cordaid also supports partners to shift mindsets to influence policy and programme towards CMDRR, continue promoting basic rights as a foundation of safety, and build awareness to adopt a change in lifestyle that reduces carbon and CFCs emission. A paradigm shift will lead to more resilient communities that are less vulnerable to the effects of climate change.

29
Enhancing Coping Mechanisms in Resilience Building

Since coastal communities are more vulnerable to floods and other natural calamities, enhancing their resilience is of paramount importance.

Source:

T.N. Balasubramanian *Independent Consultant.*

1. Concept

Coping mechanisms can be described as the sum total of ways by which a community deals with minor to major changes in ecological and social systems. The community is required to learn from previous experiences and apply them to cope with future climatic changes, including surprises. The coping mechanism includes strategies and measures that help in alleviating or containing the impact and bring efficient relief and resilience. Resilience helps to modify behaviour or activities in order to circumvent or avoid damaging effects and remain functional during the disaster and completely recover from it.

It is estimated that 60 per cent of total human population lives in the coastal ecosystem which is projected to increase to more than 75 per cent in the coming decades. Thus, enhancing resilience of coastal communities is of paramount importance. Coastal ecosystems face additional threats such as sea water intrusion, high velocity winds, cyclones and heavy rainfall in addition to floods and droughts that are commonly seen in other ecosystems.

Coastal Livelihoods:
- 60 per cent in fishing,
- 25 per cent in agriculture, and
- 15 per cent in trade associated with coastal ecosystem (barges, export, sand and salt extraction).

The livelihood vulnerability of the community living in any ecosystems depends on the community's capital investment and its flow, literacy and poverty level, food security and strength of village institutions. It also depends on the nature of degradation of natural resources, adaptive and coping capacity of the community and optional livelihood opportunities.

Floods caused by cyclonic heavy rain and inland inundation by tidal surge are annual catastrophes that disrupt the livelihood of Indian coastal communities. The intensity and frequency of these has increased over the years.

The livelihood of coastal communities of South Asian countries was severely affected by the December, 2004 tsunami. Tsunami caused greater loss to life, property and crops. This happened due to lack of disaster related information, less self-adaptive capacity of both individual and community and absence of past experience. There was no coping support extended from the external institutions.

The impact of disaster depends not only on the nature of disaster, vulnerability and exposure, but also on the coping capacity and the resilience of the community.

2. Enhancing adaptive and coping capacity for managing coastal disasters

Adaptation refers to any adjustment that takes place in natural system or human society in response to a disaster. It has the potential to reduce substantially many of the adverse impacts of disaster and enhance beneficial effects. Experience of disaster over years has evolved certain amount of adaptation. The coping capacity is adaptation plus external institutional support to bring resilience in the affected area as fast as possible. UN-ISDR (2004) defines coping capacity as the level of resources and the manner in which people or organisation use the resources and abilities to face the adverse consequence of disasters. Coping mechanisms reduce the vulnerability and build up resilience in coastal communities. The mechanisms include:

- Minimising the coastal conflicts between countries of South Asia through coastal integration. A new understanding of sovereignty wherein the coasts do not symbolise control or power but become spaces for interdependence.
- Providing salt and flood tolerant permanent infrastructures using modern engineering techniques.
- Strengthening afforestation by establishing either two or three-tier tree shelterbelts along the coastal line.
- Restoring degraded natural resources by implementing bio-village concept and community-based coastal resource management and development.

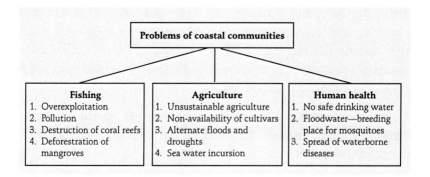

- Ban on the usage of plastic.
- Encouraging the community-based mangrove conservation.
- Providing enough drainage based on flood maps drawn for the area.
- Constructing well connected concrete cement roads.
- Establishing value addition cottage industries both for fisheries and agricultural products.
- Planning livelihood activities based on weather forecast.
- Education on coastal disaster and its management at the school level.
- Developing skills to enhance the adaptive capacity of the community.
- Monitoring overexploitation of marine resources using modern technology such as remote sensing.
- Recharging ground water by establishing simple recharge structures.
- Contingency plan to meet the disaster from floods and droughts to agriculture; and
- Imparting training/skills on disaster management to the community.

A small team of adaptive strategies can help communities build their adaptive capacity.

3. Conclusions

Coping mechanisms that build resilience among coastal communities will help to reduce their vulnerability to natural disasters. However, coping mechanisms will be successful only if there is a willingness to adapt. Adaptive capacity, therefore, depends on the ability of a society to act collectively, and to resolve conflicts among its members. Adaptive capacity confers resilience to perturbation, giving ecological and human social systems the ability to reconfigure themselves with minimum loss of function. The communities should involve a small team of adaptive strategists specialised in capacity building of the community who can also bring innovation at all levels across the system.

References

Gupta, C. and M. Sharma (2008). *Contested coastlines, fisherfolk,nations and borders in South Asia*. Routledge. *The Hindu*, Coimbatore Edition, October 22, 2009. p.8.

Kumar M. and D. Krishnan (2008). "The risk of sea level rise and submergence of resources", Proceedings of National conference on *Climate Change in India*, 28-29 August,2008, Environmental Research Centre, J.J. college of Engineering and Technology, Trichirapalli, Tamil Nadu, India. pp.113-117.

Thywissen, K. (2006). "Core terminology of disaster reduction: A comparative glossary", in J. Birkmann (ed.), *Measuring Vulnerability to Natural Hazards*. New Delhi, India: TERI Press. p.524.

UN-ISDR (United Nations-International Strategy for Disaster Reduction) (2004). *Living with risk: A global review of disaster reduction initiatives*. Geneva:UN Publications. DRR.

30
Supporting Local Capacities: From Jargon to Impact

Disaster push vulnerable people further into insecurity.

Source:

Supporting Local Capacities: From Jargon to Impact, southasiadisasters.net, Special issue 18, August 2006, All India Disaster Mitigation Institute, Ahmedabad, India.

Only a handful of emergency relief approaches produces long-term impact. Despite standards in humanitarian sector and learning opportunities, experiences show that impacts of assistance are not suitable for local communities. It should be taken into consideration that disasters push vulnerable people further into insecurity, destroying assets that are central to human security and the ability to provide security for their families. Thus, what donor agencies can do is to help remove barriers to recover assets such as supplies, finance and education for the affected.

1. The need to focus on local capacity

Capacity building is referred to as activities geared towards sustainably serving the poor. While development assistance focussing on service delivery efforts help disaster victims get back on their feet, the service delivery model is often unsustainable, making it an unviable endeavour. This is further improved with the recognition that long-term development needs are met by investing on building capacities of local people and organisations to address risk reduction.

Stakeholders of disaster management agree that there is indeed a need for capacity building. Yet, agencies fail to live up to the commitment

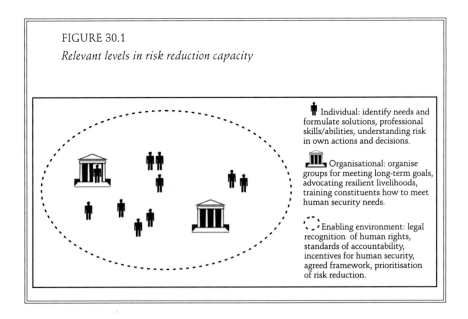

FIGURE 30.1
Relevant levels in risk reduction capacity

Individual: identify needs and formulate solutions, professional skills/abilities, understanding risk in own actions and decisions.

Organisational: organise groups for meeting long-term goals, advocating resilient livelihoods, training constituents how to meet human security needs.

Enabling environment: legal recognition of human rights, standards of accountability, incentives for human security, agreed framework, prioritisation of risk reduction.

to build the capacities of their local partners bringing about the call for further investigation of disaster risk reduction (DRR) capacity. Some initiatives in the development sector supporting capacity building at the local level include the Capacity Development Group of the United Nations Development Programme (CDG-UNDP) and the Development Assistance Committee (DAC) Network on Governance and Capacity Development (Govnet). The International Federation of Red Cross and Red Crescent Societies (IFRC) recognises and demonstrates the importance of supporting local capacities in Principles 6 and 7 of their Code of Conduct for actors in disaster response programmes. The Hyogo Framework for Action, as one of its three strategic goals, emphasises on strengthening capacities to build resilience to hazards. 'rethinking capacity for risk reduction' is also a focus of the UN Disaster Management Training Programme (UNDMTP).

Why do donors get little response on sustainable country capacity?

Despite few countries doing well, donor agencies' efforts in some countries produce little response on sustainable country capacity. This is a result of a skewed and disproportionate focus of capacity development on the organizational and enabling environment and too little emphasis on the viability of local - individual, family, community - initiatives.

Coping should be a highlighted aspect in disaster management as it points clearly to people's agency, ingenuity and abilities to help themselves individually and collectively. Without focus on coping, dependency on aid and other detrimental and unexpected outcomes will be the result.

Disaster risk reduction takes into consideration proactive attempts to prevent damage caused by natural hazards requiring political, financial and personal motivation. Although these attempts are intended to be carried out before any hazard, attempts to address disaster risk capacity mostly come after disaster occurrence. The lack of will and resources prevents more comprehensive incentives for mitigation. It is glaringly an issue of public policy after repeated cases of unfulfilled post-disaster promises to invest in risk reduction.

Sustainable DRR requires good institutions, investment and methods. Sustainability in this arena will require finding better ways to ensure capacity building investments targeting 'good' institutions. Moreover, approach to institutional capacity building is essential. Methods are yet to be developed to determine where, when and how and which capacities should be built.

2. Risk reduction capacity

Within risk reduction, we refer to the capacity to withstand the shock of disasters—the capacity to cope. Coping capacity in the context of development and DRR is the ability of an individual or group to avoid, reduce, withstand, and recover from the impact of a natural or man-made hazard, e.g. All India Disaster Management Institute (AIDMI) and the manner in which existing resources are used reactively to limit the losses brought about by disaster (UNDP, 2004).

What are the existing risk reduction capacities? Do all capacities contribute to human security? Are existing capacities well utilised? Are linkages made between levels, dimensions and phases? Are existing capacities robust, rapid, redundant and resourceful? Is room incorporated for dynamic growth? How many new capacities be generated for disaster risk reduction?

3. Coping capacity identification and support

Coping capacity development includes a multitude of activities and strategies. To come up with an identification framework, it is dissected into level, phase and dimension. After capacities are identified, strategies for support are also designed for each level, phase and dimension. The support framework associates a particular support strategy with a capacity to support for risk reduction.

4. Levels of capacities

Capacities are identified at three levels—individuals, organisations and enabling environments. Each level influences the behaviour of the actors at the lower level. Thus support to capacity at any level without regard for the other levels results to unsustainable impact. At the individual level, capacities are about dealing with direct threats to individual security on a daily basis. At the organisational level, capacities are about creating, sharing, and distributing resources, information and capabilities of individuals. On a systemic level, capacities are about creating incentives and enabling environments for organisations and individuals to develop their own capacity.

It is essential that relief support focusses on building the capacities of the national institutions and the environments. Nonetheless, it is equally important to look after the capacities of vulnerable communities themselves. Relief and recovery practice pays little attention to the needs of the ordinary people. The result is usually wasted resources, missed opportunities, and further erosion of coping skills.

Mitigation and preparedness activities may support a community's capacities for action in any time (phases)—before (to minimise), during (to survive) and after (to recover in a long-term sustainable and human development).

5. Dimensions in identifying coping capacities

Three dimensions are looked at in identifying coping capacities. Physical capacities include material or productive resources, information, technology, geographic location, cash and goods that will help people meet their needs in disaster situations. Organisational capacities include relationships, networks, organisations and systems which shall help people access necessary resources. Lastly, soft capacities include abilities to organise, participate, and manage relief, development and recovery activities. These are more on attitudinal or motivational and functional capacities useful in progressive risk reduction initiatives.

6. Four 'Rs' of disaster coping capacity

In making use of the capacity framework and assessing the resilience of each capacity, 'four R's' are to be used—especially applicable to disaster coping capacity. Robustness refers to strength or ability of elements, or systems to withstand a given level of stress without suffering degradation or loss of function. Rapidity is the capacity to meet priorities and achieve goals in a timely manner to contain losses and avoid future disruptions. Redundancy is the extent to which the elements or systems exist that are substitutable and capable of satisfying requirements in the event of disruption, degradation or loss of functionality. Lastly, resourcefulness is the capacity to identify problems, establish priorities and mobilise resources at existing threatening and disruptive conditions.

7. Conclusion

Following the assessment of the Tsunami Evaluation Coalition (TEC) after the Indian Ocean Tsunami in 2004, to a significant extent, local ownership of (tsunami) response was undermined by the actions of international agencies. Recognition and engagement with the local capacity was insufficient particularly where capacities were not visible in the form recognised by international agencies. In some cases, local capacities were depicted as even more vulnerable by the response. This brings about the clear need for relief actors to understand what capacities are relevant for supporting during recovery. The focus on local—what people do and what people need—come into place. To bring about resilience, relief agencies may support community-based management which backs the community's voices and the organisations' advocacy for protection. Furthermore, outside organisations may begin by helping community-based organisations identify the threats and needs in their specific areas.

The coping capacity framework as illustrated should help those in and outside the local administration who believe in the people's ability to reduce risk and who are dedicated towards supporting these groups. These tools shall aid local capacities at the individual and organisational levels and create an environment that enables human resilience.

References

UNDP (2004). *Disaster risk reduction: A challenge for development*. New York: UNDP.

Part 3

Building Capacities: The Path to Community Resilience

This third part of the source book focusses on importance of improved community capacities to the effects of natural disasters. This is often the best, long-term investment that development agencies can make. Resilient communities can take deliberate actions to deal with disasters. Resilience building can also serve as a useful integrating framework for integrated coastal management, community-driven development, livelihood assessments, and disaster preparedness and planning. Livelihoods have been placed at the centre of development efforts in protecting communities from the adverse affects of natural disasters. Strong social networks, social capital, habitat creation and restoration, financial mechanisms, markets linkages, and utilisation of low cost and effective tolls from information-communication technologies make significant contributions to strengthening capacities of local communities.

31
Emergence of Resilience

The concept of 'resilience' is a subject for debate.

Source:

Buddhi Weerasinghe *Communication Consultant, Disaster Management Centre, Sri Lanka, 2010.*

The concept of resilience building appears to be an intangible entity to a ground level disaster manager. The reason is that building resilience requires a multitude of stakeholders and inter-sectoral collaborations which is beyond the capability of a short-term project based disaster risk reduction (DRR) initiative. The concept of resilience itself is subject for debate. This article attempts to look at different views on resilience in search of a tangible framework which can be subjected to assessment after implementation.

How valid is the use of the term 'resilience'? Can we simply take the concept of resilience from ecology and apply it to social systems? How feasible is it for ground level disaster managers to attempt building community resilience? Can community resilience be assessed?

This paper attempts to review the theoretical basis for using the term 'community resilience'. For this discussion, 'community' is interpreted as communities of location/place (Fritzgerald and Fritzgerald, 2005), which generally consist of a network of overlapping interest groups.

1. The emergence of 'resilience' in the area of ecology, field experiments, and disaster management

1.1 Resilience in ecology

Holling (1973) first defined ecosystem resilience (L., *resiliere*, to jump back) as the measure of the ability of an ecosystem to absorb changes and persist. He also compared the concept of resilience with the notion of stability, which he defined as the ability of a system to return to its equilibrium after a temporary disturbance. Thus on one side, ecosystem resilience is a resistance to change and on another, it is the ability of the ecosystem to return to its original state.

Holling (1996) then went on to distinguish two types of resilience: engineering and ecological.

a) Engineering resilience is the time to recovery, the rate and speed of return to pre-existing conditions after disturbance.

b) Ecological resilience is the magnitude of disturbance that can be absorbed before the system changes its structure by changing the variables and processes that control behaviour.

1.2 Resilience in other fields of study

The concept of resilience has increasingly gained recognition and acceptance and it is now frequently used in many fields. In the early 20th century, economists and sociologists, such as Vilfredo Pareto began to apply J. Willard Gibbs' equilibrium criterion to the modeling of economic systems and social systems (Bailey, 1990).

Belgian chemist and Nobel laureate Ilya Prigogine (1977) however expressly stated that the Gibbs' equilibrium method is not applicable to life. He further said that evolving living systems, such as communities of people (or human ecosystems), are 'far-from-equilibrium' systems. Social resilience differs fundamentally from ecosystem resilience because of the additional factor of the capacity of humans to anticipate and plan for the future (Moberg and Galaz, 2005).

Simply put, '"community resilience" is "the ability of a community to "bounce back" and recover using its own resources' (Paton and Johnston, 2001). Adger (2000) further defines social resilience as the ability of groups or communities to cope with external stresses and disturbances as a result

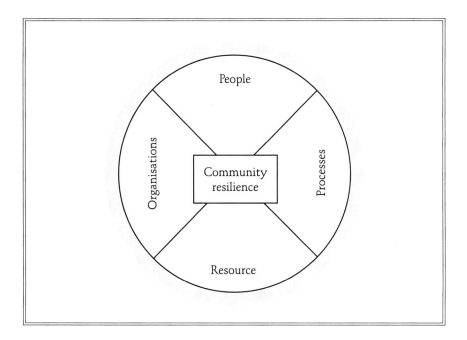

of social, political and environmental change. Such community resilience has four dimensions: people, resources, organisations and community processes (CCE, 2000).

1.3 Resilience in disaster management

Timmerman (1981) is probably the first to use the concept of resilience in relation to hazard and disaster. He defined 'resilience' as the measure of a system's or part of the system's capacity to absorb and recover from hazardous events (Figure 31.1).

While the concept could be defined, measuring it is less definite. The main challenge is how to define and develop indicators that can adequately measure it or, how it should be mapped and what unit of analysis should be used (Mayunga, 2007). When the appropriate indicators can neither be defined nor measured, the concept of resilience will not be useful for developing strategies for or sufficiently informing policies on DRR.

In relation to this, Mayunga (2007) proposes that resilience can be assessed based on five major forms of capital: social, economic, physical, human and natural (Table 31.1). These forms of relationships are necessary for development of a sustainable community economy. According to

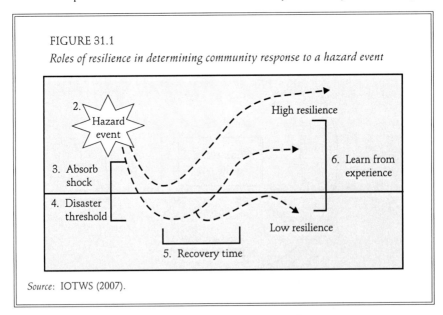

FIGURE 31.1
Roles of resilience in determining community response to a hazard event

Source: IOTWS (2007).

TABLE 31.1
Relationship between capital domains and community disaster resilience

Form of capital	Indicators of resilience	Function
Social	• Trust • Norms • Networks	• Facilitates coordination and cooperation • Facilitates access to resources
Economic	• Income • Savings • Investment	• Increases capacity e.g. insurance • Speeds recovery process • Increases well-being and reduces poverty
Human	• Education, health • Skills • Knowledge/information	• Increases knowledge and skill to understand community risks • Increases ability to develop and implement risk reduction strategy
Physical	• Housing • Public facilities • Business/industry	• Facilitates communication and transportation • Facilitates evacuation • Increases safety
Natural	• Resources stocks • Land and water • Ecosystem	• Sustains all forms of life • Increases protection to storms and floods • Protects the environment

Mayunga, determining an appropriate unit of analysis for community disaster resilience is critically important but often tricky. This is because resilience can be measured at all levels such as individual, household, group, community, or society.

A review of tsunami impact in Sri Lanka, shows that assets of all five capital domains have been destroyed but normalcy was restored through outside interventions. This means that the drivers of many of these capital domains are external and thus, beyond the control of the community. Because each domain is under the jurisdiction of many players, this clearly impedes quick recovery.

In order to protect communities from impacts of disaster, concern over gaps in the recovery process in Sri Lanka must be addressed urgently (People's Report, 2007). Many of these gaps relate to issues of people's inclusion in decision-making; justice; accountability; poverty reduction; employment/livelihood; environment; women's rights; security for women and children; sustainable development and DRR; and post-disaster rehabilitation and reconstruction. All of these are fundamental concerns for the decentralisation of DRR, local development planning and local governance. These issues are also part of Millennium Development Goals (MDGs) and South Asian Association for Regional Cooperation (SAARC) Development Goals (SDGs).

2. What is a resilient community?

2.1 It is a safe community

Geis (2000) describes a resilient community as 'the safest possible community that we have the knowledge to design and build in a natural hazard context', thus minimising its vulnerability by maximising the application of DRR measures.

2.2 It has capacity to respond to disaster

Many authors use the term capacity/ability to define the concept of disaster resilience. On the other hand, some consider it as the opposite of vulnerability.

For instance, Foster (2006) describes regional resilience as the ability of a region to anticipate, prepare for, respond to and recover from a disturbance. Moreover, Buckle et al. (2000) describe resilience as the

Disaster management focusses on, among others, preparedness and response to reduce human losses from disaster events.

'quality of people, communities, agencies, and infrastructure that reduces vulnerability'. They state further that this quality is more than 'just the absence of vulnerability'. Rather, it includes the capacities to:

- Prevent or mitigate loss;
- If damage does occur, maintain normal condition as far as possible; and
- Manage recovery from the impact.

2.3 It has control over components of resilience

Fitzgerald and Fitzgerald (2005) further state that resilience is a 'state' or 'property' of a community and must be nested within district, provincial and national level. The reason is the control level imposed from outside the community to build each component of resilience (Table 31.2).

Except for governance that is mostly controlled externally, the community exercises control over most of the components of resilience. This is especially true with risk assessment, knowledge and education, and disaster preparedness and response.

However, the component must become operational at various levels before it can become a 'state' of the community. This nested property will remain an intangible entity that will challenge assessment.

2.4 It is a community that knows and acts

Fitzgerald and Fitzgerald (2005) point out that resilient communities are 'communities that know their risks, have reduced their vulnerability, and have the ability to respond to and recover from the impacts of such risks'. Effectively echoing Foster (2006) and Buckle et al. (2000), they emphasise that the community's resilience relies heavily on the community's knowledge and abilities. Without the following abilities, the community's vulnerability increases:

- knowledge of risks faced;
- ability to anticipate and reduce potential consequences;
- ability to respond to the event; and
- ability to cope with and recover from consequences.

TABLE 31.2
Community control over components of resilience: A Sri Lankan perspective

Thematic area	Components of resilience	Community control	External control
Governance	• Policy, planning, priorities and political commitment	–	✓
	• Legal and regulatory systems	–	✓
	• Integration with development policies & planning	–	✓
	• Integration with emergency response & recovery	–	✓
	• Institutional mechanisms, capacities & structures	–	✓
	• Allocation of responsibilities	–	✓
	• Partnerships	–	✓
	• Accountability and community participation	–	✓
Risk assessment	• Hazards/risk data and assessment	✓	✓
	• Vulnerability and impact data and assessment	✓	✓
	• Scientific and technical capacities and innovation	–	✓
Knowledge and education	• Public awareness, knowledge and skills	✓	✓
	• Information management and sharing	✓	✓
	• Education and training	–	✓
	• Cultures, attitudes, motivation	✓	✓
	• Learning and research	✓	✓
Risk management and vulnerability reduction	• Environmental and natural resource management	✓	✓
	• Health and well-being	✓	✓
	• Sustainable livelihoods	✓	✓
	• Social protection	✓	✓
	• Financial instruments	–	✓
	• Physical protection; structural & technical measures	–	✓
	• Planning régimes	–	✓
Disaster preparedness and response	• Organisational capacities & coordination	✓	✓
	• Early warning systems	✓	✓
	• Preparedness and contingency planning	✓	✓
	• Emergency resources and infrastructure	–	✓
	• Emergency response and recovery	✓	✓
	• Participation, voluntarism, accountability	✓	✓

Based on these abilities, Fitzgerald and Fitzgerald (2005) identify various options that can be done by the community to reduce its vulnerability. While they identify factors that may affect successful action, they also highlight that any reduction in vulnerability entails the active and concerted action of the community concerned. The experience from the field is captured in the article Community Resilience: The Case of Wanduruppa (Article 42: 247).

3. Conclusion

Although the use of the term resilience in a social context seems inappropriate, several authors have attempted to define its use within applicable limits. The definitions are greatly useful in providing an idealistic vision. But note that frameworks proposed for resilience building can only be attempted on a national scale.

Proposed characteristics of resilience are multi-faceted and distributed across many sectors. Many actors are responsible and various aspects interconnect beyond the scope of small- and short-time scale projects of limited funding.

TABLE 31.3
Possible community action to reduce vulnerability

Component	Community level action	Comment
Knowledge of risk faced	• Decide on geographical boundary • Revive historical memory • Revive Indigenous knowledge • Do participatory hazard identification • Do participatory hazard mapping • Do identification of vulnerabilities	Within community capability facilitation may be needed
Ability to anticipate potential consequences	• Re-visit worst impact experienced • Do participatory preparedness and response planning	Capacity building may be necessary
Ability to reduce potential consequences	• Decide on acceptable risk • Implement local level mitigation	Funding may be an issue
Ability to respond to an event	• Identify evacuation routes and safe places • Establish early warning dissemination • Carry out simulation drills • Organise shelter and relief	Capacity building may be necessary
Ability to cope with situations	• Involve community leadership • Build social networking • Initiate resource pooling • Encourage volunteerism	• Capacity building may be necessary • Outside intervention may be necessary
Ability to recover from consequences	• Reconstruction of houses, infrastructure and critical facilities • Re-building livelihoods • Strengthen lobbying capacity	Outside resources necessary

Aha! Increased resilience = reduced vulnerability

In the face of different and more pressing priorities, it is possible that far-reaching policy changes needed for inter-sectoral coordination to build the resilience of communities will not become a priority in the near future in Sri Lanka. This is especially true for most developing nations. With demand on scarce financial resources for government priorities other than DRR, government commitment for upscaled holistic disaster management interventions appear unrealistic.

Another difficulty arising from the use of the term resilience is how it can be satisfactorily assessed. Resilience is an intangible asset and quality to work with. It is an entity with too many interlinks which is beyond the hands of the community and the project. In a country where much community level interventions rely on donor funding, project evaluations on assessing resilience capacities in the area would play a major role on continuity of funds. Using a framework with measurable discreet actions is therefore more prudent to show successful implementation. Comprehension of outcomes by the stakeholders would be easier and therefore, motivation to move forward would be healthier.

It appears more judicious then to depend on projects, which aim towards achieving discrete and tangible targets such as hazard-specific vulnerability reduction through mitigation, measurable objectives in capacity building, awareness creation and hazard-specific preparedness within manageable geographic areas. A bottoms up accumulation of results could gradually create more interconnected community achievements that may strengthen community safety. The ground level disaster manager could accomplish more by keeping things simple.

References

Adger, N.W. (2000). "Social and ecological resilience: Are they related?", *Progress in Human Geography* 24(3): 347-364.

Bailey, Kenneth D. (1990). *Social entropy theory*. State University of New York Press. pp. 59-61.

Buckle, P., G. Marsh and S. Smale (2000). "New approaches to assessing vulnerability and resilience", *Australian Journal of Emergency Management*: 8-15. (Winter Issue).

Fitzgerald, G. and N. Fitzgerald (2005). "Assessing community resilience to wildfires: Concepts and approach", Paper Prepared for SCION Research. Fitzgerald Applied Sociology. June.

Foster, K.A. (2006). "A case study approach to understanding regional resilience", A *Working Paper* for building resilience network Institute of Urban Regional Development, University of California.

Geis, D.E. (2000). "By design: The disaster resistant and quality-of-life community", *Natural Hazards Review* 1(3): 152.

Holling, C.S. (1996). "Engineering resilience *versus* ecological resilience", in P. Schulze (ed). *Engineering within ecological constraints*. Washington, D.C.: National Academy Pages. pp. 31-44.

————. (1973). "Resilience and stability of ecological systems", *Annual Review of Ecology and Systematics* 4: 2-23.

International Strategy for Disaster Reduction (ISDR) (2005). "Hyogo Framework for Action 2005-2015: Building the resilience of nations and communities to disasters", Extract from the final report of the *World Conference on Disaster Reduction* (A/CONF.206/6) held on 18-22 January 2005 at Kobe, Hyogo, Japan.

————. (2009). "Terminology on disaster risk reduction", Posted at *http://www.unisdr.org/eng/terminology/terminology-2009-eng.html*

Mayunga, J.S. (2007). "Understanding and applying the concept of resilience: A capital-based approach", *Working Paper*. Summer Academy for Social Vulnerability and Resilience Building, Munich, Germany, 22 - 28 July.

Moberg, F. and V. Galaz (2005). "Resilience: Going from conventional to adaptive freshwater management for human ecosystem compatibility", *Swedish Water House Policy Brief* 3. The Stockholm International Water Institute (SIWI).

Paton, D. and D. Johnston (2001). "Disasters and communities: Vulnerability, resilience and preparedness", *Disaster Prevention and Management* 10 (4): 270-277.

People's Report (2007). *People's Report series*:

————. Disaster risk reduction in the post tsunami context: India, Maldives, Sri Lanka and Thailand.

————. Violence against women in the post tsunami context: India, Maldives, Somalia, Sri Lanka and Thailand.

———. Homestead land and adequate housing in the post tsunami context: India, Maldives, Sri Lanka and Thailand.

———. Disaster risk reduction, Sri Lanka.

Prigogine, I. (1977). "Time, structure, and fluctuations", *Nobel Lecture* (in chemistry). Dec 08.

Timmerman, P. (1981). "Vulnerability, resilience and the collapse of society", *Environmental Monograph* 1, Institute for Environmental Studies, Toronto University.

US Indian Ocean Tsunami Warning System Program (IOTWS) (2007). *How resilient is your coastal community? A guide for evaluating coastal community resilience to tsunamis and other hazards*. Bangkok, Thailand: United States Agency for International Development and partners.

32

Understanding Resilience in Coastal Areas

Highly dense population along waterways and coasts sets the stage for more frequent and severe disasters that could put coastal communities in a perpetual state of response and recovery from one disaster event to another.

Source:

Adapted from: How Resilient is Your Coastal Community? A Guide for Evaluating Coastal Community Resilience to Tsunamis and Other Hazards, October 2007. Prepared by the US Indian Ocean Tsunami Warning System Program.

1. Introduction

Increased vulnerability of coastal communities to potential hazards is partly due to the constantly increasing coastal population (Adger *et al.*, 2005). Currently, an estimated 23 per cent of the world's population (1.2 billion people) live within 100 kilometres of a shoreline and 100 metres of sea level (Small and Nicholls, 2003). By the year 2030, an estimated 50 per cent of the world's population will live in the coastal zone.

Coastal habitats such as reefs, mangroves, wetlands and tide lands provide nursery and feeding areas for many marine species and serve as buffer areas for storm protection and erosion control. These coastal habitats are being destroyed by a wide range of human uses, including shoreline development, land reclamation, mining, and aquaculture. Runoff, wastewater discharges, and oil spills pollute coastal waters and endanger marine life. Overfishing and the use of destructive fishing practices are causing the decline of fishery resources and changes in marine ecosystem structure and function. The degradation of the coastal environment from chronic human-induced actions threatens food security, livelihoods, and the overall economic development and well-being of coastal communities.

What is resilience?

'Resilience determines the persistence of relationships within a system and is a measure of the ability of these systems to absorb change of state... and still persist'(Holling 1973).

'...Resilience for social-ecological systems is often referred to as related to three different characteristics: (a) the magnitude of shock that the system can absorb and remain within a given state; (b) the degree to which the system is capable of self-organization; and (c) the degree to which the system can build capacity for learning and adaptation'(Folke *et al.*, 2002).

'The capacity of a system to absorb disturbance and re-organize while undergoing change so as to still retain essentially the same function, structure, identity and feedback.' (Walker *et al.*, 2004).

Most of the coastal population lives in densely populated rural areas and small to medium cities, rather than in large cities. In these relatively rural communities, basic services and disaster warning and response mechanisms are limited (Figure 32.1). Limited capacity of a community to plan for and respond to coastal hazards makes coastal populations increasingly vulnerable to disasters.

Economic development pressures along the coast, population density and distribution, and human-induced vulnerabilities, coupled with increasing frequency and duration of storms, sea level rise, and other chronic coastal hazards, increase risk. These conditions set the stage for more frequent and severe disasters and reduced time and capacity to recover, with some coastal communities finding themselves in a state of perpetual response to and recovery from one disaster event after another. The assessment of risk is an important element of coastal community resilience (CCR). Communities must identify their exposure to hazard impacts to proactively address emergency planning, response, recovery, and implement hazard mitigation measures (Figures 32.1 and 32.2).

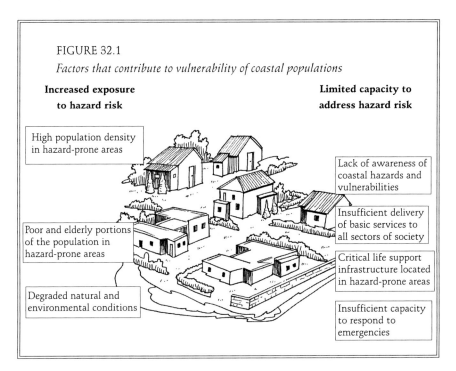

FIGURE 32.1
Factors that contribute to vulnerability of coastal populations

Increased exposure to hazard risk
- High population density in hazard-prone areas
- Poor and elderly portions of the population in hazard-prone areas
- Degraded natural and environmental conditions

Limited capacity to address hazard risk
- Lack of awareness of coastal hazards and vulnerabilities
- Insufficient delivery of basic services to all sectors of society
- Critical life support infrastructure located in hazard-prone areas
- Insufficient capacity to respond to emergencies

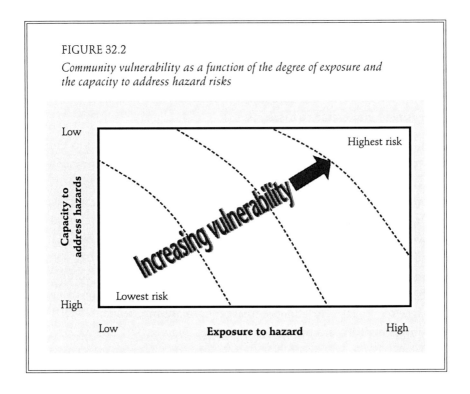

FIGURE 32.2
Community vulnerability as a function of the degree of exposure and the capacity to address hazard risks

2. Why is resilience critical to coastal communities?

Community resilience is the capacity of a community to adapt to and influence the course of environmental, social and economic change. The resilience of socio-ecological systems is often described as a combination of three characteristics: the magnitude of shock that the system can absorb and remain within a given state; the degree to which the system is capable of self-organisation; and the degree to which the system can build capacity for learning and adaptation (Folke et al., 2002). Common characteristics of resilient systems include redundancy, diversity, efficiency, autonomy, strength, interdependence, adaptability, and collaboration (Godschalk, 2003). Resilience provides the capacity to absorb shocks while maintaining function. When change occurs, resilience provides the components for renewal and reorganisation (Gunderson and Holling, 2002; Berkes and Folke, 2002).

Vulnerability is the flip side of resilience: when a social or ecological system loses resilience, it becomes vulnerable to change that previously could be absorbed (Kasperson and Kasperson, 2001). In a resilient system, change has the potential to create opportunity for development, novelty and innovation. In a vulnerable system, even small changes may be devastating.

3. Characteristics of resilient coastal communities

Resilient coastal communities take deliberate action to reduce risk from coastal hazards with the goal of avoiding disaster and accelerating recovery in the event of a disaster. They adapt to changes through experience and applying lessons learned.

CCR serves as a unifying framework for community-based plans and programmes. Enhancing CCR requires integrating and maintaining an optimal balance of three community-based frameworks typically viewed as independent and separate domains: community development, coastal management, and disaster management (Figure 32.3).

FIGURE 32.3
Resilience as an integrating framework for community development, and disaster and coastal management

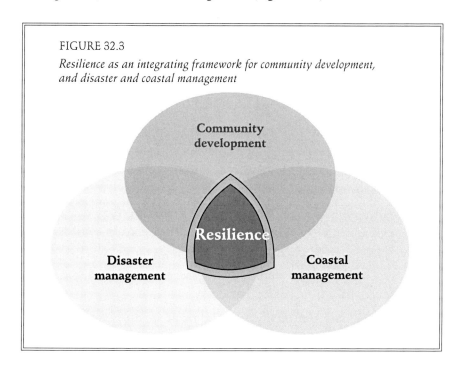

Community development provides the enabling governance, socioeconomic and cultural conditions for resilience (IMM, 2003; CED, 2000). Coastal management provides the framework for managing human uses of coastal resources and the coastal zone in order to maintain environmental and ecosystem resilience (White *et al.*, 2005; Chua, 1998; DENR, 2001). Disaster management focusses on preparedness, response, recovery and mitigation to reduce human and structural losses from disaster events (ADPC, 2005; 2004).

4. Resilience elements and benchmarks

Eight elements of resilience were identified as essential for CCR (Figure 32.4). These elements incorporate long-term planning and implementation such as society and economy, coastal management, land use and structural design. Hazard event-oriented resilience elements focus on contingency planning and preparedness for warning and evacuation, emergency response, and disaster recovery. Governance as a resilience element provides the enabling framework for resilience in all other elements. Risk knowledge is a cross-cutting requirement within each resilience element. Enhancing resilience in all of these elements is considered essential to reduce risk from coastal hazards, accelerate recovery from disaster events, and adapt to changing conditions in a manner that is consistent with community goals.

The desired outcome or overarching vision for each element of CCR can be described as follows:

a. *Governance*: Leadership, legal framework, and institutions provide enabling conditions for resilience through community involvement with government.

b. *Society and economy*: Communities are engaged in diverse and environmentally sustainable livelihoods resistant to hazards.

c. *Coastal resource management*: Active management of coastal resources sustains environmental services and livelihoods and reduces risks from coastal hazards.

d. *Land use and structural design*: Effective land use and structural design that complement environmental, economic, and community goals and reduce risks from hazards.

e. *Risk knowledge*: Leadership and community members are aware of hazards and risk information is utilised when making decisions.

f. *Warning and evacuation*: Community is capable of receiving notifications and alerts of coastal hazards, warning at-risk populations, and individuals acting on the alert.

g. *Emergency response*: Mechanisms and networks are established and maintained to respond quickly to coastal disasters and address emergency needs at the community level.

h. *Disaster recovery*: Plans are in place prior to hazard events that accelerate disaster recovery, engage communities in the recovery process, and minimise negative environmental, social and economic impacts.

Benchmarks for each resilience element evaluate the resilience condition or status of a community (Table 32.1). Each benchmark represents desired conditions against which to evaluate the resilience status of a coastal community. The benchmarks for each resilience element characterise desired conditions in four core capacities: (i) policy and planning, (ii)

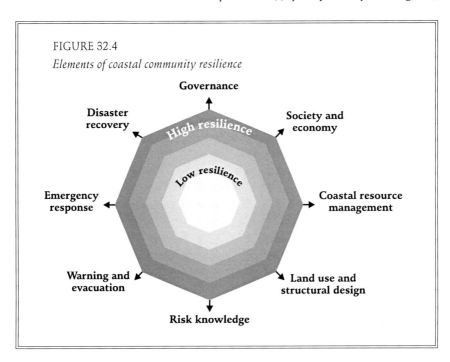

FIGURE 32.4
Elements of coastal community resilience

TABLE 32.1
Benchmarks for each resilience element by core capacities

	Benchmarks			
Resilience element	Policy and planning capacity	Physical and natural capacity	Social and cultural capacity	Technical and financial capacity
Governance	A1. Community development policies, plans and programmes are implemented and monitored in a participatory and transparent manner.	Physical and natural capacity A2. Basic services (i.e., water, transportation, security, etc.) are accessible to all sectors of society.	Social and cultural capacity A3. Participatory collaboration mechanisms among different sectors and various levels of government are established and used to manage for resilience.	Technical and Financial Capacity A4. Technical and financial support mechanisms are transparent, accountable and available to support planned community actions.
Society and economy	B1. Development policies and plans build social capital and skills for economic diversity and self-reliance.	B2. Local economies are characterised by diverse and environmentally sustainable livelihoods.	B3. Social and cultural networks promote self-reliant communities and have the capacity to provide support to disaster-stricken areas.	B4. Technical and financial resources are available to promote stable and robust economies, reduce vulnerability to hazards, and aid in disaster recovery.
Coastal resource management	C1. Policies and plans are implemented and monitored to effectively manage natural coastal resources.	C2. Sensitive coastal habitats, ecosystems, and natural features are protected and maintained to reduce risk from coastal hazards.	C3. Communities are actively engaged in planning and implementing coastal resource management activities.	C4. Communities and local governments value and invest in management and conservation to sustain their natural resources.

contd...

Understanding Resilience in Coastal Areas

...contd...

Resilience element	Benchmarks			
	Policy and planning capacity	Physical and natural capacity	Social and cultural capacity	Technical and financial capacity
Land use & structural design	D1. Land use policies and building standards that incorporate measures to reduce risks from hazards and protect sensitive habitats are established, monitored and enforced.	D2. Critical infrastructure are located outside high-risk areas and constructed to address risks from priority hazards.	D3. Developers and communities incorporate risk reduction into the location and design of structures.	D4. Education, outreach and training programmes are established to improve compliance with land use and policies and building standards.
Risk knowledge	E1. Coastal hazard risk assessments are completed at a scale appropriate to the community and routinely updated.	E2. Coastal hazard risk assessments are comprehensive and incorporate risks to all elements of resilience (e.g., livelihoods, coastal resources, land use, etc.).	E3. Community participates in the hazard risk assessment process.	E4. Information from risk assessment is accessible and utilised by the community and government.
Warning and evacuation	F1. Community warning and evacuation systems, policies, plans, and procedures are in place and capable of altering vulnerable populations in a timely manner.	F2. Community warning and evacuation infrastructure is in place and maintained.	F3. Community is prepared to respond to hazard warnings with appropriate actions.	F4. Technical and financial resources are available to maintain and improve warning and evacuation systems.
Emergency response	G1. Pre-defined roles and responsibilities are established for immediate action at all levels.	G2. Basic emergency and relief services are available.	G3. Preparedness activities (drills and simulations) are ongoing to train and educate responders.	G4. Organisations and volunteers are in place with technical and financial resources to support emergency response activities.

contd...

...contd...

Resilience element	Benchmarks			
	Policy and planning capacity	Physical and natural capacity	Social and cultural capacity	Technical and financial capacity
Disaster recovery	H1. Disaster recovery plan is pre-established that addresses economic, environmental and social concerns of the community.	H2. Disaster recovery process is monitored, evaluated and improved at periodic intervals.	H3. Coordination mechanisms at international, national, and local levels are pre-established for disaster recovery.	H4. Technical and financial resources are available to support the recovery process.

physical and natural resources, (iii) social and cultural, and (iv) technical and financial (Bruneau *et al.*, 2003 and IMM, 2003). Benchmarks on policies and plans describe enabling conditions for community resilience. Physical and natural resource benchmarks describe infrastructure or coastal resource capacity to support resilience. Benchmarks on social and cultural capacity highlight self-reliance of the community achieved through networks, cultural norms, and education and outreach. Finally, the technical and financial resource benchmarks characterise the support needed to sustain resilience efforts. The benchmarks provided in Table 32.1 are the foundation of the CCR assessment tool, which will help identify the strengths, weaknesses and gaps in community resilience.

5. Conclusion

The vulnerability of coastal communities to disasters has been highlighted and made more evident as economic development and population pressures bear down on these high-risk areas. Aside from preventing more

An important resilience benchmark is that when developers, communities, and the government incorporate risk reduction into the location and design of structures.

people from residing in these areas as well as encouraging residents to transfer to less risky locations, there is a need to build resilience among those who are already living there. Resilience is not only the ability to respond to disaster events but the conscious will to anticipate and plan for such events. Resilience is proactive and forward looking. It demands the will power of leaders to exhibit responsible governance and to show the ability to fully appreciate the situation in coastal communities, and implement coastal resource management and land use decisions that would not only reduce the time and increase the capacity to recover from disasters but that would also benefit the larger society.

References

Adger, W.N., T.P. Hughes, C. Folke, S.R. Carpenter and J. Rockström (2005). "Social-ecological resilience to coastal disasters", *Science* 309(5737): 1036-39. 12 August.

Asian Disaster Preparedness Center (ADPC) (2004). *CBDRM field practitioners' handbook*. Thailand: Asian Disaster Preparedness Center.

———. (2005). *Disaster risk management in Asia: A primer*. Prepared under USAID Cooperative Agreement No. DFD-A-00-03-00077-00.

Berkes, F. and C. Folke (2002). "Back to the future: Ecosystem dynamics and local knowledge", in L.H. Gunderson and C.S. Holling (eds.), *Panarchy: Understanding transformations in human and natural systems*. Washington, DC: Island Press. pp.121–146.

Bruneau, M., S. Chang, R.T. Eguchi, G.C. Lee, T.D. O'Rourke, A.M. Reinhorn, M. Shinozuka, K. Tierney, W. Wallace and D. von Winterfeldt (2003). "A framework to quantitatively assess and enhance the seismic resilience of communities", *Earthquake Spectra* 19(4): 733-52.

Chua, T.E. (1998). "Lessons learned from practicing integrated coastal management in Southeast Asia", *Ambio* 27: 599-10.

DENR (2001). *Philippine coastal management guidebook series* (8 volumes). DA-BFAR, and DILG (Department of Environment and Natural Resources, Department of Agriculture-Bureau of Fisheries and Aquatic Resources, and Department of the Interior and Local Government). Coastal Resource Management Project of the DENR. Cebu City, Philippines.

Folke, C., J. Colding and F. Berkes (2002). "Synthesis: Building resilience and adaptive capacity in social-ecological systems", in F. Berkes, J. Colding and C. Folke (eds.), *Navigating social-ecological systems: Building resilience of complexity and change*. Cambridge University Press.

Godschalk, R. David (2003). "Urban hazard mitigation: Creating resilient cities", *Natural Hazards Review* 4(3): 136-43.

Gunderson, L.H. and C.S. Holling (eds.) (2002). *Panarchy: Understanding transformations in human and natural systems*. Washington, D.C., USA: Island Press.

Holling, C.S. (1973). "Resilience and stability of ecological systems", *Annual Review of Ecological Systems* 4: 1-23.

Integrated Marine Management Ltd. (IMM) (2003). "The sustainable livelihoods approach in the coastal context", *SCL Working Paper* 1. Sustainable Coastal Livelihoods South Asia. University of Exeter, United Kingdom. *http://www.innovation.ex.ac.uk/imm/SCL.htm*

Kasperson, J.X. and R.E. Kasperson (2001). *SEI risk and vulnerability programme report 2001-01*. Stockholm: Stockholm Environment Institute.

Small, C. and R.J. Nicholls (2003). "A global analysis of human settlement in coastal zones", *Journal of Coastal Research* 9 (3): 584-99.

Walker, B., C. Folke, S. Carpenter, M. Scheffer, T. Elmqvist, L. Gunderson and C.S. Holling (2004). "Regime shifts, resilience, and biodiversity in ecosystem management", *Annual Review of Ecology, Evolution and Systematics* 35: 557-82.

White, A.T., P. Christie, H. d'Agnes, K. Lowry and N. Milne (2005). "Designing ICM projects for sustainability: Lessons from the Philippines and Indonesia", *Ocean and coastal management* 48: 271–96.

33

Towards Creating a Resilient Community

Communities require guidance on DRR to progress towards resilience.

Source:

John Twigg *University of London. 2010.*

1. Introduction

Resilience is the ability of groups or communities to cope with external stresses and disturbances as a result of social, political and environmental changes. Building resilience into both human and ecological systems is an effective way to cope with environmental changes characterised by future surprises or unknowable risks.

The government and civil society organisations require disaster risk reduction (DRR) initiatives at local and community levels to create a resilient community. They require guidance notes/processes to progress towards resilience.

A structure of thematic areas is essential for identifying resilience issues (Figure 33.1), but, as all frameworks, this imposes artificial distinctions between different aspects of the subject when there is actually much more connection and overlap operation. Governance is a cross-cutting theme that includes: planning, regulation, integration, institutional systems, partnerships and accountability, because it contains issues likely to affect any initiative in DRR, development or relief.

The processes help users to visualise the widest possible range of options from which to make their own choices. However, they do not set DRR

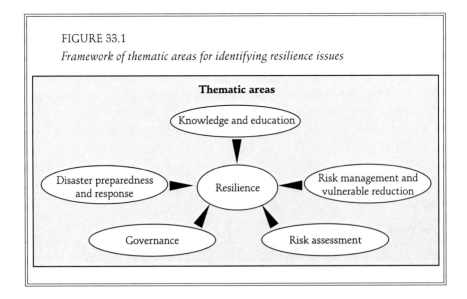

FIGURE 33.1
Framework of thematic areas for identifying resilience issues

priorities, because their framework and components are more general, for all contexts. Each group or organisation has to decide its own priorities and methods of intervention according to needs, operating contexts and capacities. Thematic areas aim to give a comprehensive picture relating to resilience and DRR of communities.

The thematic areas are grouped into three sub-sections: components of resilience, characteristics of a disaster-resilient community, and characteristics of an enabling environment. National governments and many NGOs have accepted these components and characteristics that have been generated from UN's Hyogo Framework for Action 2005-2015.

2. Components of resilience

These are still quite broad categories but they start the process of breaking disaster resilience down into more precise and comprehensible features. The Table 33.1 sets out the components of resilience for each thematic area.

2.1 Characteristics of a disaster-resilient community

Characteristics of a disaster-resilient community bring users closer to reality on the ground. Therefore, these characteristics are widely used at field level.

> *Characteristics of a disaster-resilient community*
>
> - Communities have a comprehensive picture of all major hazards and risks including potential risks.
> - It has a participatory process that includes representatives of all sections of community and sources of expertise.
> - Assessment findings are shared, discussed, understood and agreed among all stakeholders, and incorporated into community disaster planning.
> - Continuous monitoring of hazards and risks and updating the assessments are done.
> - Capacity building to carry out community hazard and risk assessments through support and training are accomplished.

TABLE 33.1
Components of resilience per thematic area

Thematic area	Components of resilience
1 Governance	• Policy, planning, priorities and political commitment • Legal and regulatory systems • Integration with development policies, planning, emergency response and recovery • Institutional mechanisms, capacities, structures and allocation of responsibilities • Partnerships, community's accountability and participation
2 Risk assessment	• Hazards/risk data and assessment • Vulnerability and impact data and assessment • Scientific and technical capacities and innovation
3 Knowledge and education	• Public awareness, knowledge and skills • Information management and sharing • Education, training and research • Cultures, attitudes and motivation
4 Risk management vulnerability reduction	• Environmental and natural resource management • Health and well-being, sustainable livelihoods • Social protection, financial instruments • Physical protection; structural and technical measures • Planning régimes
5 Disaster preparedness	• Organisational capacities and coordination • Early warning systems • Contingency planning and preparedness • Emergency resources, response, recovery and infrastructure • Participation, voluntarism and accountability

2.2 Characteristics of an enabling environment

The enabling environment possesses the following characteristics:

- political consensus on importance of DRR;
- DRR policy priority at all levels of government;
- strategy and implementation plan with clear vision, priorities, targets and benchmarks;
- local government DRR policies, strategies and implementation plans in place;
- official (national and local) policy and strategy of support to CBDRM; and

- local-level official understanding of and support for community vision.

They identify the importance of wider institutional policy and socioeconomic factors in supporting community level resilience. Some NGOs have found the characteristics of an enabling environment a helpful tool in their advocacy work, especially in developing an agenda for mainstreaming DRR at higher levels to complement grassroots projects.

3. Applications

The characteristics can be used at various stages of the project. They serve purposes such as:

- Indication of gaps in existing DRR work and highlighting the potential for new or stronger collaboration on specific issues;
- Identification of threats, vulnerabilities and capacities. (The characteristics ensure that the baselines are sufficiently coherent and wide-ranging to cover all relevant issues);
- Sharpening the analysis of vulnerability by identifying areas of inquiry before vulnerability and capacity analysis (VCA) by providing a framework for organising and interpreting data.
- A tool in project planning, especially in designing indicators for logical and results-based frameworks; and
- Monitoring and evaluation (of individual projects and for comparative analysis of projects).

4. Relevance

Organisations working in DRR should select those elements that are most relevant to:

- the needs and capacities of the communities they are supporting;
- the hazard and threats the communities face;
- the type of DRR work the organisations are expert in and their capacities to deliver; and
- the wider operational and policy environment.

The outcome should create positive attitudes amongst operational users. This is the result of the characteristics being solution-focussed rather than problem-driven.

5. Conclusions

The characteristics provide full and clear criteria to evaluate progress and give a systematic overview of work done and identify the gaps. Characteristics of a disaster-resilient community are not the same as indicators instead they should be seen as signposts for indicator development, describing attributes and elements that contribute to the resilience of communities. They have to be translated into measurable indicators for project purposes.

The characteristics can be used by researchers as a conceptual framing tool for designing their projects and directing their questioning. The findings of reviews and assessments carried out using the resource may have value in advocacy work at local and higher levels.

The characteristics are not a model for every situation. It is a resource, not a checklist to be ticked off. It should stimulate and facilitate discussion about appropriate approaches to creating more resilient communities. It must always be adapted to the context in which it is being used and the needs and capacities of those who use it.

Adaptation entails rewriting individual components of resilience and characteristics of a disaster-resilient community, adding new ones or even arranging the framework differently with additional thematic areas. Such 'customising' is encouraged, because it makes the characteristics more relevant to different circumstances. This should be through participatory processes of discussion and validation at local level.

Resilient communities are capable of bouncing back from adverse situations. They can do this by actively influencing and preparing for economic, social and environmental change. When times are bad, they can call upon the myriad of resources that make them a healthy community.

World Resources Institute (2008) argues that properly designed enterprises can create economic, social and environmental resilience that cushions the impacts of climate change, and help provide needed social stability.

References

BEDROC (Building and Enabling Disaster Resilience of Coastal Communities) (2009). *Building local capacities for disaster response and risk reduction: An Oxfam-BEDROC study*. Nagapattinam, Tamil Nadu, India. p.141. BEDROC. *www.bedroc.in*

Twigg, J. (2007). *Characteristics of a disaster-resilient community: A guidance note*. Version 1 (for field testing). DFID Disaster Risk Reduction Interagency Coordination Group. London, UK. p.36.

34
Learning to Respond to Disasters: Community Empowerment

Community-based disaster risk management reduces vulnerabilities and enhances capacities.

Source:

Adapted from *Learn to Manage Your Own Risk: Tips from Asian Disaster Reduction Training,* southasiadisasters. net, Issue 35, August 2007, All India Disaster Mitigation Institute, Ahmedabad, India.

1. Introduction

Development and disaster management experts have realised the importance of empowering communities and thus increased efforts to involve relevant communities in their activities. The involvement of the most vulnerable social groups—those often excluded from public efforts—in the process is critical for a successful implementation.

Disaster risk reduction (DRR) includes any and all measures that can reduce disaster related risks to avoid loss of life, property or assets by reducing the vulnerability (or hazard) of the elements at risk and increasing the coping capacities. In community-based DRR programmes, the community's participation in the process from prevention to the recovery from a disaster is emphasised. This approach considers the local people and the necessity to increase coping mechanisms of everyday life and the need to reduce vulnerability to disasters.

Community-based disaster risk management (CBDRM), is a process in which at-risk communities are actively engaged in the identification, analysis, treatment, monitoring and evaluation of disaster risks, in order to reduce vulnerabilities and enhance capacities. Governments and non-government organisations (NGOs) play a supportive role by providing information, strengthening community groups, giving financial and technical assistance, building linkages, and supporting complementary between groups. When dealing with CBDRM, two key concepts must be taken note of: vulnerability and resilience and coping capacity.

2. Vulnerability

Vulnerability is a set of prevailing or consequential conditions, which adversely affect the community's ability to prevent, mitigate, prepare for and respond to hazardous events. It can be computed by using the following equation:

$$R = H \times V$$
where R represents risk; H, hazard; and V, vulnerability

This political economic approach shows that disasters are influenced also by human actions, politics and history.

To be able to respond to a disaster efficiently, communities and households need capacities which enable them to cope with, be prepared for, and quickly recover from disaster. The coping capacity is determined by the following equation:

$$\text{hazard} \times \text{vulnerability/capacity}$$

Community-based approaches for disaster risk management are essential because it promotes a culture of safety by effectively removing local vulnerabilities and building capacities. Practicing CBDRM involves handling the organisational mandates of the concerned organisations; understanding the social, economic and political development of the communities in a country; and awareness in the funding cycles of donor organisations. Another key trend in the recent years has been the efforts by NGOs, the United Nations and other international organisations to mobilise government support for CBDRM in policy, planning and programming.

One of the features of CBDRM is to provide solid outcomes such as a community-based organisation(CBO); community disaster risk reduction fund; community hazard, vulnerability, capacity map (HVCM); community disaster risk management plan; CBO training system; community drills system; community learning system and a community early warning system.

The success of CBDRM lies on the proper dissemination of its goals clearly and simply to all stakeholders. The methodology must also be simple. Convincing local and national governments of the importance of preparedness and their role in providing a holistic vision to the communities is essential. CBDRM aims to reach a level of resilience, beginning with regular training programmes to ensure belongingness.

3. Resilience from CBDRM

Resilience can be described as the capacity of a system, community or society potentially exposed to hazards that can adapt. Rooting from a successful CBDRM implementation, resilience pertains to resisting or changing in order to reach and maintain acceptable level of function and structure. This is determined by how well the social system is capable of increasing its capacity for protection by learning from past disasters. Resilience depends on being informed and aware of the situation in the community, as well as being prepared to respond to a disaster effectively. All members of a community must be involved in DRR programmes be it through formal or informal institutions.

4. Characteristics of a resilient community

Resilience can be measured by some indicators such as an established community organisation; a community early warning system; trained manpower for risk assessment, search and rescue, medical first aid, relief distribution, masons for safer house construction, fire fighting; physical connectivity with local authorities and NGOs; safer houses to withstand local hazards; and safe sources of livelihoods (Table 34.1).

TABLE 34.1
Disaster phases and resilience indicators

Phase	Capacities of a resilient community	Result
Pre-disaster	The ability to absorb the shocks of hazard impact, so that they do not become disasters	Reducing the probability of failure
Post-disaster: immediate relief	The capacity to bounce back during and after a disaster	Reducing the consequences of failure
Post-disaster: Long-term recovery	The opportunity for change and adaptation following a disaster	Reducing the time needed for recovery as well as patterns of vulnerability

Source: Asian Disaster Preparedness Center (ADPC) (2006). *Critical guidelines: Community-based disaster risk management*. Bangkok, Thailand.

5. Conclusion

The aim of the CBDRM approach is achieved when the community, involving all social groups in the process of disaster risk management can exhibit resilience at all disaster phases. Maintaining an appropriate level of resilience will be an ongoing process, requiring links between governments, NGOs and community members establishing institutional structures. Aside implementing disaster risk management, communal activities must adapt to the changing contexts of today's uncertainties.

Advocacy empowers people by supporting them to assert their views and claim their entitlements and, where necessary, representing and negotiating on their behalf. It aims change in policy to meet the needs of the people. CBDRM advocacy revolves around communities who are disadvantaged. The campaign seeks to become part of the planning and programming of risk reduction that may affect local groups. Because of focussed advocacy efforts, one can influence both policy and social change.

During an advocacy initiative the following steps may be adopted:

1) identify policy issues;
2) select an advocacy objective;
3) research audiences;
4) develop and deliver advocacy messages;
5) understand the decision-making process;
6) build alliances;
7) make effective presentations;
8) fund raising for advocacy; and
9) evaluate and improve the advocacy.

Advocating for CBDRM's integration in government planning will improve the efficiency of DRR programmes. CBDRM will enable more people to plan for themselves and access local funding. Since it is long term, CBDRM advocacy is a commitment against poverty because the programme aims to reduce disaster risks before they happen, with the participation of the community.

The following example shows how communities may be involved in the activities of the local government. It presents how advocacy can influence policy and social change.

Dagupan City, the Philippines: cooperation between local authority and communities
Flooding is a common problem in Dagupan City, the Philippines which was exacerbated by an earthquake in 1990. The city government recognised the importance of addressing the issue of flooding and actively initiated different phases of disaster management. Recently, the government shifted to a preventive focus by addressing issues of vulnerability. The city was selected to become the Philippines' demonstration city for the Programme for Hydro-Meteorological Disaster Mitigation in Secondary Cities in Asia, named PROMISE.

The programme promotes the adoption of specific disaster mitigation measures in city and community levels. To achieve this, Dagupan City's eight pilot barangays were involved in the process mostly tackling the goal by organising disaster risk management workshop as well as an early warning and evacuation workshop in respective communities. The Technical Working Group coordinated with the legislators of the city council and developed a resolution to implement a Dagupan City Disaster Preparedness Day. Disaster safety is now a large part of the city's culture while effectively raising awareness.

35
Community Resilience: A Social Justice Perspective

Social vulnerability occurs when unequal exposure to risk is coupled with unequal access to resources.

Source:

Betty Hearn Morrow Adapted from: Florida International University, USA, September 2008, Resiliency, Sustainability, Social Justice, and Environmental Justice. CARRI Research Report 4.

Resilience has been referred to as the ability to absorb changes or disturbances (Handmer and Dovens, 1996; Adger *et al.*, 2005). It requires: (1) knowledge of the hazard; (2) accurate perception of the risk; (3) understanding of available alternatives; and (4) the resources and flexibility to respond successfully.

A resilient community is defined as 'one that anticipates problems, opportunities, and potentials for surprises; reduces vulnerabilities related to development paths, socioeconomic conditions, and sensitivities to possible threats; responds effectively, fairly, and legitimately in the event of an emergency; and recovers rapidly, better, safer, and fairer' (Wilbanks, 2008).

1. Types of resilience

Resilience can be categorised into several types:

Physical resilience refers to the strength to deal with an impact (such as the ability of a house to withstand high winds or the physical health of an individual to survive a disaster).

Economic resilience is the robustness and diversity of the economy to survive and recover from a disaster.

Ecological resilience is the biological ability of natural ecology to survive environmental pressures.

Social resilience describes abilities within human societies to adjust to change, particularly 'to absorb recurrent disturbances such as hurricanes and floods so as to retain essential structures, processes and feedbacks' (Adger *et al.*, 2005). At the community level, it is closely tied to the economic and political circumstances of a community, as well as to the strength of its social institutions and social networks.

2. Social vulnerability

Social vulnerability occurs when unequal exposure to risk is coupled with unequal access to resources (Bolin and Stanford, 1998). Social structures systematically discriminate against the socially, culturally, and economically marginalised (Mustafa, 1998). The effects increase

Community Resilience: A Social Justice Perspective **395**

vulnerability, or potential for loss, at all levels of society—from individuals to communities to nations.

The following risk factors influence vulnerability:

2.1 Economic status

The extent to which people have sufficient resources to meet their basic needs and to anticipate and respond to inevitable change and disruption is a core factor in resiliency and varies considerably among communities and nations. When significant segments of the population are poor and live with daily risk and insecurity, they cannot be expected to anticipate and respond to external changes and threats effectively without outside assistance.

2.2 Political power

The power to affect decisions, such as those related to economic development, the use of public resources for infrastructure development and services, and the location of environmental and technological

The rich have sufficient resources to prepare and respond to impending crisis.

hazards, largely determines which communities and households are most vulnerable. Some individuals and groups will garner greater respect and thus more influence in the community. Any individuals or groups that are marginalised, whether by poverty, gender, minority status, or disability, are likely to be more vulnerable when it comes to dealing with unexpected events.

The importance of political power becomes most obvious during the recovery period after a disaster. Political power is closely associated with economic status (Logan and Molotch, 1987). More affluent, and thus politically powerful, neighbourhoods are the first to receive public services after a disaster. The interests of businesses tend to be served before social services are restored. Poorer neighbourhoods are more likely to flood and less likely to have debris cleared and services restored in a timely manner.

During reconstruction, political and economic power determines what is rebuilt and where (Vale and Campanella, 2005). Former residents of poorer areas will have less power over the process of determining what gets rebuilt.

3. Factors influencing social vulnerability

A number of vulnerability factors have been well documented as being related to differential exposure and impact, as well as slow or inadequate recovery.

3.1 Poverty

The difficult position of poor people, neighbourhoods, communities and states becomes painfully obvious after every disaster. The case of Hurricane Katrina in New Orleans (2005) provides the most recent and dramatic example of the effects of poverty on vulnerability. According to census data, 27 per cent, or about 125,000 people, did not have access to a car. Given the lack of public transportation for evacuation, it was not surprising that over 100,000 were in the city when Katrina made landfall.

During recovery, stark examples of the effects of social class become obvious. Similar circumstances occurred after the Northridge Earthquake in California (Bolin and Stanford, 1998). Poor communities have fewer resources to devote to recovery, and, surprisingly, in one case, households in a poorer community actually received less federal assistance than those in a more affluent neighbouring town (Dash et al., 1997).

3.2 Minority status

Patterns of discrimination against racial and ethnic minorities have resulted in groups being located in highly segregated neighbourhoods that is geographically located in the least desirable, most hazardous areas.

Minorities are also rarely part of the influential power structure in either the private or public sector. There is substantial evidence that the households of racial and ethnic minorities, irrespective of income, tend to be more vulnerable at all stages of disaster response (Fothergill et al., 1999; Bolin, 2006).

Culture plays an important role in how people assess risk, weigh options, make decisions, and carry them out (Peacock et al., 1997; Bolin and Stanford, 1998; Bankoff, 2002). Cultural representation of events plays a role in transmitting knowledge and can influence how people prepare and react (Webb et al., 2000). The foreign born and those for whom English is a second language are likely to encounter greater difficulty when interpreting warnings, understanding alternatives and seeking information and assistance.

3.3 Gender

Gender-based socialisation affects attitudes, behaviours, and status in ways that become particularly important in crisis. Women tend to be more risk aversive and more likely to respond to warnings. Men are much more likely to volunteer to assist with rescue, security, early cleanup, and other more hazardous, community activities. In addition to assessing risk differently, women and men often respond to disasters and to losses in different ways. It has been noted that even couples who do not follow traditional gender roles tend to revert to them in a crisis (Hoffman, 1998).

3.4 Age and disabilities

While there is a great deal of difference in the physical health and circumstances of the elderly, age is often a factor in determining one's ability to respond and recover from a disaster. Disabilities, regardless of age, can restrict individual and household options.

Even when in good health and with sufficient resources, older people are less likely to heed hurricane evacuation orders (Gladwin *et al.*, 2001). They are reluctant to leave the comforts of home for the discomforts they will likely to encounter during transport, and in shelters and other places of refuge.

3.5 Human capital

In the disaster context, human capital can be defined as the personal abilities and skills that promote resilience. The following factors contribute to the individual's capacity in mitigating and responding to disaster:

- *Education and literacy*: At every stage of response from interpreting warnings to negotiating the complex processes associated with

Age is often a factor in determining one's ability to respond and recover from a disaster.

recovery, it helps to be literate, educated and experienced in dealing with authorities and bureaucracies.

- *Culture*: Being from a different culture can hinder understanding of the process involved in anticipating and responding to an event.
- *Relevant skills and knowledge*: Lack of skills and understanding related to building construction can be a tremendous handicap during home reconstruction. Lack of knowledge about topics such as insurance, building codes, assistance programmes, and government can hinder recovery. Understanding natural hazards and emergency procedures is important to resilience.

Minorities often rely on kin for help.

3.6 Social capital

The importance of social connections and social networks to social resilience is obvious. Family and friends can be important sources of information, advice and assistance. This is particularly true in times of crisis. Minorities, including immigrants, often rely on kin networks as their source of guidance and help.

Membership in social networks such as churches, social clubs, parent groups, and other community organisations provides connections that can be called upon when needed, whether for job opportunities, recommendations, guidance, or resources.

To be isolated is to be far more vulnerable in everyday life, and when serious problems arise. Groups that tend to have less social capital include recent immigrants, new residents in the community, elderly who live alone, and the homeless. Tenants are less likely to be connected to the community than homeowners. Adults without children in the home tend to have fewer outside connections.

4. Intersecting vulnerabilities

Vulnerabilities are likely to intersect. Obvious examples are that ethnic minorities are more likely to be poor, as are women, the elderly, the disabled, and the least educated. Women-headed or women-alone households are more likely to be poor. Women live longer and thus are over-represented in the elderly population, especially the poor elderly. The tragedy of Hurricane Katrina provided vivid evidence of the compounding vulnerabilities associated with gender, age, class and disability.

5. Building resilient communities

5.1 Commitment to change

- *Recognition and acknowledgement of inequities and vulnerabilities*: The community vulnerability mapping that is becoming routine in emergency management operations should include a component that denotes where high-risk or socially vulnerable residents tend to be clustered (Morrow, 1999; Hill and Cutter, 2001).

- *Addressing the needs*: Once vulnerabilities are identified, there need to be a consensus and commitment to address the needs of vulnerable groups and communities through public and private initiatives, including, but not limited to, emergency management.

The following methods can help address community needs:

- *Use of social marketing techniques*: This must be done to reach key groups with culturally sensitive educational materials, written at appropriate reading levels and in other languages when needed.
- *Civic participation at all levels*: Community organising around safety, such as the Community Emergency Response Teams (CERTs), is an important step toward building more resilient neighbourhoods.
- *Economic development paradigm change*: At some point it becomes clear that more and more development, particularly in environmentally sensitive areas, is not likely to result in long-term gain, either economically or in the quality of life.
- *Funding source*: Government programmes related to emergency management and disaster response at all levels need permanent and sufficient funding sources, as more areas prone to disaster are denied of property insurance.

5.2 Building on strengths

An important asset in the reduction of vulnerabilities lies with the people and groups themselves. The use of indigeneous strengths at every level from administration and management through on-the-ground response will result in more effective policies and programmes. For example, community women are most likely to know what's going on in their neighbourhoods and can be valuable informants, facilitators, and leaders; while extended family networks are common to ethnic minorities and are an important source of information and assistance in emergencies. Finding ways to tap into social resources is essential to building resilience at the grassroots level.

5.3 Essential resilience

Social justice should form an interpretative framework to guide planning at all levels of government. Hazards mitigation is not just about building stronger buildings or levees 'but equity and justice in resource management' (Mustafa, 1998). People need to have faith in their government and trust that they are being represented fairly. This requires community involvement. Top-down paternalistic official activities do not lead to meaningful resilience (Murphy, 2007). It requires hard work at the grassroots level to build strong community social structures and for government and officials to earn public trust.

Addressing basic social and economic problems is a major step toward building resilient communities.

References

Adger, W., T. Neil, P. Hughes, C. Folke, S.R. Carpenter and J. Rockstrom (2005). "Social-ecological resilience to coastal disasters", *Science* 309: 1036-39 (August 12).

Bankoff, G. (2002). *Cultures of disaster: Society and natural hazards in the Philippines*. London: Routledge.

Bolin, R. (2006). "Race, class, and disaster vulnerability", in H. Rodriguez, E. L. Quarantelli and R. Dynes (eds.). *Handbook of disaster research*. New York: Springer.

Bolin, R. L. Stanford (1998). *The Northridge earthquake: Vulnerability and disaster*. London: Routledge.

Dash, N., W.G. Peacock and B.H. Morrow (1997). "And the poor get poorer: A neglected black community", in W.G. Peacock, B.H. Morrow and H. Gladwin (eds.), *Hurricane Andrew: Ethnicity, gender and the sociology of disasters*. London: Routledge. pp.206-25.

Fothergill, A. (2004). *Heads above water: Gender, class, and family in the grand forks flood*. Albany: State University of New York Press.

Fothergill, A., E.G.M. Maestas and J.D. Darlington (1999). "Race, ethnicity and disasters in the United States: A review of the literature", *Disasters* 23(2):156–173.

Gladwin, C.H., H. Gladwin and W.G. Peacock. (2001). "Modeling hurricane evacuation decisions with ethnographic methods", *International Journal of Mass Emergencies and Disasters* 19 (2):117-443.

Handmer, J.W. and S.R. Dovens (1996). "A typology of resilience: Rethinking institutions for sustainable development", *Industrial & Environmental Crisis Quarterly* 9(4): 482-511.

Hartman, C. and G.D. Squires (eds). (2006). *There is no such thing as a natural disaster: Race, class, and hurricane katrina*. New York: Routledge.

Hill, A.A. and S.L. Cutter (2001). "Methods for determining disaster proneness," in Susan L. Cutter (ed.), *American hazardscapes: The regionalization of hazards and disasters*. Washington: Joseph Henry Press. pp.13-36.

Logan, J.R. and H.L. Molotch (1987). *Urban fortunes: The political economy of place*. Berkeley: University of California Press.

Morrow, B.H. (1999). "Identifying and mapping community vulnerability disasters", *The Journal of Disaster Studies, Policy and Management* 23 (1):1-18.

Mustafa, D. (1998). "Structural causes of vulnerability to flood hazard in Pakistan", *Economic Geography* 74 (3):289-305.

Peacock, W.G. (2003). "Hurricane mitigation status and factors influencing mitigation status among florida's single-family homeowners", *Natural Hazards Review* 4 (3): 1-10.

Peacock, W.G., B.H. Morrow and H. Gladwin (1997). *Hurricane andrew: Ethnicity, gender and the sociology of disaster*. London: Routledge.

Vale, L.J. and I.J. Campanella (eds.) (2005). *The resilient city. How modern cities recover from disaster*. New York: Oxford University Press.

Webb, G.R., T. Wachtendorf and A. Eyre (2000). "Bringing culture back in: Exploring the cultural dimensions of disaster", *International Journal of Mass Emergencies and Disasters* 18 (1): 5-19.

Wilbanks, T.J. (2008). "Enhancing the resilience of communities to natural and other hazards: What we know and what we can do", *Natural Hazards Observer* XXXII (5). May 2008.

36

Boosting Community Resilience

Appropriate resilience strategies and tools keeps the community afloat during disasters.

Source:

Michelle Colussi, Mike Lewis, Sandy Lockhart, Stewart Perry, Pippa Rowcliffe and Don McNair *The Community Resilience Manual: A Resource for Rural Recovery and Renewal.*

Many marginalised communities largely depend on available resource base for their livelihood and subsistence. But social, economic, and physical volatilities, (i.e., changes in the market), technology, policies, and the resource base itself, compel the affected communities to find means on how they could adapt and appropriately respond to such changes. Community resilience is the order of the day. A resilient community is one that takes intentional action to enhance the personal and collective capacity of its citizens and institutions to respond to and influence the course of social and economic change.

In the face of increasing levels of volatility, the ability to assess and specify their level of resilience allows communities to identify areas of weakness, and select and implement strategies proven to target those difficulties. In order to help local communities avoid errors and replicate successes, the Communities Committee of Forest Renewal British Columbia (FRBC) funded a team from the Centre for Community Enterprise (CCE) to develop a conceptual framework and process through which resource-dependent communities can strengthen local resilience. A *Community Resilience Manual: A Resource for Rural Recovery and Renewal* is the offshoot of all their efforts.

1. A new understanding of resilience

A new conceptual model in strengthening community resilience starts with four dimensions:

a. *People*—includes residents' beliefs, attitudes and behaviour in matters of leadership, initiative, education, pride, cooperation, self-reliance and participation.

b. *Organisations*—the scope, nature and level of collaboration within local organisations, institutions and groups.

c. *Resources*—the extent to which the community builds on local resources to achieve its goals, while strategically drawing on external resources.

d. *Community process*—the nature and extent of community economic develop-ment planning, participation and action.

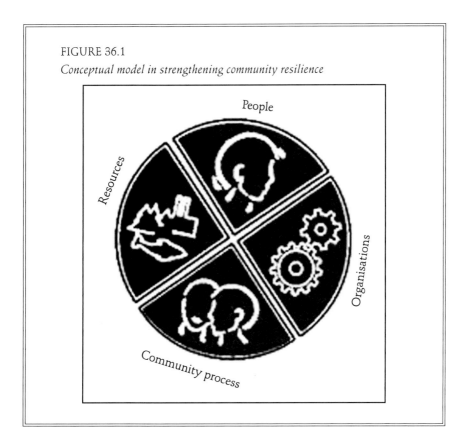

FIGURE 36.1
Conceptual model in strengthening community resilience

These are the core components of a community's social and economic structure. They form a foundation for a detailed analysis of community functions and for the creation of a Portrait of Community Resilience. Each of the four dimensions is described by a series of characteristics of resilience. Though not exhaustive, the identified characteristics have been proven to be strongly predictive in assessing resilience (Table 36.1).

2. Three steps towards resiliency

The three-step process attempts to help communities through a structured and focussed process of prioritising and planning.

TABLE 36.1

Characteristics of a resilient community

People	Organisations	Resources	Community process
1. Leadership is representative of the community.	10. A variety of community economic development (CED) organisations in the community serve well the key CED functions.	12. Employment in the community is diversified beyond a single large employer.	18. The community has a CED plan that guides its development.
2. Elected community leaders are visionary, shares power, and builds consensus.		13. Major employers in the community are locally-owned.	19. Citizens are involved in the creation and implementation of the community vision and goals.
3. Community members are involved in significant community decisions.	11. Organisations in the community have developed partnerships and collaborative working relationships.	14. The community has a strategy for increasing independent local ownership.	20. There is ongoing action towards achieving the goals in the CED plan.
4. The community feels a sense of pride.		15. There is openness to alternative ways of earning a living and economic activity.	21. There is regular evaluation of progress towards the community's strategic goals.
5. People feel optimistic about the future of the community.		16. The community looks outside itself to seek and secure resources (skills, expertise and finance) that will address identified areas of weakness.	22. Organisations use the CED plan to guide their actions.
6. A spirit of mutual assistance and cooperation is present in the community.			23. The community adopts a development approach that encompasses all segments of the population.
7. People feel a sense of attachment to their community.		17. The community is aware of its competitive position in the broader economy.	
8. The community is self-reliant and looks to itself and its own resources to address major issues.			
9. There is strong belief in and support for education at all levels.			

Step 1: Draw a portrait of a resilient community

The data collected for each of the 23 characteristics of a resilient community are documented and analysed in a portrait of community resilience. The portrait is important in identifying current resilience strengths and weaknesses. Collected data are streamlined by using indicators that specify the exact information needed. The model uses both quantitative indicators (e.g. population data) and qualitative indicators (e.g. survey of people's perceptions) to identify the extent to which each characteristic is present in a given community.

Step 2: Establish community priorities

The portrait provides a new perspective on the social and economic structure of a community. Analysing these can help communities gain new insights into factors that can increase their capacity to adapt and influence the course of change. At this point, the community is helped in a decision-making process to further analyse the significance of the portrait to their community resilience. The process suggests ways of involving community members, applying their insights and knowledge and prioritising resilience weaknesses.

Step 3: Select appropriate strategies and tools

Once community priorities have been established, it is time to decide what actions a community can take to best improve its resilience. A wide range of promising strategies and tools have been created by gathering information about CED 'best practice.' Details of the strategies and tools, who applied them effectively and how, together with contact information and additional recommended resources, can be found in the *Community Resilience Manual*.

All these considerations address one or more of the six functions essential to community's economic well-being:

Boosting Community Resilience

- Access to equity capital;
- Access to credit;
- Building human resource capacity;
- Capacity for research planning and advocacy;
- Creating partnerships within and outside the community; and
- Infrastructure.

3. Conclusion

The information drawn from the community resilience is neither a quick fix nor a mere research methodology. Instead, it opens up a way of

'The Tool framed things differently for us. The way questions were asked and the way data was presented triggered discussion that did not happen before. It showed us that we could have stronger communities by addressing the characteristics of resilience and taking a more holistic approach to CED.'
— Nadina Community Futures

'Even being a participant in the research stage as one of the test communities we see the benefits. It has been a great reality check. Even for some of us with a long history of CED practice, the results have been instructive. It is a good educational tool and it produces results that can help strengthen communities.'
— Revelstoke Economic Development Commission

thinking and helps focus community dialogue on key aspects of healthy community functioning that seldom find their way into a community strategic plan.

Early on, several benefits have already been gained with the conception of this manual:

- Existing planning methodologies and processes in communities were enhanced.
- Communities discovered the salient benefits of their local economic development planning process.
- By applying the process in the manual, the communities were able to gather new information about local attitudes and organisations which provided a framework for local decision-making and priority setting.
- The process engaged a broad cross-section of the community in thinking about resilience and its economic impact, thus, creating new energy for local initiatives.
- The manual provides communities with a means to systematically strengthen their capacity to steer towards the future which they choose.

37
Community Resilience and Community-based Management

Communities have always responded very creatively to risks and disasters recognising the effectiveness of CBM.

Source:

Stephen Tyler *Independent Consultant, Adaptive Resource Management, Ltd. Canada 2010.*

1. Introduction

Poor coastal communities are especially vulnerable to natural disasters for a number of reasons. Their livelihoods are tied mainly to the ocean, the resources of which are severely depleted. Tropical storms and the accompanying high tide and storm surge threaten their modest assets on shore and at sea, and the frequent damage and loss only pushes them further into poverty. While devastating tsunamis are rare, the severity and frequency of tropical storms are expected to increase due to global climate change, which threatens key coastal ecosystems, such as wetlands and reefs, further eroding livelihoods and natural storm defenses.

Communities have always responded, often very creatively, to risks and disasters. But there is now increasing formal recognition of the effectiveness of community-based management (CBM) mechanisms to avert, prepare for and respond to natural disasters. There is also recognition of the need to build community initiative and management capacity for natural resources and for alternative livelihoods. Disaster risk reduction (DRR), livelihood diversification and ecosystem management all build community resilience.

2. Community resilience and CBM

Resilience is the capacity to respond to and recover from long-term stresses or sudden shocks without losing the ability to function effectively.[1] Resilience is normally considered as a characteristic of a system, such as a human being or a household or a community. One important aspect of systems is that they include many interlinked components whose elements function together to create outcomes that might not be obvious from a simple cause-effect assumption. In the case of coastal communities, the system includes the various component households and individuals, the ecosystem, as well as infrastructure, organisation, social relations, and governance (decision-making, regulatory processes and structures).

---

1. This definition is based on the pioneering work of the Resilience Alliance, a loosely-organised group of multi-disciplinary researchers with strong roots in ecology. Its work on the application of systems theory and resilience thinking to socio-ecological systems is an important resource for underlying concepts and theory: *www.resalliance.org*

In CBM, decisions about planning, allocation of scarce resources and implementation are made by, or in close collaboration with the women and men who will be most affected by them. In managing natural resources or natural disasters, there is also a role for senior levels of government and policy, so CBM does not mean excluding government. It means recognising that top-down programming and policies must support bottom-up initiatives, in order to build on the people's knowledge and capacities. It means recognising that local resilience comes from the ability to organise, prepare for and respond to local shocks and stresses.

3. CBM key principles and mechanisms

CBM is a process that places local people at the centre of problem diagnosis and response. Its starting point is the recognition that even poor communities may have extensive knowledge and capacity to manage complex local systems for resilience. It values and actively engages

Damaged coastal ecosystems such as wetlands have led to the erosion of livelihoods and natural storm defenses. CBM involves the people in finding solutions to such problems as this.

local knowledge. It recognises that community members have different knowledge depending on the diversity of their experience, and the extent to which they are able to cope with change and shocks.

Social networks and exchange relations are part of the community asset base and can serve to provide valuable resilience in times of stress. The strength of these networks comprises an important element of community capacity to adapt and respond to stress. Capacities and vulnerability should be assessed from local and technical evidence of hazards as well as ecological, social and political processes.

Adaptive capacity, which is closely related to resilience, depends in part on characteristics of individual households. But it also crucially depends on collective arrangements based on collaborative social relations at multiple scales. This is true at the household level, where adaptation and resilience depend on links to external resources (e.g., information, credit, political support, remittances etc.); so too at the community or national scale, where external linkages such as flows of information, trade and investment may reinforce or undermine local adaptation decision-making and economic options, depending on their structure (Vincent, 2007).

This reliance across scales means that new kinds of collaborative mechanisms become important, allowing scientific researchers to interact respectfully and supportively with local resource managers, government authorities to work with local resource user groups rather than at cross-purposes, and local governments to improve representation, transparency and accountability. CBM organisations also benefit from networking with external organisations to provide new information and to support policy change and innovation (Nong and Marschke, 2006).

In order to ensure that community management mechanisms include the interests of marginalised groups, these should be specifically targeted for engagement and inclusion (Beck and Fajber, 2006). Local leadership of resilience initiatives may easily be captured by elites and fail to serve those in the community who are most at risk. Engagement of marginal groups is not easy, and requires organisational structures that are both transparent and able to be held accountable for their performance. Where local governments are democratic and responsive, efforts should be made

Resilience is exemplified by the people who created vegetable gardens on top of clumps of water hyacinth that grow on the waterways of Bangladesh.

to improve their performance rather than substitute completely with new organisations.

4. Applying CBM to build resilience

CBM can be applied through community-based disaster management (CBDM), community-based ecosystem management, alternative livelihoods, and knowledge generation. All of these areas are crucial to building the resilience of coastal communities.

4.1 Community-based disaster management

Disaster management not only recognises that local disaster responses must always be the first responses, but that local experience and perceptions of vulnerability and hazards are crucial to understanding and reducing risks. If they have experience with natural hazards, local people are often well aware of sources of vulnerability and the measures needed to reduce it. Their coping mechanisms can be recognised and improved with better training, warning, equipment and risk avoidance (Shaw and Okazaki, 2004).

4.2 Community-based ecosystem management

Coastal communities are often excluded from the management of the resources and ecosystems on which they depend. Without having a strong voice in the use of natural resources that are crucial to coastal protection, livelihoods, and habitat conservation, coastal communities will be unable to implement ecosystem measures to boost resilience. Some of the key success factors include collective tenure systems for common pool resources, fair and inclusive mechanisms for decision-making and conflict management, and basic monitoring to be able to recognise emerging ecological problems (Tyler, 2006). Ecosystems that are healthy are themselves resilient to shock and stress.

4.3 Collaborative relations

Effective CBM provides much of the 'glue' of collaborative relations to strengthen the ability of community members to develop alternative livelihoods based on fair exchange. Collective action and economic organisation, such as cooperative societies to organise technical inputs, extension services and marketing support to small-scale aquaculture producers, have been successful in introducing better management practices, higher returns and lower risk production in India (Umesh *et al.*, 2009). This community-based structure shows how collaborative organisation and management increases community resilience.

4.4 Knowledge generation

A central role in all of these community-based initiatives is played by knowledge generation. While local knowledge and current practice are important bases on which to build community resilience, they are generally insufficient. New knowledge, i.e., source and intensity of hazards, techniques for restoring natural storm barriers, culture techniques and inputs for healthy fish or shrimp in aquaculture ponds, is essential to strengthening CBM for resilience. It allows community leaders and organisations to make better decisions about risk reduction,

Cyclone shelters in Bangladesh, a delta nation, reduces risks from flooding.

adaptation, ecosystem management and productivity, and introduce new livelihoods. Knowledge and mechanisms for learning and sharing knowledge are essential elements of resilience building.

4.5 Community-based resilience: What would it look like?

CBM for resilience provides a framework for focussing on processes that assess local vulnerabilities, and that builds increased capacity for managing environmental shocks or stresses. It is vital to have a varied menu of appropriate technologies and methods to respond to diverse coastal stresses. But these technologies and methods can be selected and implemented more effectively under a CBM framework that connects

TABLE 37.1
Community-based resilience: What would it look like?

	Disaster management	Ecosystem management	Alternative livelihoods	Generating new knowledge
Organisation	Vulnerable groups including women provide input, share knowledge, agree to key decisions and participate in implementation through accountable and transparent leadership using existing local government institutions where suitable	Key resource user groups including women provide input, share knowledge, agree to key decisions and participate in implementation through accountable and transparent leadership using existing local government institutions where suitable	Poor women and men organised in self-help groups with accountable and transparent leadership	Disaster management, ecosystem management or livelihood groups identify need for new knowledge about risks ecological systems, production systems, value added or markets
Diagnostics	Vulnerable groups and individuals share experience of hazards, as well as adaptation, compared with technical experts and best practices to assess vulnerability and risks	Local knowledge and scientific assessment combined to assess ecological and institutional factors causing degradation	Community leaders, entrepreneurs and self-help groups undertake strategic assessment of skills, resources, value-added, and market prospects	Research to explore relevant best practices and technical knowledge on hazards, risk assessment, ecological degradation, sustainable use and market opportunities undertaken by local or external experts.
Planning	Previous adaptive and coping mechanisms inform risk assessment to help identify local training, early warning, and risk avoidance requirements	User groups (including women and men) develop plans for ecosystem rehabilitation, conservation and sustained use based on evidence from assessment and best practices	Market feasibility and production system assessment undertaken by self-help groups with external support as required. Development addresses underlying causes of vulnerability	Develop shared vision and values to drive planning options, and identify potential new solutions and associated risks

contd...

...contd...

	Disaster management	Ecosystem management	Alternative livelihoods	Generating new knowledge
Implementation	Local leadership and voluntary organisation plus external resources (expertise, funds, equipment, warning information systems)	Government officials review and sanction plans and help to implement, monitor and enforce conservation and productivity efforts with user groups	Support from credit programmes, marketing networks, extension and quality advisors	Monitoring of key success indicators is undertaken at the local level and reported to all stakeholders
Networking	Local leaders and community workers share experiences with other communities and to exchange views with national officials. Local warning systems are tied into regional and national information networks	Local resource decision-makers and government co-managers share knowledge and experiences in local forums	Producer networks exchange non-proprietary information on markets, inputs, prices, production techniques and processing in order to improve returns	Researchers support access to latest relevant technical and financial innovations with local leaders and user groups. Extension services provide up-to-date information
Policy development	Disaster management and risk reduction policies commit regular budget to building local capacities and in providing fast response and external support in recovery	Policies enable and support collective tenure systems, resource co-management and conflict management procedures	Small business and micro-credit policies, along with entrepreneurship extension programmes, support livelihood diversification	Support for applied research and extension services to match national research resources with diverse local needs and bridge gaps between research and practice
Shared learning	Response and preparation are evaluated after each disaster, with a view to improving practice and building skills	Government resource managers and local resource users evaluate field experience together to exchange knowledge and lessons from innovations in practice	Production and processing innovations raise standards and improve skills, and are taught to other local producers	Collaboration between researchers and local people researchers contribute methodological knowledge and scientific assessment to local experimentation, share and interpret results jointly

Note: Community-based management processes (rows) can provide the framework to support resilience in ways that are mutually reinforcing. Resilience can be fostered through different domains of intervention (columns). Some or all of these activities may be facilitated by external support organisations (e.g., CBOs, NGOs, research groups etc.).

local knowledge and capacities to external resources. The framework, not the individual techniques on their own, will help assure that community capacities and resilience are strengthened.

5. Conclusion

CBM is an approach towards resilience to disasters and risks in our increasingly dangerous and unstable global climate. It puts the local people at the heart of problem diagnosis and response because they are primarily the ones who would know the best solutions to the problems that they experience due to disaster risks created by climate change. However, CBM does not eschew the importance of social networks, external linkages, governments, the ecosystem, social relations, and governance. It recognises the fact that the lack of resources and poverty are linked to vulnerability. It sees the problem from a system's perspective that considers all interlinking components to be important and to have an impact on the community.

References

Beck, T. and E. Fajber (2006). "Exclusive, moi? Natural resource management, poverty, inequality and gender in Asia", in S. Tyler (ed.), *Communities, livelihoods and natural resources: Action research and policy change in Asia*. Reading and Ottawa, ITDG Publishing and International Development Research Centre. pp.297-320.

Nong, K. and M. Marschke (2006). "Building networks of support for community-based coastal resource management in Cambodia", in S. Tyler (ed.), *Communities, livelihoods and natural resources: Action research and policy change in Asia*. Reading and Ottawa, ITDG Publishing and International Development Research Centre. pp.151-168.

Oxfam (2008). *Viet Nam: Climate change, adaptation and poor people*. Oxford and Hanoi, Oxfam International. http://www.oxfam.org.uk/publications

Shaw, R. and K. Okazaki (2004). *Sustainable community based disaster management (CBDM) practices in Asia: A user's guide*. Kobe, Japan: United Nations Centre for Regional Development, Disaster Management Planning Hyogo Office. http://www.hyogo.uncrd.or.jp/publication/pdf/Guide/UsersGuidePDF/index.html

Umesh, N.R., A.B. Chandra Mohan, G. Ravi Babu, P.A. Padiyar, M.J. Phillips, C.V. Mohan and B.V. Bhat (2009). "Shrimp farmers in India: Empowering small-scale farmers through a cluster-based approach", in S.S. De Silva and F.B. Davy (eds.), *Success stories in Asian aquaculture*. Ottawa, International Development Research Centre and Springer Verlag.pp.41-66.

Vincent, K. (2007). "Adaptive capacity and the importance of scale", *Environmental Change* 17: 12-24.

38

Information and Communications Technology for Disaster Risk Reduction[1]

ICT plays an important role in establishing effective early warning systems.

Source:

Adapted from: A Policy Brief on ICT Applications in the Knowledge Economy, Issue No. 4. September 2009. Economic and Social Commission for Asia and the Pacific (ESCAP). United Nations.

---

1. This article is a remake from the Policy Brief on ICT Applications in the Knowledge Economy. It was prepared by the Information and Communications Technology and Disaster Risk Reduction Division of UN Economic and Social Commission for Asia and the Pacific (ESCAP) to provide a brief introduction on selected ICT applications, identify issues for implementation, and provide policy direction for the promotion of the applications.

1. Introduction

Extreme weather conditions and natural disasters are taking an increasing toll, in both human and economic terms. The international community, having realised the gravity of this ominous trend early on, adopted the Hyogo Framework (2005-2015). In the Framework, countries agreed on actions to reduce the loss of life and the socioeconomic and environmental impacts of disasters, by identifying, assessing, and monitoring disaster risks, enhancing early warning, as well as strengthening disaster preparedness. Globally, 2008 was the third most expensive year on record in regard to disaster related economic damage. In the Asia and the Pacific region, 2008 was marred by Cyclone Nargis, which devastated the Irrawaddy Delta in Myanmar and killed an estimated 130,000 people, and the earthquake that struck Sichuan Province in China, which affected millions and left more than 85,000 dead. In terms of deaths, Cyclone Nargis ranks among the worst 5 cyclones, and the Sichuan earthquake among the worst 10 earthquakes, since 1900.

It is widely recognised that information and communication technology (ICT), including space based technology, plays an important role in establishing effective early warning systems and successfully conducting emergency preparedness and response activities. Aimed at policy- and decision-makers from developing countries working on disaster risk reduction in the Asia and the Pacific region, this policy brief discusses useful infrastructure and applications and recommends actions on how to mainstream and enhance their use in disaster risk reduction (DRR) efforts, in particular with respect to early warning systems.

2. Information collection and sharing

A number of ICT tools are available to help systematically collect data and undertake risk assessments so that the behaviour of hazards and the socioeconomic vulnerabilities of communities can be better understood. These tools can be offshore (tsunami buoys), ground-based (automated hydro-meteorological observing systems, broadband seismometers, portable digital cameras and electronic handheld devices), airborne (lidar) or space-based (optical and radar satellite remote sensing, global positioning systems); all are used to acquire data for various types of

Information and Communications Technology for Disaster Risk Reduction

ICT tools for early warning systems, emergency preparedness and response

The main functions supported by the tools presented in the present policy brief include the following:

- Information collection and sharing,
- Decision support systems, through the integration of geo-spatial data,
- Communication and dissemination, and
- Emergency preparedness and response.

A simplified model of an early warning system is provided in Figure 38.1.

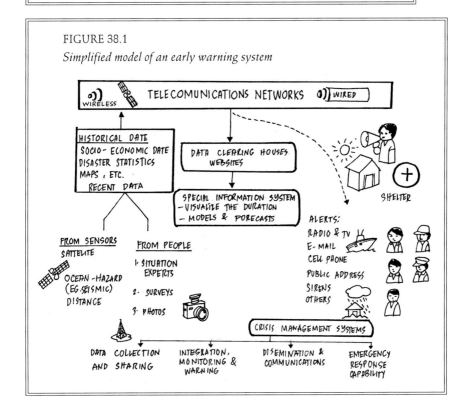

FIGURE 38.1
Simplified model of an early warning system

hazard monitoring and at different stages of disaster risk management. For example, the World Meteorological Organization collects data through more than 20,000 manned and automatic weather stations in order to monitor hazards. Processed satellite images of affected areas for damage assessment and disaster response can be provided by other global and regional initiatives, such as the International Charter on Space and Major Disasters and Sentinel Asia, in the event of major impending disasters and upon the request of countries.

Stakeholders in DRR began collaborating with each other through national and regional multi-sectoral mechanisms, with the objective of sharing knowledge and tools for improving access to information and implementing measures of DRR. This is increasingly done through the Internet, using data clearing-houses, informational websites, document depositories, discussions forums and communities of practice (for example, PreventionWeb). It is important to establish international standards for describing data (metadata) to enable the practical usage of the data that are made available.

3. Decision support systems

Promoting synergies in hazard monitoring and risk identification through the functional integration of scientific and technical organisations working in meteorology, geology and geophysics, oceanography and environmental management, among other fields, has been identified as a key action towards reducing disaster risks. It is equally important to reflect socioeconomic perspectives in the process. Relevant organisations should have the capability to integrate hazard and risk information, as well as to identify and monitor the parameters influencing the hazards, and should have access to decision-support tools.

Decision-support tools include spatial information systems designed to assist in integrating and analysing vast amounts of historical and real-time data and in displaying the data in user-friendly ways. Geographic information systems can super impose multiple layers of spatial information derived from the processing and interpretation of remotely sensed data (such as land use and geomorphology) with other geographical and cartographic information (such as elevation and slope). This information can be linked to statistical databases (for example, those with figures on population density), resulting in maps on which high-risk areas can be matched with the socioeconomic features of the society.

> *Communication and dissemination*
>
> Voice and data communication continue to be of crucial importance in the context of disaster management. Some tools, such as traditional radio and television, are ideal for one-way mass communication, as they have high penetration rates in most countries. Other types of radio tools, such as community, amateur, shortwave and satellite broadcasting, are also suitable for transmitting information for universal coverage. The Internet, e-mail and mobile telephones are becoming increasingly important broadcasting tools. Cellular phones provide mobility, two-way communication, location based services and privacy. As more poor people in many developing countries of Asia and the Pacific obtain access to such phones, special attention must be given to making new content and early warning alerts suitable for these devices.

Countries in the Indian Ocean receive international tsunami warnings from the Pacific Tsunami Warning Center, US and the Japan Meteorological Agency, which use decision support systems to forecast whether an earthquake may cause a tsunami. These tools are also useful to policy and decision-makers addressing issues such as unplanned and poorly managed urban growth, rural poverty and vulnerability, and declining ecosystems, which together have been identified as the drivers underlying the increase in disaster risk.

4. Emergency preparedness and response

National and community disaster preparedness and response capabilities should be developed by all levels of government as well as by communities. In addition to planning for adequate evacuation routes, emergency shelters, and emergency stockpiles of food, water and medicines, efforts must be made to ensure redundant and reliable communication systems, as well as efficient operating procedures in and reports originating from disaster-affected areas. Such critical infrastructure, facilities and communication systems can be developed using ICT tools, including those that are space-based.

Risk assessment maps, generated by geographic information systems, play a critical role in determining safe locations for emergency shelters and evacuation routes. Disaster management systems and field reporting mechanisms are key tools for understanding and managing relief and recovery activities, including the relocation of people and the management of the logistics for food, fuel, water, medicines and other critical assets. This logistical information should also be linked to spatial information systems, enabling disaster managers to visualise on a map the correlation of disaster damage and casualties together with the available and needed facilities and supplies.

Another essential tool for disaster response is emergency communication. Emergency communication operating procedures and equipment must be established and made available to various actors involved in disaster response. In an emergency situation, these actors must have reliable and redundant communication channels, as some forms of telecommunications infrastructure may be damaged by extreme disasters. Emergency communication should consider such requirements as: (a) enabling the relevant actors to continuously report from within the disaster areas; and (b) maintaining their access to information sources, such as meteorological reports, at all times. Pre-established crisis situation centres should be equipped with crisis management systems to handle the command and control of rescue teams.

5. Challenges in the use of ICT for disaster risk reduction

5.1 Lack of data and information sharing

The data and information needed for DRR come from a wide variety of sources which often are not shared or integrated in a way that facilitates timely and accurate decision-making in a disaster situation. This is further complicated due to the differences in standards used for data collection and classification within national boundaries, as well as between neighbouring countries in the case of trans-boundary disasters, creating difficulties when users attempt to access and analyse data. Some countries lack historical records about hazards and the quality of the data may vary. Frequently, historical data is not available in an electronic format, and it lacks proper classification and descriptive information (metadata), which makes it difficult to compare data among regions.

5.2 Insufficient human and institutional capacity

Policymakers working on DRR may be aware of the potential that ICT tools may hold for their work; however, there may be a lack of skilled staff to analyse and interpret data for evidence-based policy and decision-making.

Also, national DRR entities may lack the institutional arrangements that would enable them to mobilise sufficient human and material resources to benefit from ICT, or to obtain such resources from other countries or institutions in the region through cooperation or assistance mechanisms.

5.3 Lack of connectivity and unreliability of telecommunication networks

The flow of information in an early warning system often originates from global or regional sources (eg., meteorological and seismic data), and needs to reach, through a national centre, local authorities and ultimately people in their communities. Therefore, it is critical that all the stakeholders in an early warning system have access to communication tools that enable them to fulfill their role in a timely and cost-effective way.

In the event of severe disasters, connectivity in many countries fails. Moreover, developing countries, particularly the least developed countries and small island developing states, have frequent power outages and possess unreliable telecommunications networks that offer only low connection speeds, both nationally and internationally. Outdated or insufficient equipment supporting the telecommunications backbone as well as the user terminals, such as television, radio and personal computers, contribute to this problem. These conditions may result in average citizens and emergency response teams being incapable of receiving early warnings or achieving their objectives, if the communication infrastructure on which they depend is affected or rendered unavailable by a natural hazard or other reasons.

> *Policy recommendations*
>
> The successful implementation of ICT applications, including those that are space-based, for disaster risk reduction requires an enabling environment which fosters the development of ICT infrastructure, capacities and institutional arrangements. Despite the increasing awareness and availability of resources for disaster risk reduction, and the affordability and reach of ICT to previously unconnected communities, it would be unrealistic to expect everything to be put in place soon. In the meantime, it is important for developing countries to leverage the available ICT resources and services, while prioritizing and planning the mainstreaming of ICT in plans, efforts and initiatives related to disaster risk reduction.

The ESCAP secretariat recommends the following areas of policy intervention pertaining to the issues and challenges raised above. Information collection and sharing. National data collection, standardisation and sharing procedures and guidelines should be established for collecting environmental and social data that are needed for risk assessments, hazard monitoring and disaster forecasting. Data should be appropriately classified and made widely available electronically for use by national and international stakeholders. To obtain urgent space based information for response and early relief activities, countries could seek support from global and regional initiatives, such as the International Charter on Space and Major Disasters, and the Sentinel Asia project.

5.4 Capacity-building

Countries should develop human resources and strengthen the capacities of institutions to integrate and make use of ICT applications, including those that are space-based, for effective DRR and disaster management. Countries should seek to leverage available ICT resources and services that are available regionally and globally by joining cooperative mechanisms and capacity-building programmes for disaster risk management, such as: (a) the ESCAP Regional Space Applications Programme for Sustainable

Development (RESAP); (b) the Asian and Pacific Training Centre for Information and Communication Technology for Development (APCICT), an ESCAP regional institution; and (c) the United Nations Platform for Space-based Information for Disaster Management and Emergency Response (UN-SPIDER), executed by the United Nations Office for Outer Space Affairs (UNOOSA). Tsunami-prone countries can seek support from sources such as the Tsunami Regional Trust Fund in order to strengthen their multihazard early warning capacities, as part of the efforts towards a regional early warning system for tsunami and other hazards.

Additionally, countries should establish or strengthen agreements with other countries and institutions in their subregion to provide institutional support and share ICT resources (such as equipment, including the expertise required to use the equipment), which may be too costly for a single developing country to acquire and maintain.

5.5 Reliable connectivity

Governments should implement policies and regulations to provide universal access to ICT services, reaching users located in unserved and underserved areas. Policymakers should identify the hazard-prone areas that are home to the highest numbers of unconnected people (lacking radio, television, telephones and/or Internet), and take immediate action to provide these areas with at least the minimum service required to communicate awareness campaigns, early warnings, and emergency response information. In particular, policymakers may consider encouraging investment in wireless voice and data networks, as they offer opportunities for poor rural areas to achieve connectivity rapidly and cost-effectively, as can be demonstrated by the continuous growth in the adoption of mobile telephones. In the immediate term, policy and decision-makers should maximise the use of the most widespread technologies in their country, such as radio, television, satellite television and mobile phones, and establish partnerships with the operators of those networks in the private sector, as well as with civil society and mass media, to ensure that they cooperate fully with government in all DRR and disaster management activities.

As emergency response teams and early warning systems rely on telecommunication networks, such networks should be designed to be available at all times and should have a backup service in case of failure. They should be tested frequently and treated as critical national infrastructure. However, funding the large investments required by such activities can be difficult, particularly so in the midst of the current economic crisis, given competing development priorities. Insufficient investment in early warning capacities is a significant challenge. As recommended by the Global Platform for Disaster Risk Reduction (GPDRR) at its second session, governments should implement policies which enable the mobilisation of national development funds to risk reduction measures. This could help to: (a) extend the reach of communication networks to underserved areas; (b) ensure the reliability of such networks; and (c) reduce the price of telecommunications equipment and user terminals, such as computers and cellular phones. Such policies may include establishing Universal Service Obligation Funds (USOF), providing tax waivers for the import of equipment, and promoting the use of free and open-source software.

Policymakers should also support the implementation of standby systems for regional and subregional emergency communications, which can be shared among participating countries during emergencies.

Reference

UNISDR (2007). *Hyogo framework for action 2005-2015: Building the resilience of nations and communities to disasters.* The World Conference on Disaster Reduction (WCDR), Kobe, Hyogo, Japan. January.

39
Management of Coastal Resources through Village Level Planning

Source:

B.H.J. Premathilake *Coast Conservation Department, Sri Lanka, 2010.*

1. Introduction

Coastal resource management (CRM) is one of the key pillars of community resilience for reducing coastal hazards. Livelihood component in post-tsunami project sites reveal that a significant portion of the local community are dependent on coastal resource based livelihoods mainly fisheries, coastal tourism, and lagoon resources extraction. On the other hand, coastal environment is subject to continuous socioeconomic and ecological changes (human and natural) mounting to enormous pressure and degradation to the available resources. Pollution, over exploitation, and habitat degradation are to be avoided through legal or community actions so as to reverse coastal environment destruction. The massive destruction such as the Indian Ocean tsunami in 2004 awakened policymakers and practitioners to take all possible measures to prepare for such disasters.

As a centralised government regulatory authority, the Coast Conservation Department (CCD) focussed on top-down planning approach which fails to ensure enough community involvement to their field level

Coastal zone management (CZM) in Sri Lanka

Sri Lanka is a pioneer country in the Asian region having a national CZM programme. The Coast Conservation Department established in 1983 under the legal provisions of the Coast Conservation Act no. 57 of 1981, is responsible government agency for managing and conserving coastal resources in the country. The department prepared the country's first generation national CZM plan in 1990 which addressed three common coastal issues namely; coastal erosion, coastal habitat degradation and protection of archeological and scenic sites in the coastal zone. The plan was revised in 1997 and 2005 to incorporate new challenges that arose as a result of changing socioeconomic and ecological conditions. The national CZM plan provides policies, strategies and actions for solving key coastal environmental issues in the country. Although Sri Lanka has progressed considerably in this direction, weaknesses such as inadequate enforcement capacity, weak institutional coordination and insufficient funds exist in implementation of regulations and plans.

implementation programmes. On the other hand, as a national level agency, the CCD has no capacity to undertake village level resource management. Yet, the CCD understood the importance of community participation and community ownership of sustainable management of coastal resources. Therefore, the village level coastal resource management approach was adopted in collaboration with Sarvodaya and Practical Action. The experiences spread over 12 coastal villages belonging to the four districts of Ampara, Hambantota, Matara, and Galle are documented in this article.

2. Significance of coastal resource management (CRM)

Sustainable community development, disaster management, and CRM are the key issues in developing community resilience to reduce coastal hazards. Nearly 60 per cent of the four districts is dependent on coastal resources such as fisheries, tourism, salt and sand extraction and lagoon resources extraction. Others depend indirectly on coastal resources. Fast changing socioeconomic and ecological (human and natural) are mounting enormous pressure on the available resources. Pollution, overexploitation, and habitat degradation have to be reduced through legal or community actions.

Identified Problems:

- Over extraction of fishery resources
- Coral breaking, sand extraction
- Destruction of mangrove and non-mangrove vegetation
- Alteration of land use forms
- Absence of actions to reverse the resource degradation process

3. Village level CRM

The approach adopted for sustainable management of coastal resources in village sites was new in Sri Lanka, though the country has a good reputation for national scale coastal planning. The project was aimed at capacity building of communities in each village to undertake their own initiatives for coastal resource management. The community took responsibility of the entire process beginning from identifying issues to monitoring and implementing strategies such as mangrove rehabilitation in Andaragasyaya, Medilla, and Pathegama; sand dune protection in Wanduruppa; and coastal area greening. The project facilitated and provided enabling environment to the communities.

4. Challenges for village level CRM planning

The project faced a number of challenges in establishing a sustainable programme for conserving and managing village level resources in the initial stages. These were:

- Lack of organisation within the community for resource management due to the lack of knowledge on coastal resource and their management.
- Difficulty in mobilising the community to undertake the CRM programme as their own initiative.
- There was no incentive for the national level agencies to participate in planning due to the small geographical coverage and less number of beneficiaries.
- There were inadequate financial and technical resources and guidance for implementing their own initiatives.
- The existing administrative system of the country did not encourage community based initiatives.

To overcome these challenges, the implementing organisation undertook the following measures:

- Initially, project field staff was made aware of coastal resources, their benefits and management. Later they were used as facilitators and coordinators in the village level processes.
- Community was organised and mobilised by the project field staff under the guidance of a hired CRM consultant.
- Linkages between the community and relevant local government agencies such as CCD, forest department, divisional secretariat,

and disaster management centre were established to ensure the implementation in more collaborative manner.

- Awareness about the concepts of CRM was created among local community and field officers from various organisations who were engaged in disaster management, community development, and environmental programmes in the area.
- Village development committee (VDC)—a local village institution—was formed to implement various thematic areas concerned with improving communities' resilience. A community group was selected to work as environmental committee which assigned the responsibility of village level coastal planning.

5. CRM plan preparation

'Community-driven coastal resource management initiative' for each village was a novel experience for the community of each project site. The project facilitated the community to prepare a management plan to address their coastal resource issues. A CRM consultant was hired by the project to carry out this facilitation role in more effective manner. The Strengthening Resilience in Tsunami Affected Communities (SRTAC) project noticed that the communities had a clear picture of declining resources that had affected their livelihoods and scenic beauty even before the tsunami 2004. They also had noticed the link between coastal resources and tourism.

People at each site identified three broad issues: (i) coastal habitat degradation, (ii) coastal disaster risk reduction, and (iii) livelihood development. Although nature and severity of these issues differ from site to site, solving them generates both short-term/long-term benefits to local people. The communities at each site identified root causes and threats responsible for degradation of their coastal resources and proposed more practical solutions to reverse the degradation process.

People also understood the role of coastal ecosystems as natural shields against coastal hazards during tsunami disaster 2004. Once the communities were mobilised, they felt the need to have a series of actions to transform resource exploitation into sustainable management of

resources. The plan also introduced bio-shield development as a disaster risk measure. Furthermore, local people at each site identified sustainable coastal resource based livelihoods and introduced interventions to promote the identified livelihood options.

6. Plan implementation

The VDC was responsible for implementing the prepared plan. It was required to ensure and identify required funding and other resources for the implementation through coordination with external donor agencies and relevant government/non-government agencies. Resource scarcity for plan implementation led the community to discuss among themselves and prioritise activities proposed by the CRM plan. In the meantime, the implementation of number of some activities in each project was financially supported by the project as a means of mobilising the local community.

7. Achievements of the VDC

- The VDC proved to be a good learning experience of the community as it involved different stakeholders such as government, coastal managers, practitioners, policymakers, students and academics.
- The project ensured community involvement and implementation of their resource conservation and management programmes.
- The 12 communities are now aware of the coastal resources and have come forward with their own initiatives to protect coastal resources.
- Wanduruppa and Andaragasyaya communities became good examples to other communities by coming forward to protect coastal sand dunes from mineral sand extraction.
- Local people became strongly convinced that sand dunes protected them during tsunami disaster in 2004 and are taking measures to conserve them.

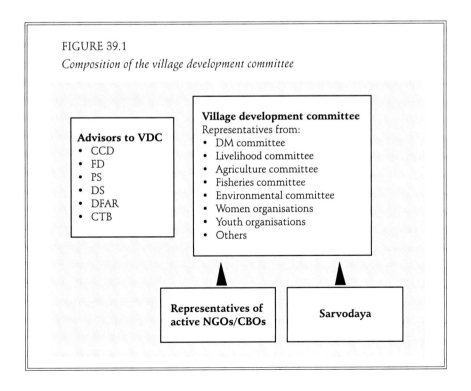

FIGURE 39.1
Composition of the village development committee

The community have come forward with their own initiatives and implement resource conservation and management programme to protect coastal resources.

- Each community at each project site came up with its own organisation and a mechanism for working together for resource conservation and management.
- They continue to have dialogues with government and non-government agencies to strengthen the community resilience.
- VDCs at Talalla, Pathegama, Wanduruppa, Andaragasyaya and Karathive maintain contacts with government agencies and their comments and suggestions are sought through *grama niladhari* (village administrator) for divisional development plan prepared by the divisional secretariat.
- Karathive VDC is working with district disaster management centre to strengthen community disaster management capacity in the village.

- The committees at Andaragasyaya, Wanduruppa, Medilla, Talalla, Pathegama and Karathive show high potential for further development.

8. Challenges ahead

The capacity of each VDC at each site has to be strengthened further to undertake successful CRM initiatives. There are many good project proposals that require coordination with external donor agencies to find resources for village level implementation. The VDCs, however, are faced with limited or no resources for the implementation of the identified CRM activities. They require external financial and technical assistance for another three years to gain stability within the VDC. The VDC will be institutionalised only when it is able to integrate sustain activities. CRM initiatives should be pursued for both short-term and long-term benefits.

40

Development Planning using Spatial Data

GRS and RS are important tools to improve understanding of coastal management needs and in supporting rehabilitation efforts.

Source:

Gnanappazham Lakshmanan and V. Selvam *M.S. Swaminathan Research Foundation, Tamil Nadu, India, 2010.*

1. Introduction

The area of interaction between sea and land which is influenced by terrestrial and marine environments is known as a coastal zone. It is very dynamic in terms of its physical processes, endowed with a variety of ecosystems with a high population density with varying livelihood needs, trades and transports. The coastal ecosystems, which individually support their own unique assemblages of plants and animals, are also complexly linked with and dependent on each other. Hence, sustainable management of coastal zone becomes essential to safeguard the coastal environments to ensure its productivity upon, which the livelihood of millions of people depend.

In the past, there was a gap in understanding the environmental and anthropogenic factors acting on coastal ecosystem. This was due to the lack of updated spatial information of the ecosystem available to management authorities. Only limited site-specific information on coastal resources, interrelationships between the physical and biological components and interactions between ecological processes and human needs was available. The coordination among management, scientific community and the user community was not adequate for purposes of conserving and managing the coastal environment. After the development of remote sensing (RS), Global Positioning System (GPS) and Geographical Information System (GIS) technologies, key information on coastal resources management became transparent to all. Every stakeholder can now view the area using RS data. Monitoring coastal resources and its environment becomes easy with satellite images. Manual surveys of entire ecosystems is a difficult and tedious task. Satellite data also provides accurate and periodic information. Field information such as biophysical and socioeconomic data can be integrated with maps derived from RS data. These are now being used as decision-supporting tools in coastal zone conservation and management.

In India, coastal regulations were formulated in 1991 with coastal regulation zones (CRZ) I, II, III and IV defining the limits for ecologically sensitive zones and zones for development. Later in 2007, coastal management zones (CMZ) I, II, III and IV have been formulated with modifications by considering the benefits for both the coastal community and for the protection of coastal environments. In order to regulate these activities in coastal areas, the foremost step is to acquire knowledge about current coastal land use conditions, changes in the coastal zone over a period of time and to precisely delineate the high tide line (HTL) and low tide line (LTL). RS helps in deriving this information from satellite data in hands with GIS while the societal influence on coastal zone and their

---

Components of remote sensing

1. Energy source or illumination—the first requirement for remote sensing is to have an energy source to illuminate or provide electromagnetic energy to the earth's surface. Enormous amount of energy is available from Sun.

2. Radiation and the atmosphere—as the energy travels from its source to the earth surface, it will interact with the atmosphere it passes through. This interaction will take place second time as well as the energy travels from the earth to the sensor.

3. Interaction with the target—once the energy reaches the earth through the atmosphere, it interacts with the objects on the earth depending on the properties of the objects and the solar radiation. That is the light is either reflected or absorbed (then emitted) or transmitted.

4. Recording of energy by the sensor—after the energy has been reflected by, or emitted from the object, we require a sensor (remote—not in contact with the object) to collect and record it.

5. Transmission, reception, and processing—the energy recorded by the sensor has to be transmitted, often in electronic form, to a receiving and processing station where the data are processed into an image (hardcopy prints and/or digital).

6. Interpretation and analysis—the processed image is interpreted, visually and/or digitally, to map the area of the earth which was acquired by the satellite.

7. Application—the final element of the remote sensing process is achieved when we apply the information about the area derived from map (prepared using satellite image) in order to better understand, reveal some new information, or assist in decision-making process of planning, monitoring and management of that area or locality. These elements comprise the remote sensing process from the beginning to end.

interaction is effectively and efficiently done when participatory rural appraisal (PRA) is integrated with GIS recently recognised as Participatory GIS (PGIS). With the advent of high resolution satellite images and development in the field computer technology, these tasks have now become easier than before and help planners and decision-makers in the periodic CZM activities.

2. What is RS?

The literal definition of RS is 'the science and to some extent, art of acquiring information about the Earth's surface without actually being in contact with it. This is done by sensing and recording reflected or emitted energy and processing, analysing, and applying that information.' The process of RS can be exemplified by the basic components as given in Figure 40.1.

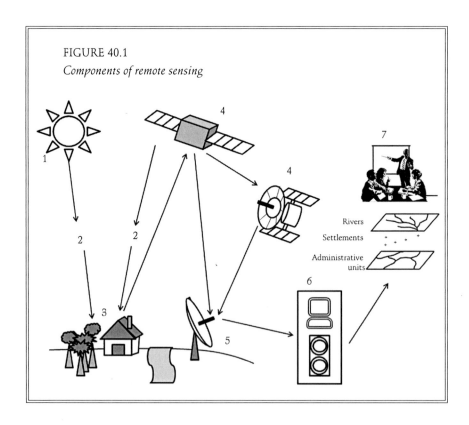

FIGURE 40.1
Components of remote sensing

3. What is GIS?

GIS is broadly defined as a system for data input, storage, manipulation and output of geographic information and a practical instance of a GIS combine's software with hardware, data, a user, etc., to solve a problem, support a decision, help to plan. Nowadays GIS also includes the people involved in database development and decision-making process using GIS.

RS and GIS are used in coastal environment either individually or together in many aspects. Some of them are listed below:

1. Mapping of the inaccessible areas of coastal ecosystems through a synoptic view.

2. Monitoring of the changes in coastal land use/ land cover, shoreline changes and the coastal geomorphology.

3. Monitoring of the offshore dynamics like suspended sediments, coastal currents, near shore bathymetry, oceanic circulation etc.

4. Mapping of the density and zonation of ecologically sensitive areas such as mangroves, coral reefs, sea grass, seaweed, sand dune, dune complex, estuarine system etc.

5. Quantifying the impact of developments and pollution to the coastal environment; and

6. Supporting tool in decision-making process of CZM, coastal hazard mitigation and management.

Example: Application of RS and GIS in mangrove conservation, restoration and management of Pichavaram mangrove wetland, Tamil Nadu.

Satellite data of different periods (1977, 1986 and 1996) and Survey of India toposheet map (surveyed in 1970) were used to assess and monitor the mangrove forest area in the selected periods. Anthropogenic pressures such as diversion of fresh water from the catchment area of the river basin to agriculture purpose by construction of dams since independence, dependency of the local community for their livelihood are integrated with satellite data derived maps such as mangrove changes and shoreline changes. Changes showed that there was a reduction of mangrove area from 615 hectares to 325 hectares from 1977 to 1986. These maps were analysed by the mangrove specialists to find the actual causes of

> *Components of GIS*
>
> 1) Data input—Spatial information in the form of maps and charts are entered in GIS by digitising or scanning them. Qualitative and quantitative data are entered in the tables, which are linked with the digitised or scanned maps. The spatial data of the location of our interest is entered with its geographic coordinates either in terms of latitudes/longitudes or in projected coordinates such as Universal Transverse Mercator (UTM) projection or Conical projection.
>
> 2) Database management—This deals with the editing, spatial correction and geocoding the spatial data entered and managing them with their tabular attributes of the records such as the identity number, name, values, prices, etc.
>
> 3) Manipulation and analysis—The database developed in the GIS environment is ready for the user to work on, to compose the maps required for the analysis, to make a query out of a single map or the combination of different maps etc. Analysis includes overlay analysis, buffer analysis, extraction, multiple layer interactive query and many more.
>
> 4) Output—Once the GIS data is entered, linked with the attributes and analysis is done, the results of the analysis can be presented in the form of maps, charts, tables or the combination of either or all of them for documentation.

reduction in mangrove area over these periods. The major causes for degradation of mangroves are development of high soil salinity due to changes in the biophysical condition, which is in turn due to clear felling system of management followed in the past, reduction in fresh water flow in core mangroves and cutting trees for fire wood and grazing cattle in the peripheral areas of mangroves. To overcome these causes, proper management measures such as ensuring proper tidal water flushing in the hypersaline degraded areas by establishing artificial canals and plantation of mangroves, proper cattle management practices, alternate income generation activities to local communities etc., were taken up. The satellite image of 2006 (Figure 40.3) shows the increase in mangrove area which can be compared with that of satellite image of 1986. Thus, remote

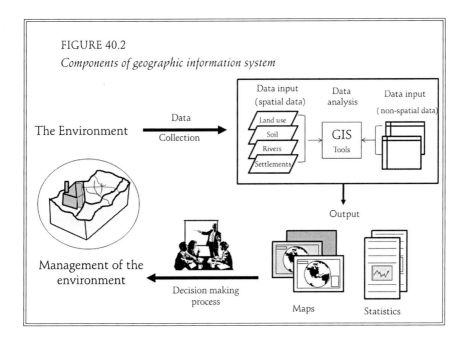

FIGURE 40.2
Components of geographic information system

sensing data integrated with GIS helped in planning and decision-making process of mangrove forest management and monitoring results of various activities periodically.

4. RS and GIS in coastal zone management methods

The development and implementation of Integrated Coaztal Zone Management Plan (ICZMP) includes zoning, vulnerability mapping, and defining of setback zone.

Zoning of the coastal area is done by delineating the zones of habitats, coastal hazards, suitable areas for aquaculture, tourism, industries and recreation. Thus, RS plays a major role in assessing and monitoring the current land use and land cover, soil types, geomorphology of the coast. Integration of GIS by adding information about other spatial data such as types of industries, settlements, road/railway network leads to the zoning of the coastal area.

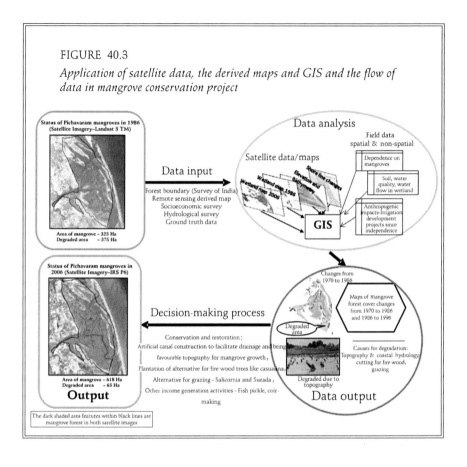

FIGURE 40.3

Application of satellite data, the derived maps and GIS and the flow of data in mangrove conservation project

Vulnerability levels are mapped by integrating the details and maps of elevation, geology, geomorphology, sea level trends, shoreline displacement, tidal ranges and wave heights using GIS analysis (National Technical Service, US. GIS helps in the overlay of the maps (prepared using RS data and field survey) to calculate and map the vulnerability of the coastal zone and community.

Setback zones are mapped by drawing buffer around different features such as houses, road networks, drainage lines, habitats, infrastructure facilities such as hospitals, industries, cyclone shelters etc. Overlaying different buffer zones of varying features is easily accomplished by GIS analysis once the database is generated with accuracy. Figure 40.4 shows a coastal area with critical habitat—mangrove vegetation and location of

Development Planning using Spatial Data 451

FIGURE 10.4
Buffer analysis used in delineating the setback zones for development activities by considering three criteria: (a) 1000 m away from the critical habitat, (b) 500 m away from villages and c) 500 m away from high tide line

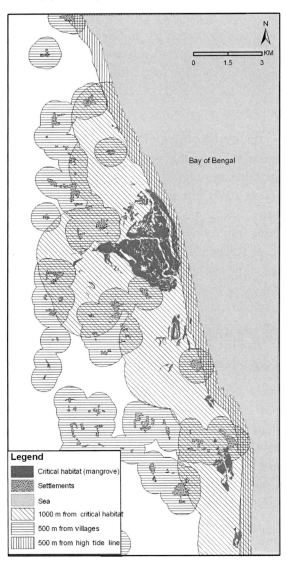

Note: The shaded portions are restricted and the zones outside the shades are defined as development zones.

hamlets along the coast. For example, if the zones for development sector has the criteria of: a) not within 500 metres from high tide line, b) not within 500 metres from settlements, and c) not within 1,000 metres from critical habitats, then the maps of coastal habitats, high tide line, and the hamlets are used in GIS environment to create the buffer zones for each of these criteria. These maps are then overlaid to identify the areas that are restricted and not restricted for the development. GRS and RS are important tools for improving the understanding of coastal management needs and in supporting rehabilitation efforts in these areas.

5. Conclusion

GIS, PGIS, and RS are becoming more accessible tools for storing and analysing information, mapping, visualising and modelling development scenarios for monitoring progress and change in rural development and rehabilitation efforts. While it still tends to be expert-driven and not so very cost-effective tool, the concepts and the application of GIS is still being utilised for improving the understanding of CRM. GIS can be applied at various scales and levels of complexity for specific contexts. With such a flexibility, the application of participatory GIS becomes more relevant in rural development, especially for land and resource mapping. Development planners and decision-makers will have better access to a collection of both primary (PRA, GPS, field survey, and baseline survey) and secondary (satellite data, survey maps, census data, and published maps) data for their immediate action.

Local and scientific spatial knowledge, community-based natural resource management, community-based disaster risk reduction, area planning, and environmental and coastal resource mapping can be integrated, thereby giving three-dimensional view to the critical state of the environment and communities vulnerability. The purpose of PGIS is also to improve networking and critical analysis among various stakeholders such as community, development practitioners, government officials, etc. The next article captures the application of PGIS for village development planning in greater detail.

41
Towards a Better Assessment Framework: The Case of Wanduruppa

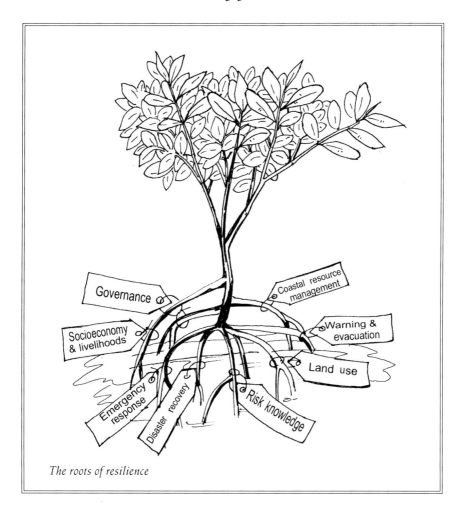

The roots of resilience

Source:

Buddhi Weerasinghe *Disaster Management Centre, Sri Lanka, 2010.*

In Sri Lanka, the project Strengthening Resilience in Tsunami-affected Communities (SRTAC), co-funded by Canadian International Development Agency (CIDA) and the International Development Research Centre (IDRC), was implemented at 12 sites in Wanduruppa and Karaitivu. This site-specific case study specifically looks at Wanduruppa. Partners in Sri Lanka were Sarvodaya and Practical Action. The project was designed to build community resilience to natural disasters through:

- bioshield establishment;
- livelihood development and diversification;
- disaster management planning; and
- strategic information for communities.

This case study looks comparatively at two assessment frameworks for measuring resilience. The aspect of resilience types and assessment is covered in Part 3 which revisits the concept of community resilience.

Wanduruppa Grama Niladhari (GN) division is situated in Hambantota district bordering Walawe river estuary, South of Sri Lanka, with a population of 2,597. The village is semi-urban with 435 families. Major income sources are farming, fishing and small scale businesses.

Key issues in Wandurruppa are flood risk to paddy lands, infrastructure development needs, markets for producer groups, management of household wastes and degrading of the coastal environment.

Assessment of resilience has been conducted using selected characteristics for components of resilience formulated for coastal community resilience (CCR) by Indian Ocean Tsunami Warning System (IOTWS) and Asian Disaster Preparedness Centre (ADPC). Table 41.1 categorises them under the Hyogo Framework for Action, 2007 (HFA) themes and capital domains.

1. Governance

Government is one of the actors in governance. Other actors involved in governance vary depending on the level of government that is under discussion. In rural areas, for example, other actors may include: influential landlords, associations of peasant farmers, cooperatives, non-

TABLE 41.1
Criteria used to assess community resilience

Characteristic used	HFA theme	Capital domain
Governance	Governance	Human capital
Socioeconomy and livelihoods	Risk management and vulnerability reduction	Economic capital
Coastal resource management	Risk management and vulnerability reduction	Natural capital
Land use	Risk management and vulnerability reduction	Natural capital
Risk knowledge	Knowledge and education	Human capital
Warning and evacuation	Disaster preparedness and response	Physical capital
Emergency response	Disaster preparedness and response	Physical capital
Disaster recovery	Disaster preparedness and response	Physical capital

government organisations (NGOs), research institutes, religious leaders, finance institutions, political parties, the military, etc. The situation in urban areas is much more complex.

At Wanduruppa, there are about 13 community-based organisations (CBOs) and several NGOs working at ground level, which offers partnership for Wanduruppa village activities. The local government mandated with maintenance of drains and other infrastructure apparently does not function at accepted levels.

Disaster risk reduction (DRR) policy planning, priorities, and political commitment at local government level are currently inadequate and are compounded by the scarcity of government funding for mitigation. However, there is acceptance of funded interventions by outside agencies.

At the time of the project commencement, 2,000 acres of land in Wanduruppa has been flood-affected for over two decades. The local government was unable to mitigate this flood vulnerability. Sand bars that form across the local river mouth impede floodwater drainage. The project intervention supported the establishment of a trust fund (Rs. 600,000) with the accruing interest being ear-marked for maintenance of the canals and the river mouth into which they drain.

Catalysed by this intervention, the community now collects money on a voluntary basis (Rs. 11,000 per season) to cut sand bars without using available funds and saving for future contingencies. Reduction in paddy yield losses due to the initiative is estimated at Rs. 50 million benefiting about 1,000 beneficiaries.

Although sand bar is now cut in time, the water subsidence is slow due to sedimentation of existing drainage channels with tsunami debris. Clearing this rests with local government but has not been taken care of, even today. A probable reason may be lack of financial allocation. Whether the current arrangements would hold in a high-return-period event is to be seen. Community initiative has gained high recognition from the district and divisional authorities and has now been invited to the divisional and districts development forum as a stakeholder. This intervention is an excellent example of a community based flood preparedness and mitigation effort rather than a governance initiative.

2. Socioeconomy and livelihoods

Most of the livelihoods of Wanduruppa villagers depend on the natural resources. The community livelihoods are therefore very vulnerable to disaster impact. Of the total population, 45.7 per cent form an employable sector aged between 19-60 years. However, only about a third of this work force find regular employment. Table 41.2 provides the employment profile of the community. According to Practical Action assessment, 31.5

TABLE 41.2
Employment profile of the 19-60 year old age group

Type of employment	Per cent employed
Self-employed including agriculture	11.0
Fishing	8.5
Masons/carpenters	4.2
Daily labour	3.3
Garment factory workers	1.2
Private sector	1.0
Government	0.5

per cent of the families belong to the poor category including poverty alleviation beneficiaries and daily wage earners. Ninety-five per cent of families live in their own houses. Housing typology varies. Family savings are minimal.

Overall framework used within project implementation was the Disaster Resistant Sustainable Livelihoods (DRSL) framework by Practical Action. The formulation of a village development committee (VDC) led to a need of an analysis and the development of a village development plan through a participatory approach.

VDC was the village level boundary partner and the divisional secretariat office was selected as the second boundary partner of the project. Various capacity building training programmes were carried out with the identified producer groups. These are on financial and book keeping and technical trainings to improve the quality of the final products and productivity.

Social capital of the producer groups was enhanced through networking. This has enabled them to work together to solve the common problems of the sector concerned. Basket weaving has been a traditional craft industry and the range of raw materials used is very wide: pandanus, palmyra, coconut, water hyacinth and reeds. Training was provided to weavers in Wanduruppa and a sales centre was established allowing producers to market their products directly. Lack of raw materials was a major constraint of the handicraft production group. As a long-term solution, plants that could be used as raw materials in the industry were incorporated into the bioshield. As a short-term strategy, technical trainings were conducted after studying the demand and other factors to manufacture handicrafts from easily available materials. Dress-making, home gardening, and food processing are the other livelihood options encouraged.

According to the results of the market system analysis, existing problems were: weak linkages with the market, lack of negotiation skills, and poor business orientation of the producers. These were also the constraints in reaching out for credits from microfinance organisations, which had presence in the project locations.

Financial entrepreneurship of the groups was enhanced through building their capacity to develop business proposals and linking the producers with microfinance institutions. Markets for products and microfinance for producers thus rely on outside entities. Forging linkages with entities outside the community and sustaining them again requires outside intervention such as International (INGO) or NGO partners working on a project. The linkages forged by the project between the groups and the Export Development Board, Industrial Development Board and Rural Enterprise Network have been initiatives of Practical Action and the longevity of such linkages beyond the project's lifetime is a concern.

On disaster impact, loss of raw materials, disturbed links between the value chain actors and the service providers are challenging and would delay economic recovery time.

3. Coastal resource management

Another major focus of the project has been the bioshield development through mangrove and non-mangrove plantations at suitable sites near project villages. Where there is substantial width of mangrove forests as for instance in India, their shielding capacity to break the tsunami wave is understandable.

According to Dahdouh *et al.* (2005), despite the popular and widely accepted view that mangroves act as living dykes against tsunami waves, there is surprisingly little data available to test that hypothesis. Alongi (2008) suggests that mangroves may, in certain circumstances, offer limited protection from tsunamis; some models using realistic forest variables suggest significant reduction in tsunami wave flow pressure for forests at least 100 metres in width. The magnitude of energy absorption strongly depends on tree density, stem and root diameter, shore slope, bathymetry, spectral characteristics of incident waves, and tidal stage upon entering the forest.

Mangroves flourish in the ecological niche of the intertidal zone. Since the tidal amplitude is very low in Sri Lanka, the distribution of mangrove forests are confined to a narrow intertidal belt. All these are associated with some of the lagoons and estuaries in the country.

Towards a Better Assessment Framework: The Case of Wanduruppa

Crossed purposes: Planting a bioshield versus 'harvesting' the sand.

Only species that can grow beyond niche conditions can be planted in developing mangrove belts outside the intertidal zone. Although there is a survival rate of about 80 per cent in the planted mangrove areas under this project, the effectiveness of the artificial belt without typical mangrove species and tree forms is debatable.

Indirect benefits of bioshield plantation include capacity building on nursery management, raising and selling nursery plants, income received by planting and maintenance. Low-income family members have been involved in these activities.

The Wanduruppa project has also established village committees for coastal ecosystem management. Although this initiative has had the concurrence of the CCD, the committees have no jurisdiction to prevent degradation apart from people power. If this is enduring, it will indeed be a very positive outcome. However, end of project activities, which provides an incentive for tree planting and maintenance, may reduce the level of people participation. Even during implementation, community participation has been low. More information on land availability, suitability of lands for bioshield development, relationship between rural livelihoods and bioshield resources etc., at specified project locations may have helped during project implementation. Only time will tell whether the community resource management will be internalised in community behaviour. Until then this effort will remain a stand-alone mitigation effort.

4. Land use

Practical Action has pioneered a new planning tool named Disaster Risk Sensitive Land Use Planning (DRSLUP) at Wanduruppa and has been successful in reaching the Urban Development Authority to pilot the tool in the project area and its administrative umbrella—the Ambalantota divisional area. The tool provides reference to overall hazard situation, to categorise land with regard to:

- Hazard-free areas (safe areas).
- Areas prone to specific hazards where development should take care of mitigation, construction standards etc., for prospective risk reduction and as areas that would need further study prior to designing of development projects.
- Critically sensitive areas due to hazards, archeological or cultural reasons where development should be avoided; and
- Areas where development is already in and where mitigation measures have to be planned to safeguard community as well as development investments for corrective risk reduction and others as relevant.

Practical Action has also pursued possibility of incorporating DRR into the Environmental Impact Assessment (EIA) process with the Central Environmental Authority (CEA) and the discourse is ongoing. Yet, how well this will be integrated into planning remains unsolved.

In Sri Lanka, there is currently no national land use policy in place and land falls under the control of a land commissioner, many ministries and line departments with an equally large number of legal statutes with overlapping jurisdiction. It is therefore outside the sphere of community influence to control land use.

5. Risk knowledge

The project has established village knowledge centres (VKC) or village information centres (VIC) to increase the quality, quantity and local relevance of information content to improve conditions for both men and women. These were provided with computer equipment and Internet. In most cases, information centres were utilised for communication and for informal education purposes of children. Apparently, the services offered are not considered unique and community ownership seems to be lacking.

Risk communication can only become successful when the right channel is chosen to match the target audience with relevancy of content. Livelihood and disaster related information sharing was not up to the expected level as envisaged in the project formulation phase.

6. Early warning and evacuation

At community level, warning relates to the dissemination of hazard-specific messages generated by the national level technical agencies. Community is the last mile of the early warning. The Director General of the Department of Meteorology currently issues evacuation orders for widespread disasters such as Tsunami and cyclones in consultation with the Director General of the Disaster Management Center (DMC) and the Secretary to the Ministry of Disaster Management and Human Rights. The District Secretary in consultation with District Irrigation Engineer and National Building and Research Organisation (NBRO, MoDM) issues flood and landslide evacuation orders in concurrence with divisional secretaries. The district disaster management unit affiliated to the district secretariat carries out capacity building of communities for evacuation. The national early warning dissemination system is multi-faceted and adequate.

Community involvement occurs through the village level disaster management committees centred on the *grama niladhari* (GN) who is the administrative officer. They assist in dissemination of early warning and evacuation.

Learning together to map hazards and identify safe routes and safe areas.

The DMC and ground level NGO collaboration has carried out a fair amount of capacity building at community level to strengthen these community inputs. Nearly all tsunami vulnerable communities have carried out participatory hazard mapping, identification of safe routes and safe areas and practiced mock-drills for evacuation.

Early warning and evacuation process is a preparedness measure driven by national level policy. The Wanduruppa project has facilitated the preparation of the community disaster preparedness and response plan, which is under the supervision of the preparedness planning division of the DMC. Sarvodaya is capable of linking up with the VKCs through the satellite hub at its headquarters and is therefore capable of strengthening the last mile dissemination.

7. Emergency response

The Ministry of Disaster Relief Services is mandated with providing emergency relief and restoring normalcy to the lives of affected persons. Government funding is channelled through the Ministry. The National Disaster Relief Services Centre (NDRSC) is its implementing arm and has a network of staff affiliated to divisional secretariats. NGO and volunteer collaboration is vital. Establishment of divisional and village level disaster management committees has added an orderly relief distribution and response mechanism coordinated by divisional secretaries. Shelter for the displaced and camp management capacity has improved. However, emergency funding available to the district secretary is an issue, which needs addressing.

The activities of the Wanduruppa project have created awareness and community networking which would strengthen response capacity.

8. Disaster recovery

Sri Lanka has mandated disaster recovery implementation to the NBRO, MoDM. Disaster impacts beyond a threshold value disrupt assets and processes of human, social, economic, physical and natural domains in communities and recovery becomes a dependent process of outside interventions. Events of the tsunami magnitude overwhelm national capacities and the assistance of the international community becomes critical to developing nations such as Sri Lanka. The Wanduruppa project has not been able to contribute in any significant way towards this assessment criterion.

9. An overall assessment in relation to resilience building

The project has been implemented in a community low in economic capital. Livelihood interventions have attempted to build capacity for diversification of income avenues. It has also attempted to address gaps in product value chains.

Assessment of success will need further time. However, it would be inappropriate to suggest that economic resilience of the community has been significantly improved. To achieve this, many more facets of the economic status in Wanduruppa may need to be addressed.

The development of the bioshield is identifiable as a mitigation activity against storm surges, which may have additional spillovers to the community in the future. Community ownership of the intervention is recorded to be low and therefore its sustainability remains in doubt.

The flood mitigation intervention must find merit and be recognised and disseminated as a 'best practice' in community based flood mitigation.

The VKC appears to be a supply-driven rather than a demand-driven activity due to its low acceptance by the community.

Contributions to developing social capital of the community have been creditable. The social animation created due to project implementation must be recognised and long-term sustainability must be monitored.

The eight characteristics in the community resilience framework adopted by the project clearly indicate the limited scope for proper assessment of inputs. Even out of the selected characteristics, disaster recovery appears to be beyond the capability of a project of this nature.

10. Assessing vulnerability reduction plus capacity building *vs.* resilience

Many authors use the term capacity/ability to define the concept of disaster resilience. Some consider it as the opposite of vulnerability.

11. Wanduruppa Project in a new light

The definition of resilient communities is given as 'communities that know their risks, have reduced their vulnerability, and have the ability to respond to and recover from the impacts of such risks'. Fitzgerald and Fitzgerald (2005) has translated these concepts to a graphical presentation given in Figure 41.2 relating resilience to vulnerability. Table 41.3 provides possible community action to achieve this reduction of vulnerability.

The focus is the reduction of direct vulnerability to the hazard. A re-look at Wanduruppa in this framework provides a more positive picture of inputs and outputs as shown in Table 41.3.

Participatory hazard mapping and community preparedness planning have been carried out. The flood mitigation endeavour is a success and its ownership is with the community. Ability to respond is built into early warning and evacuation capacity by the DMC. Last mile early warning dissemination has been strengthened by the VKCs. Both mitigation and response will depend on the social capital, which has been enhanced at

ISDR defines capacity, capacity building, and vulnerability. As follows:

Capacity

The combination of all the strengths, attributes and resources available within a community, society or organisation that can be used to achieve agreed goals.

Capacity building

The process by which people, organisations and society systematically stimulate and develop their capacities over time to achieve agreed goals, through improvement of knowledge, skills, systems and institutions.

Vulnerability

The characteristics and circumstances of a community, system or asset that make it susceptible to the damaging effects of a hazard.

Source: ISDR (2009).

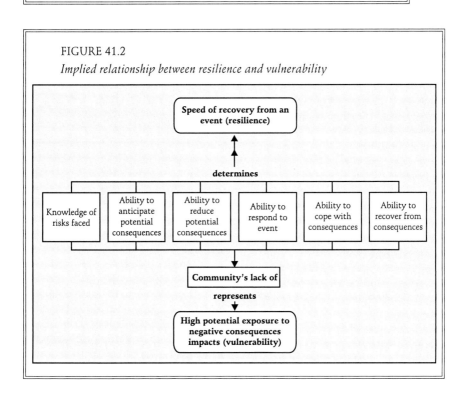

FIGURE 41.2
Implied relationship between resilience and vulnerability

TABLE 41.3
Possible community action to reduce vulnerability

Component	Community level action	Comment	Project achievement
Knowledge of risk faced	Decide on geographical boundary Revive historical memory Revive indigeneous knowledge Do participatory hazard identification Do participatory hazard mapping Do identification of vulnerabilities	Within community capability, facilitation may be needed	✓
Ability to anticipate potential consequences	Re-visit worst impact experienced Do participatory preparedness response planning	Capacity building may be necessary	✓
Ability to reduce potential consequences	Decide on acceptable risk implement local level mitigation	Funding may be an issue	✓
Ability to respond to an event	Identify evacuation routes and safe places Establish early warning dissemination Carry out simulation drills Organise shelter and relief	Capacity building may be necessary	✓
Ability to cope with situations	Involve community leadership Build social networking Initiate resource pooling Encourage volunteerism	Capacity building may be necessary Outside intervention may be necessary	✓
Ability to recover from consequences	Re-building livelihoods Strengthen lobbying capacity	Outside resources necessary	✓

Wanduruppa through project activities. Livelihood interventions, although too early to measure success have put in capacity building to enhance the economic capital. Ability to recover will depend on external sources, as the economic capital is weak in the community.

This case study therefore shows that it is more judicious to aim towards achieving discrete and tangible targets such as hazard-specific vulnerability reduction through mitigation, measurable objectives in capacity building, awareness creation and hazard-specific preparedness within manageable geographic areas. Assessment of outputs becomes more realistic. The ground level disaster manager could find more comprehensible directions for intervention by keeping things simple.

References

Adger, N.W. (2000). "Social and ecological resilience: Are they related?", *Progress in Human Geography* 24(3): 347-64.

Alongi, D.M. (2008). "Mangroves as protection against tsunami", *Estuarine, Coastal and Shelf Science* 76(1): 1-13, January.

Ashby, W.R. (1963). *An introduction to cybernetics*. New York: John Wiley & Sons, Inc.

Bailey, K.D. (1990). *Social entropy theory*. State University of New York Press. pp. 59-61.

Bruneau, M., S. Chang, T. Eguchi, G.Lee, T. O'Rourke, A. Reinhorn, M. Shinozuka, K. Tierney, W. Wallace and D. von Winterfeldt (2003). "A framework to quantitatively assess and enhance the seismic resilience of communities", *Earthquake Spectra* 19(4): 733-52.

Buckle, P., G. Marsh and S. Smale (2000). "New approach to assessing vulnerability and resilience", *Australian Journal of Emergency Management*: 8-15.

Carney, D. (1998). *Sustainable rural livelihoods-What contribution can we make?* DFID.

CARRI Research Report 5 (2009). *Comparing ecological and human community resilience*. January.

CCR (2006). "Concepts and practices of 'resilience': A compilation fro various secondary sources", *Working paper* CCR (IOTWS), May.

———. (2000). *Community Resilience Manual*. www.cedworks.com

Dahdouh, D., L.P. Jayatissa, D. Di Nitto, J.O. Bosire, D.L. Seen, K. Koedam (2005). "How effective were mangroves as a defense against the recent tsunami", *Current Biology* 15(12): R443-R447, June.

Fitzgerald, G. and N. Fitzgerald (2005). "Assessing community resilience to wildfires: Concepts and approach", Paper Prepared for SCION Research, June.

Foster, K.A. (2006). "A case study approach to understanding regional resilience", *Working Paper* for building resilience network. Institute of urban regional development, University of California.

Geis, D.E. (2000). "By design: The disaster resistant and quality-of-life community", *Natural Hazards Review* 1(3): 152.

Goldsmith, E. (1981). "Thermodynamics or eco dynamics", *The Ecologist* 11(4). July/August.

Holling, C.S. (1973). "Resilience and stability of ecological systems", *Annual Review of Ecology and Systematics* 4: 2-23.

Holling, C.S. (1996). "Engineering resilience *versus* ecological resilience", in P. Schulze (ed.), *Engineering within ecological constraints*. Washington, D.C.: National Academy Pages. pp. 31-44.

ISDR (2009). "Terminology on disaster risk reduction", Posted at *http://www.unisdr org/eng/terminology/terminology-2009-eng.html*

Kasperson, J.X., R.E. Kasperson and B.L. Turner (1996). Regions at risk: Exploring environmental criticality. *Environment* 38(10): 4-15, 26-29.

Klein, R.J.T., R.J.Nicholls and F. Thomalla (2003). "Resilience to natural hazards: How useful is this concept?", *Environmental Hazards* 5: 35-45.

Mayunga, J.S. (2007). "Understanding and applying the concept of resilience: A capital-based approach", A draft working paper prepared for the summer academy for social vulnerability and resilience building. 22-28 July, Munich, Germany.

Odum, E.P. (1969). "The strategy of ecosystem development", *Science* 164: 262-70.

Oxfam America (2006). *Report on disaster management policy and practice: Sharing lessons among government, civil society & private sector*. Institute of Policy Studies in collaboration with Oxfam America. October.

Thims, L. (2007). *Human Chemistry* (Volume One) Morrisville, NC: LuLu.

Paton, D. and D. Johnston (2001). "Disasters and communities: Vulnerability, resilience and preparedness", *Disaster Prevention and Management* 10 (4): 270-77

People's Report (2007). *Disaster risk reduction in the post tsunami context, India, Maldives, Sri Lanka & Thailand*.

———. *Violence against women in the post tsunami context, India, Maldives, Somalia, Sri Lanka & Thailand*.

———. *Homestead land & adequate housing in the post tsunami context, India, Maldives, Sri Lanka & Thailand*.

———. *Towards a people centered and sustainable fisheries policy*.

———. *Disaster risk reduction, Sri Lanka*.

Practical Action (2008). *Decentralized disaster risk management training manual*.

Prigogine, I. (1977). "Time, structure, and fluctuations", Nobel Lecture (in chemistry). December 8.

Solangaarachchi, D.T. (2008). *Assessment of geographical risk and vulnerability of coastal communities of the galle municipal area for tsunami*. M.Sc Thesis, Post Graduate Institute of Science, Peradeniya University, Sri Lanka.

UNISDR (2007). *Hyogo framework for action 2005-2015: Building the resilience of nations and communities to disasters*. Geneva: ISDR.

USIOTWS (Indian Ocean Tsunami Warning System Program) (2007). *How resilient is your coastal community? A guide for evaluating coastal community resilience to tsunamis and other hazards*. Bangkok, Thailand: IOTWS/USAID. p.144.

Wickremaratne, H.J.M. and A.T. White (1992). "Concept paper on special area management for Sri Lanka coasts", *Coastal Resources Management Project Working Paper* No.10-92. Colombo.

42

Building Resilient Communities: Lessons from Cordaid

Cordaid ensured and supported a platform for learning and sharing of CMDRR between partners and communities.

Source:

Rustico Binas *Global Advisor- Cordaid, Philippines. 2010.*

1. Introduction

Cordaid has funded several projects on Community Managed Disaster Risk Reduction (CMDRR) in Uganda, Kenya, Ethiopia, Malawi, Zambia, India, Bangladesh, Indonesia, El Salvador, and Honduras. The project activities included participatory community risk assessment and analysis, identifying and implementing risk reduction measures like water related projects in drought areas, and implementing drought early warning systems and flood coping strategies. The Cordaid projects may have been varied but one common thread defined their purpose: to shift mindsets from relief to building resilient communities.

One very important initiative of Cordaid to support its capacity building effort and disseminate its approach was the publication of its training manual on CMDRR in collaboration with the International Institute of Rural Reconstruction (IIRR). It was produced mainly to provide supporting materials to all its DRR programme initiatives. The manual has been translated in Spanish, Bahasa and Bangla.

Cordaid ensured and supported a platform for learning and sharing of CMDRR between the implementing partners and the communities in countries where the DRR programme is implemented. This is to guarantee that work on the ground is properly acknowledged with their significant stories of social change. It highlights learning processes, offers solutions to enhance and strengthen the programme, and notes significant changes at the institutional and community levels and among the development workers.

2. Lessons learned

Since CMDRR started, both Cordaid and its local partners shared successes and failures in implementation. Cordaid believes that in surviving the waves of climate change, there is a need to have a strong vessel named CMDRR. For this vessel to survive the treacherous waves of climate change, the captain who is the local community partner, needs to be equipped and knowledgeable in navigating the impending hazards of the sea.

The following is a summary of lessons learned:

2.1 Accompaniment

Cordaid accompanied partners to further understand the CMDRR. Better understanding empowers communities and makes them more effective and efficient in their work. The community becomes the centre of gravity in identifying disaster risk and providing solutions, and this gives the people a sense of ownership of the programme. In Indonesia, after the menacing tsunami in 2004, Cordaid built earthquake-resistant buildings the designs of which evolved from consultations with the communities.

In Kenya and Ethiopia, communities developed adaptation measures like rainwater harvesting technologies, and the construction of sand dams and water ponds as a response to drought problems. They have also shifted from using cattle to camels, since camels can survive during the dry season.

Cordaid projects espouse the following principles:
- Communities have accumulated local knowledge in addressing hazard events;
- Communities are survivors, not victims;
- Basic rights are foundations of safety;
- Community organisations are mechanisms for successful DRR initiatives and the government is a major player;
- Communities have to take responsibility for their most at risk members;
- It is the communities that decide if they are in a state of disaster; and
- Resiliency is not merely accumulated assets or secured livelihood. It is also the will to survive and to stake a claim to justice and equity as members of society.

The same can be seen in Karamoja and Teso in Uganda where communities have diversified livestock and agricultural production to improve their income, again as a response to drought. They also saw the need to develop peace building and conflict resolution initiatives to address the situation between tribes in the region.

2.2 Linking experiences to policies

Innovative and successful DRR projects are scaled up through linkages that influence government programmes and policies. This is evident in Honduras and El Salvador wherein local community partners worked hand-in-hand with the local government to ensure that their adaptation strategy will be implemented and legislated within the local policy framework. Management of water sources and river systems were also integrated in the policy framework. Cordaid's local community partners in Indonesia together with the Ministry of Health helped each other to

Making the community the centre of gravity in identifying disaster risk and providing solutions gave the people a sense of ownership of the project.

promote the use of clean water and proper hygiene. On the other hand, local authorities in Bangladesh were involved in the conduct of training and risk analysis.

2.3 Resource training materials

DRR is now a 'hot concept' in development. Cordaid is obviously ahead and leading because of its experiences on the ground and clear understanding on what it takes to do DRR at the community level. Cordaid made a strategic choice in promoting CMDRR. CMDRR is a package with philosophy, principles and practices, and Cordaid is utilising its leadership in this field by influencing and engaging partner funding agencies in both international and 'in country' operations. The DRR measures and programmes of the partner communities of Cordaid is testament that the CMDRR training manual has transformative results.

2.4 Learning and sharing

DRR is a young thematic area. Many who are dealing with DRR are still confused. This is why the mantra for all the DRR projects is 'to learn what works and what does not work in terms of appropriate technology, tools for risk assessment and social processes for building resilient communities, and ensuring that learning is contributing to the global pool of knowledge'. Cordaid and its local partners regularly share the knowledge generated from the ground to further develop the strategies so it can benefit the world.

In November 27, 2008, the Cordaid staff held a Programme Review: 'Harvesting Life Changing Experiences in the Ground' to discuss lessons learned, achievements, innovative strategies, and innovative social processes. Also discussed were implementation bottlenecks at the country, partners', and community levels. In one of the sharing venues of Cordaid and its partners, Binoy who is the director of UNNATI, Gujarat exclaimed, 'I find this learning and sharing exercise valuable and indispensable. We learned from each other. We are enriched because we get ideas on how to strengthen our work. I thank Cordaid for this'.

> The CMDRR Manual is my new bible. It helps me in work with my communities. I even sent copies to other development workers of Caritas and they are happy about this new knowledge resource.
>
> — Norma of Caritas Honduras

Cordaid conducts water related projects especially in drought-stricken countries of the world.

2.5 Monitoring and evaluation

In the realm of DRR, the baseline data are risk assessment and analysis. The same risk analysis should be used in monitoring and evaluation to measure if the risk was reduced or not. If the risk was reduced, the probability of disaster is low. In this area, although many partners are still learning to make their risk assessment more robust, there are those who are already making headway in monitoring and evaluation work in the community through a more effective, efficient, and a clear link with the government setup.

2.6 Advocacy and lobbying

Partner communities of Cordaid have replicated their know-how of the DRR process in response to the threats of climate change. Community leaders initiated DRR facilitation and trained people in adjacent communities. They encouraged communities to lobby with their government in using the DRR process in addressing human induced hazards and risks posed by climate change. In Bangladesh for example, six

The road to disaster resilience is long and tortuous. It helps that communities have partners in organisations like Cordaid that has implemented CMDRR.

months after the implementation of CMDRR in 2008, core groups and task forces were formed in 56 selected villages, and these communities are conducting risk assessments and maintaining updated information. Government linkage and lobbying has also been a significant result of the CMDRR in El Salvador and Honduras in Central America wherein Cordaid local partners ensured the legislation of climate change adaptation. In Ethiopia, the local government adopted a drought early warning system and subsequently, this was implemented in other communities.

2.7 Learning from experiences and making headway

As early as the year 2000, Cordaid was 'mixing and matching' 'what works' and 'what did not' at the community level. It started linking relief and development and now has a complete holistic perspective from situational analysis to a vision. DRR was integrated in the Strategic Goal and Programme for 2007-2010, and since then, it grew from concept into practice. Climate change adaptation did not come as an add on but was placed as part of the bigger framework of DRR. Many organisations were interested and trained in Cordaid's CMDRR. Implementing partner organisations became champions in this theme too. Many of them reported that they were recognised by both government and peer organisations on their work with the communities and their capacity to clearly explain the CMDRR concept, philosophy and practices.

3. Conclusion

The Cordaid experience exemplifies an approach towards DRR that is steered by the people. The people themselves identify the disaster risks, and as a community, manage the implementation of solutions that they have helped identify. Coupled with policy support and complemented by education and information, continuous knowledge generation, and monitoring and evaluation, this approach has had much success, thus it is already being replicated in many countries all over the world.

43

Community Disaster Resilience Fund: Early Insights and Recommendations

Source:

Rustico Binas *Global Advisor- Cordaid, Philippines. 2010.*

Workshop discussions, baseline assessment reports, in-field discussions with women's groups and communities, and learning notes from facilitating organisations have generated early insights from the implementation of the Community Disaster Resilience Fund (CDRF). These insights reveal key findings on development priorities in grassroots communities and on the implementation and effects of the CDRF:

> In disaster-prone communities, the most vulnerable groups are not only the income poor, but the socially marginalised, those without stable livelihoods and those who live in vulnerable habitats.

Women's groups identify the most vulnerable groups as women-headed households and families with alcoholics. Their vulnerability stems from the fact that they do not have a secure livelihood source in non-disaster times, which decreases their ability to cope once calamity strikes. Reducing poverty and vulnerability is hence linked to building disaster resilience and vulnerability, not solely to raising incomes.

> The current disaster and climate change trends are affecting many traditionally agro-based communities.

Many agro-based communities, especially in the states of Assam, Bihar and Rajasthan, are turning to wage labour because water scarcity and/or salinity, and droughts and floods have made agricultural livelihoods vulnerable. Building resilience in these communities will centre on strengthening existing livelihoods as well as diversifying livelihood sources. In Assam, where communities are inundated by river floods, the CDRF is used to give interest-free loans to women's groups for restoring women-managed weekly markets. In Bihar, the fund is being used to set up small businesses and vegetable producer groups. The sustainability of new livelihoods is crucial to the long-term development of the regions.

> Government and media early warning mechanisms are disconnected from grassroots communities.

Both private and public early warning mechanisms are disconnected from grassroots communities and in some cases, are constantly transmitted on the news and the radio, which desensitises communities to disaster. Also, grassroots communities rarely know how to interpret cyclone severity ratings and understand all warnings as being the same. This illustrates a critical gap in knowledge and information which leads to a failure of early warning systems. Early warning mechanisms need to be brought closer to communities for these to benefit them.

> Women's groups identify access to drinking water, sanitation facilities, and health services as the key factors in decreasing vulnerability and alleviating disaster risk.

Most grassroots women's groups identify access to clean drinking water, sanitation (especially toilets) and health care as their biggest priorities during disaster. Other short term or immediate post-disaster priorities include livelihoods security, availability of irrigation water, connectivity, public transport, and fodder security.

> Sustainable livelihoods are critical in building long-term resilience to disaster and lessening the impact of disaster after they occur.

Communities invariably point to livelihoods as the area of life most affected by disaster and a good economic base as a necessary precursor to other resilience building strategies. Other long-term developments that reduce disaster vulnerability are accessibility to higher education, availability of sociocultural infrastructure, fertile land, economic stability, high universal literacy and socioeconomic equality. The CDRF is about linking poverty reduction with DRR to build community resilience. The creation of local institutions that integrate DRR with poverty reduction is the key to developing long-term resilience.

> An area-based disaster resilience fund is a powerful catalyst that can fast track the development of disaster resilience.

Women-managed bazaars have helped provide livelihoods to disaster-prone communities in India.

Activities associated with the creation of a disaster resilience fund, such as crafting of a project proposal, formation of a village level CDRF committee, mapping of vulnerabilities and surveying, bringing the men and women in a community to work together and act as a trigger for further development. Voices of the marginalised in society that are only heard for the first time contribute significantly to the design of initiatives. Community-wide discussion ensures that local development priorities and issues are met. The CDRF project is the first stepping stone. While it does not cover the full cost of the proposed initiative or contribute to later development projects, it will leverage alternative resources from the government and the private sector.

> Women's collectives are a natural choice as managers of the resilience fund.

Women's collectives and networks are natural managers of the CDRF, especially at the local level, as they bring transparency into fund management. This comes from years of experience in managing microcredit, emergency funds, and enterprise and revolving funds.

Community Disaster Resilience Fund: Early Insights and Recommendations

By being vocal about their concerns, women ensure that their needs are addressed by DRR initiatives.

Women's groups screen proposals, direct the use of funds, and ensure the proper implementation and repayment of funds. Women are also aware that long-term resilience funds can be sustained only with the mobilisation of local resources. In Bihar as well as in other states, women's groups have pooled savings to create emergency funds to meet medical expenses in times of disaster as well as in calm.

> Communities have practical knowledge, context-appropriate technology and experience.

These must be leveraged to effectively drive DRR initiatives. Involving communities in collective mapping and analysis of vulnerabilities and hazards and the dissemination of local knowledge, capacities and resources is critical to long-term disaster resilience and development. Local communities are highly aware of the complex nature of the risk situations that they are in and also of multi-pronged solutions that will best address identified problems. Locally developed solutions will often be more effective in addressing specific needs of grassroots communities in different geographic areas.

In the West Godavari district of Andhra Pradesh, women's fish-vending federations are creating shelterbelts and regenerating mangroves forests to reduce the force of cyclonic winds, prevent water logging, and contribute to their livelihoods. Using their local knowledge of the needs of plant species, they have derived a cost-effective long-term resilience plan as an alternative to more costly top-down initiatives.

Communities and families have a repertoire of preparedness and coping mechanisms to deal with seasonal changes in climatic conditions but the changing nature, frequency and intensity of disasters seen in many places, such as Assam and Bihar, is not something that community coping mechanisms alone can address. What is required are new modes of thinking and multistakeholder partnerships that recognise and build on community strategies, resources, and technical skills and knowledge to meet complex challenges.

> In order for women's needs to be addressed by DRR initiatives, women need to be vocal about their concerns and active in the design and implementation of initiatives.

The priorities of men and women, although not necessarily in opposition to each other, are different. Women prioritise health, sanitation, and food security while men emphasise income generation activities. The CDRF enables all groups in society to voice out their concerns and respect and address the needs of all people in a given locality. The CDRF, with its focus on enabling women to play a leading role in DRR, gave women a platform to define their priorities and the community can work towards finding a solution to women's priorities.

Women are vocal about the disasters that communities face, including drought, storms, cyclones, floods, and other calamities. Once given the opportunity and forum, they can identify traditional and social practices that hinder the ability to cope with disaster, such as the prohibition on them to enter an in-law's home without being properly dressed/covered. This can be a problem when women evacuate in an emergency at night and do not have time to be adequately dressed and an in-law's home offers a haven on higher ground at a time of a crisis.

> Women describe the CDRF as a means through which their awareness on disasters and disaster preparedness and response has increased.

The CDRF has engaged women in work beyond savings and credit and other general economic activities. Savings and lending practices have empowered women in villages and enabled them to take further steps to initiatives such as the CDRF.

> Sustainable livelihoods are a critical priority for building long-term disaster resilience and for lessening the impact of disasters (after they occur).

To be effective, DRR programmes must tap the knowledge and experience of women and grassroots communities.

When disaster affected communities are asked: 'What is most affected by disasters?' they invariably point to livelihoods. The CDRF is about linking poverty reduction with DRR to build community resilience, since poverty is the major cause for vulnerability. Local platforms around resilience add value by leveraging resources and knowledge networks at the grassroots level.

> The CDRF has spawned local platforms and networks that offer an ongoing space for learning to upstream resilience lessons into policy and programming at district/national levels by linking DRR priorities with goals of poverty reduction and development.

Creation of local institutions that integrate DRR with poverty reduction is the key to long-term development of resilience.

> Grassroots women's groups need to be seen as key allies in bringing DRR issues to the forefront and in decreasing community vulnerabilities.

Grassroots women are armed with evidence on how disasters affect livelihoods, health and everyday survival. They know their risks, and have developed solutions to cope with disasters. Social barriers prevent women from voicing these solutions and participating in community level decision-making. By insisting that local DRR platforms include and strengthen women's voices, the CDRF presents yet another entry point for women in poor communities and local governments to work together.

1. Recommendations

The establishment of the CDRF and initiation of multiple projects have generated the following early recommendations on community-driven DRR and sustainable development.

1. Government, NGOs, and private entities need to recognise women and community groups as key actors in DRR and use local/insider information as well as abilities present in these groups in conducting resilience building and development projects

Women and grassroots communities should not be seen solely as beneficiaries of programmes, but they should be seen as source of specific knowledge and experience in building community disaster preparedness, response ability, and resilience. Effective DRR programmes must tap into these local groups in order to supply effective and need-based solutions to risk and disaster that simultaneously contribute to sustainable development of the area. They must provide clear mandates for formal public roles that position communities and women's groups as agents of DRR and adaptation.

2. DRR programmes must be aligned with poverty reduction, income generation and sustainable development initiatives

Vulnerability is associated with the absence of a secure livelihood source, the lack of prospects, and insufficient access to basic services, such as clean drinking water, sanitation facilities and health services. The availability of public utilities and infrastructure in communities, and proliferating and diversifying livelihood sources is critical to building resilient communities. Sustainable community development and DRR go hand-in-hand. Both need to be fuelled in cooperation with the grassroots communities themselves.

3. Community level disaster preparedness and resilience strategies need to be used as resources and starting points in creating social safety nets

Traditional knowledge and existing networks need to be used as foundations for the creation of new and enhanced safety nets and capacities. They cannot be neglected to create dual structures that will likely to act in discord with each other and lead to ineffective resilience building projects.

For more details on CDRF, please contact P. Chandran at *sspinfo@gmail.com*

CDRF project

Facilitating organisations:

- Rural Volunteers Centre (RVC), Assam, *rvcassam@gmail.com*
- Institute for Development Support (IDS), Uttarakhand. *ids_pouri@rediffmail.com*
- Sanghamitra Service Society, (SSS), Andhra Pradesh, *sanghamitra.ankita@gmail.com*
- Covenant Centre for Development (CCD), Tamil Nadu, *nadarajan.vc@gmail.com*
- URMUL Trust, Rajasthan, *ajhaarvind@gmail.com*
- Saurashtra Voluntary Actions (SAVA), Gujarat, *savaahm@gmail.com*
- Kanchan Sewa Ashram (KSA), Bihar, *vijayksa@rediffmail.com*
- Udyama, Orissa, *udyama.pradeep@gmail.com*

Institutional partners:

GROOTS International, *www.groots.org; www.disasterwatch.net*

Huairou Commission, *www.huairou*

Coordination and knowledge management

Swayam Shikshan Prayog (SSP), India, *www.sspindia.org*

Knowledge Links Pvt. Ltd., India, *knowledge.links@gmail.com*

44

Capacity Building Interventions: Drawing Lessons from the Field

Capacity building played a crucial role in bridging the gap of conceptual clarity and streamline the focus among different players in strengthening resilience of tsunami-affected communities.

Source:

Nancy Anabel *M.S. Swaminathan Research Foundation, Tamil Nadu, India, 2010.*

Capacity building plays a predominant role in leading the community towards development by improving their lives and livelihoods. Through capacity development efforts, individuals, organisations and communities develop abilities to perform functions, build consensus, solve problems, and set and achieve objectives in a systematic manner to realise change. Useful experiences were gained in the project on 'Strengthening Resilience of Tsunami-Affected Community'. The need to build the resilience of coastal communities emerged due to the indelible imprints left by the tsunami on human, financial, natural and physical assets both in Tamil Nadu and Andhra Pradesh.

1. The project sites

Project sites were identified based on few criteria such as low level of socioeconomic condition, vulnerability to natural disasters, inaccessibility in terms of information, relief, rehabilitation and development, and negligence of natural resources.

The project covered 18 villages encircling 10 in Tamil Nadu (TN) and two in Andhra Pradesh (AP). Most of the villagers in TN and AP depended on fishing, agriculture and livestock. It is notable that the villagers in AP lost their main source of livelihood due to the South Asia tsunami in 2004.

2. The approach

Under this scenario, the entire components were planned to be implemented embracing 'process-oriented, people centred and science-based' approach, contemplating community as the prime movers to building their resilience with an aim to sustain the outcome. Alongside, it was also decided to implement it through partner organisations who closely work with grassroots.

2.1 Top-down approach

Convincing community about the project is always the first phase and it requires a solid top-down approach. It is always the part of implementing organisation to make the community comprehend the significance of the project and its role in bringing out a change in their life and livelihood by fulfilling some of their basic demands.

2.2 Bottom-up participatory approach

When the community has fully comprehended the significance of a project, the top-down approach stems and switches on to bottom-up or participatory approach of planning and execution.

3. The role of capacity building

Capacity building played a crucial role to bridge the gap on conceptual clarity and streamline the focus among different players; integrate scientific, technical, and ground level know-how on all the components; better planning and diversification of livelihood opportunities, alliance building among different stakeholders for delivering timely information for decision-making; and building a strong people's organisation with gender in order to implement and sustain the project initiatives in both TN and AP.

Capacity development is about change. Change involves institutionalising participation and learning.

The entire project cycle, that focusses on bringing about change in the community, requires capacity building.

4. Integrating formal and informal capacity building programmes

A systematic approach and plan was used, incorporating formal and informal programmes to build the capacity of the stakeholders involved in the project.

4.1 Formal capacity building

Formal capacity building indicates a structured approach of planning, implementing, monitoring and evaluating the programmes. It undertakes a systematic capacity assessment using appropriate methods at various levels of stakeholders. Figure 44.1 provides a snap shot of how the formal process of building capacities was implemented.

During the formal programmes, a combination of different learning techniques such as role-play, group discussions, demonstrations, field works, use of visualisation cards, short films, learning games, group work were used to capture the attention of the participants.

4.2 Informal capacity building

The informal capacity building programme helped assess the requirement of the stakeholders at different levels and plan programmes. Through informal

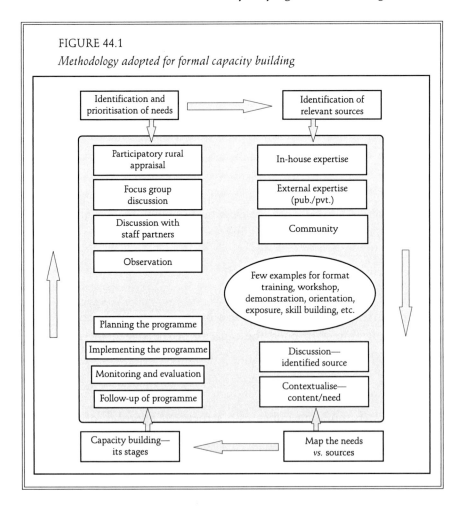

FIGURE 44.1
Methodology adopted for formal capacity building

interaction with the community, useful insights and critical opinions were gathered that helped strengthened the project implementation processes.

Anecdotal evidences also played a crucial role in streamlining methods and processes adopted in the planning stage. One-on-one or group discussions with the community and observations helped bridge the gaps between conceptual planning and reality. Field visits of the staff, for example, enabled them to observe and understand the gender issues in the community which helped them in the conduct of gender sensitisation workshops.

5. Changes due to capacity building

The whole attempt to build the resilience of the community resulted changes on various aspects. The changes realised at different levels were:

5.1 Partners' level

Capacity building yielded very positive results for the partner organisations. Not only was there a scaling-up in terms of capacity but the partners were also able to work independently with other organisations. Bioshield, village level institutions (VLIs) and the village knowledge centres (VKCs) have now become a model for other organisations. Some of the NGOs from West Bengal, Orissa, Tamil Nadu and Andhra Pradesh that are involved in the establishment of VKC visited People's Action for Development (PAD) and VKC villages and interacted for learning and strategising their implementation.

5.2 Group level

Community participation and management was realised later in the implementation process. It was processed intensively but repeated reinforcement enabled them made decisions on:

- planning and managing common and revolving funds;
- providing space for VKC as community contribution;
- identifying location for bioshield plantation;
- forming into VLI with gender participation;

- involving women in decision-making;
- finalising charges for certain courses and services delivered through VKC;
- deciding suitable business plan for enhancing their livelihood; and
- following monitoring and evaluation system.

The village communities are now escalated to the extent of organising inter-exchange programmes to share, learn and grow through their peers' sharing on success factors for harvesting the result, key learning, and challenges during the project implementation process along with innovations and techniques adopted. This widened the opportunity to have incremental learning and to seek solutions from peers. It is also significant to note that the villages that visited other villages for exposure during the inception stage of the project, have now become the hub for others to share, learn and grow.

6. Legitimate space for women in VLI

One of the most significant changes that formal and informal capacity building brought to the community is the attitudinal change with regard to gender roles. Due to culture and custom of the village, women could not enroll as members in the executive committee of VLI. A series of sensitisation workshops on gender organised for both men and women facilitated to resolve the gender issues and challenges faced by women.

Exposure visit to the model VLI and the interaction with the persons contributed considerably to make them understand the need for gender participation in development process. The shift in attitude allocated legitimate space for women in VLI, giving them equal rights as men to participate in development planning and implementation in their community.

7. Participation in disaster mitigating measures

Exposure visit by some of the community stakeholders to bioshield in Pichavaram and Chidambaram facilitated their comprehension of the framework of the Joint Mangrove Management (JMM) approach, types of bioshield, role of mangrove and non-mangrove costal vegetation in

Capacity Building Interventions: Drawing Lessons from the Field

Hands-on: Students being exposed to computer literacy at the inaugural of the village knowledge centre (VKC) of the M.S. Swaminathan Research Foundation (MSSRF) at Neduvasal near Pudukottai (www.thehindu.com).

reducing the impact of disasters, method of establishing and managing bioshield unit, and the links between livelihood and bioshield. The impact of the exposure visits on the community was so great that even women who refused to be members in the VLI in the initial stage had begun to take up tasks like mangrove plantation after undergoing practical training.

8. Constraints and challenges

- Retention of knowledge workers and staff to continuously pass on the learnings and build the capacity of new entrants/entourage.
- Much time and effort is still needed to bring about an attitudinal shift in gender issues. So that women can actively participate in VLIs and be included in the decision-making process.
- Specific problems such as the hampering of the bioshield growth due to occurrence of pest attack, sea grass deposit, shore fishing, and cattle grazing posted a challenge to the community. These

issues were later incorporated in the management plan and solved by allocating jobs to members of the community.

9. Learnings

- Exposure visit stood on top, among all other methods, in yielding the best results to bring about changes in attitude and knowledge for confidence building and awareness creation.
- Frequent visits and informal discussions consequent to formal programmes are highly essential as they play a crucial role in skill building and raising community understanding about the initiatives, leading to action and result.

Poonkothai, 35 yrs old, hailed in Kattumavadi recollects the days when women were set aside from participating in a common meeting. If there was a case filed in the traditional *panchayat* against a woman, she had to fold her hands and stand in the *panchayat* without being given chance to speak. But now, the case is different, she says with pride, as she actively takes part in the VLI during planning and decision-making. She also feels powerful managing the revolving fund and being a joint signatory for the account.

'Horizontal transfer of knowledge and skills by their own peers from within and other villagers had its own reach and effect as a source than vertical transfer of knowledge by external resource persons'

- Training and field demonstrations had its mark for enhancing skills with regard to bioshield plantation and livelihood programmes.
- Cultural programmes contributed remarkably in efforts to mobilise the community, and to emphasise the importance and need for

specific project initiatives. It also helped in creating awareness among community members.
- During the implementation of the project, integrating the key components was a good learning experience and added value to the organisational capacity by means of hands-on experience on new thematic areas.

Capacity building is a pivotal process in any developmental initiative as it has the power to trigger the stakeholders at different levels to come to a uniform understanding of a given context. It tries to tap the potentials of the individuals and channel them by chiselling their knowledge and skills as required in the development arena.

45

A Role for Customised Financial Mechanisms

Source:

S. Olagnathan, Vijay Pratap and Singh Aditya *Ekgaon Technologies, India, 2010.*

1. Introduction

Vulnerable communities need strong social networks to build resilience. To reduce their vulnerability, the short term and long-term needs of the communities must be addressed. Building resilience will benefit from the formal and informal transfer of intergenerational skill and knowledge, effective provision of public goods and services by governance structures and evolved or customised financial mechanisms.

Need-based financial mechanisms can build the resilience of the communities facing poverty traps or disaster. Often, these mechanisms take the form of customised products and services that are designed to either reduce the vulnerabilities or enhance the coping skills of the communities. To be successful, these customised products and services must be based on the perceived needs of the community and be delivered by local institutions.

Efficient and performing financial mechanisms possess the following attributes:

- The user/beneficiary holds a stake in the ownership and governance structure;
- The mechanism itself is the logical outcome of a participatory design process;
- The offerings are a mixture of both financial and non-financial services and products;
- The offerings are based on a meticulous analysis of the risk environment and life cycle needs of the user community; and
- User feedback is solicited and is factored into the product development process and the enhancement of the service delivery system.

In addition, norms on the access to and use of the financial mechanisms have to evolve through the participation of the community itself. Here, the facilitation of the non-government organisations (NGO) is important.

2. Financial mechanisms

Disasters affect community finances. To cope, communities use various financial means to smoothen income, meet contingencies, support orphaned dependents in the long term, build assets and reduce risks, etc. Various financial mechanisms can facilitate the recovery of the communities living in the disaster situations (Figure 45.1).

3. Credit

Communities that face recurrent disaster situations need affordable and non-collateral credit. With credit, the community can continue its livelihood activities.

Credit tides communities over the sudden disruption of normal livelihoods when assets or household goods are wiped out. To be effective however,

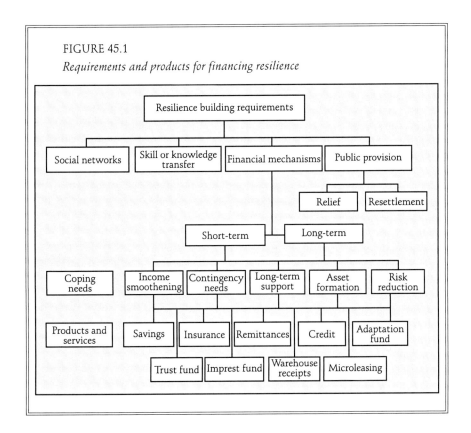

FIGURE 45.1
Requirements and products for financing resilience

Financial resilience the Association of Sawa Seva Farm (ASSEFA) way

ASSEFA runs a Cattle Promotion Scheme in Kariapatti block of Virudhunagar District in Tamil Nadu. The NGO evolved norms to administer its animal mortality fund, a community mutual fund to compensate the loss of the members' animals.

Because of the high risks in managing this fund, cattle raisers set stringent rules such as compensating only 75 per cent of the animal value and deducting for salvages of skin value, etc. Moral hazards were eliminated through monitoring mechanisms and intermediary structures such as village level committees and an apex level inspection committee. These ensured both a speedy settlement of genuine claims as well as the fund sustainability. The initial year's gains were translated into full compensations of claims in subsequent years and the fund has proved ideal in the drought and epidemic-prone area.

loan products must be customised to the needs of the community because such customised loans are better absorbed and more effectively used. A rigorous market survey involving an analysis of household cash flow and an assessment of credit needs throughout the life cycle helps in the design of effective loan products.

Credit as a service should act as growth capital instead of consumptive capital. Thus, it is essential to differentiate the products so that borrowers do not revolve 'credit' from various sources to pay for other loans.

4. Savings

Savings reduce the impact of downward economic pressures caused by seasonal household and livelihood demands and/or predictable life cycle events such as marriage, schooling of children, community celebration, etc. Savings serve as shock absorbers and as stepping-stones for asset building.

Coupled with community organisations, the pooled savings of the participating communities has a very high leveraging strength in accessing external resources using the community affinity as the collateral substitute. In drought-prone areas, the household savings play a critical role in helping tide over the severe drought years.

With credit, savings need to be customised to purpose. Communities that live in disaster situations generally consider savings as a reserve rather than an investment and its liquidity is considered very important. Communities living in difficult or contested areas, have a tendency to convert savings into income generating or safe assets with easy liquidity and prefer to broaden their assets base than reduce the risk of money loss through diversification.

The communities have customised a number of ingenious savings mechanisms. We find in most vulnerable communities, voluntary chit funds of various names that are a hybrid of both savings and loans. Some of these chit funds have instant and large markets; there are chit funds for pilgrimage, festivals, temple constructions, financing purchase of utilities such as tractor, car, motorcycle, etc.

Due to the probability of future uncertainties, communities in difficulty prefer products that have multiple functions. But though the demand for savings products is high amongst the poor, these services are still unavailable to them. The formal banking financial institutions are unable to extend the services due to infrastructure and networking constraints, while the non-banking financial institutions are not allowed to offer these services.

5. Insurance

Insurance or risk transfer is inadequately developed in disaster-prone communities. Often, the mainstream insurance service providers do not adequately cover the livelihood and life risks that many communities face. As such, the communities have evolved their own ways of coping with risk, some of which happen to be financial mechanisms.

The Kazhi Kadaimadai Farmers Federation (KKFF) in Thiruvengadu, Tamil Nadu, extends company insurance products to farmers for ensuring their crop. The insurance comes from a public sector company, the Agriculture Insurance Company of India (AIC) Limited. This insurance has picked up dramatically as the region is prone to flooding both from the Kaveri River as well as the sea. Considering the high risk to the portfolio, it is unlikely that a private sector company could offer such products in these regions.

6. Remittances

Money remittances are important particularly for those communities that migrate for work during a comparatively adverse farming season, often to cope with debt traps. Usually, only the productive group within a family migrates. The rest are left behind with essential commitments such as asset upkeep and care for dependents.

Regular flooding in the eastern region of the state of Uttar Pradesh in India has resulted in its unproductivity for the large populations living there. The people then migrate seasonally with as many as 35 per cent of the population reaching far off Punjab, Mumbai and New Delhi. This separation means that families will need a facility to remit their income back home. At present, this service is being provided by occasional migrant commuters or by the postal network. The former is always risky and undependable while the latter is comparatively costly. Also, the postal network involves issues of confidentiality and inconvenient business hours.

7. Adaptation fund

Adaptation funds facilitate the transition from vulnerability to stability or resilience. This graduation demands sustained investment in household or farm level infrastructures that either enhances productivity or buffers shock events.

Adaptation funds accommodate and adjust for the long gestation period before the supported activities bear any outcomes. Thus, they provide the much-needed flexibility for the use of the funds. For example, funds can be used to upgrade mud houses to earthquake-resistant concrete houses in high seismic areas, or convert thatched houses into fire-resistant houses in high fire risk areas. Land development banks that help fund irrigation facilities in dryland farms are in fact effectively providing adaptation funds.

8. Microleasing

Expensive machines that support some livelihoods are often beyond the purchasing power of the individual members of the community. Through microleasing, these machines could be acquired as common assets and then leased for member use.

Microleasing helps the user in two ways: it eliminates the acquisition cost for the individual and the usage cost is recurrent and so, cheap and affordable. A mechanical harvester, drying yards, and village level godowns are some of the potential investments for collective investment and micro-leasing. For instance, low value livelihood implements can be bought on installment so that the user owns the implements over time. The user pays regular installment for the price of the implements and associated costs.

Microleasing requires a sustained quality of management and user care. Critically then, the institution must have a good delivery system with

the needed internal capacity. In most developing countries however, these mechanisms remain mostly in an unorganised sector while access to support machineries remains cost-ineffective.

9. Warehouse receipts

Warehouse receipts allow owners to dispense with holding commodities in the physical form and can be used to avoid distress sales due to emergencies or disasters. In well-managed warehouses, the commodities' asset quality and shelf-life are maintained. In the owner's care, these characters may be compromised.

Warehouse receipts are important particularly in drought and flood-prone areas where people can retrieve their commodities and consume them in times of food grain scarcity. Furthermore, these receipts give owners some flexibility in the handling of their assets. Owners can use the substantial money value of the commodities in the present and realise their full value in the future, when the market is more conducive.

However, the power of this mechanism has been grossly underutilised due to bureaucratic red tape. For instance, banks require government authorisation to consider the ware-receipts as collateral property. Forwarding marketing mechanism such as spot and futures trading, also remain under-developed in most cases due to lack of access. However, access could be improved by effective use of information technology.

10. Trust funds

Trust funds are exclusively created for the benefit of a third party. Trust funds play a crucial role by transferring the much-needed resources for the recovery or development of the third party. This is especially true when beneficiary cannot cope with disasters due to age, physical condition or utter poverty. However, such funds are rarely set up in disaster-prone regions of developing countries.

11. Imprest funds

Imprest funds can be built either by community contributions or through external assistance. The fund may be designated as a source of immediate cash in sudden events such as emergency health care, accidents, or funerals. The funds can be perpetually replenished and function efficiently when they are integrated with larger community funds. While these funds are valuable in mitigating the risk of families/communities falling into debt trap, they have been rarely used as risk mitigation mechanism.

The imprest funds could be managed better at group level to cater to immediate needs while the trust funds must be administered centrally for effective deployment of the fund. However, the selection of eligible beneficiaries, the level and the terms and conditions of the support, the norms for the use, and the support duration should be decided by an appropriate forum of elected representatives from the supported communities.

> *Rarely-used financial resilience builders*
>
> - Warehouse receipts—limited by red tape and bureaucratic impediments.
> - Trust funds—limited by difficulty in setting up the funds.
> - Imprest funds—limited by difficulty in building up the fund.

12. Conclusion

The effective administration of the financial mechanism in resilience building and rehabilitation demands strong financial institutions and support systems. The financial institutions have to balance between the social objective of serving the vulnerable people and the economic objective of survival as a service provider.

The seemingly contradictory objectives can better be addressed by building layered institutional structures. Vulnerable people can organise into self-help groups at grassroot level and federate at a much higher level. These organisations serve as effective conduits for product and service delivery and intermediate structures for managing some of the funds.

However, these community-based organisations must be motivated to develop a sense of ownership by contributing towards the equity. The ownership then triggers governance efficiency and creates a demand system for management accountability. Given the nature of its clientele and the variety of their needs, financial institutions must deliver a bundle of services. Some of these services have to be subsidised by the income generated from the more popular financial products.

For financial institutions to deliver effective services and to have desired impact, the community must have a role and responsibility in the decision-making process. Only with such participatory decision-making processes can an institution share with rehabilitating communities and building their resilience to disasters.

46

Village Information Centres: Harnessing the Potential of Technology

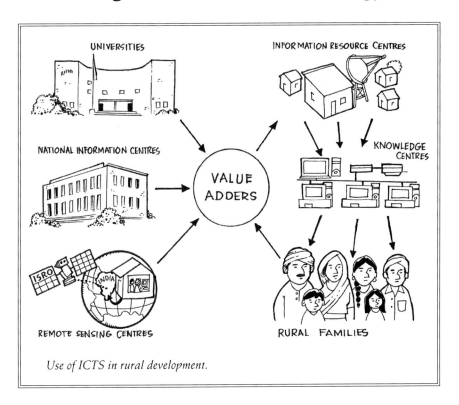

Use of ICTS in rural development.

Source:

S. Senthilkumaran, Nancy Anabel and Ganga Vidya N *M.S. Swaminathan Research Foundation, Tamil Nadu, India, 2010.*

1. Introduction

To address the needs of farm women and men for just-in-time information and instructions on farm technologies, agricultural policies and development, farmers' rights, market prices, health care and more, the M.S. Swaminathan Research Foundation (MSSRF)[1] has started a Rural Knowledge Movement (Grameen Gyan Abhiyan) to mobilise the power of partnership among the public and private sectors, academia, mass media, civil society organisations and virtual networks. The intent of the movement is to create a large shared knowledge pool by tapping the potential of information communication technology (ICT) and harnessing technologies for sustainable equitable development.

Through a series of annual interdisciplinary dialogues on 'New Technologies: Reaching the Unreached' in the early 1990s, the MSSRF realised that many laboratory research results have not reached the rural families and that likewise, as much traditional knowledge and practices tested and proven through generations have only been shared by word of mouth within the communities. Their dialogue with the other experts has led to the belief that the future of food security in the developing world, especially South Asia, would depend more on scientific, indigeneous and shared knowledge as a catalyst for growth and development. This has triggered the concept of an ICT-enabled information village in 1992 as an infrastructure needed to facilitate the movement of information and knowledge between, within, and among the different rural and scientific communities. MSSRF has piloted the concept in Puducherry and from the experience now calls the block level information resource centres as village resource centres (VRCs) and village level knowledge centres VKCs. VRC acts as the village information hub to which several VKCs connect for up-to-date information. In 2003, MSSRF strengthened the VRC and VKC with the creation of the Jamsetji Tata National Virtual Academy, JTNVA

---

1. The MSSRF is a non-profit research organisation located in Chennai, Tamil Nadu. The MSSRF research focusses on six thematic areas namely coastal systems research, eco-technology, biodiversity, biotechnology, food security and on information communication and technology for development or ICT4D. The premise of MSSRF's work is on the transition from the hybrids of traditional and modern agricultural practices of India towards an era of knowledge intensive agriculture that is pro-nature, pro-poor, pro-women and pro-sustainable livelihood orientation.

(value adder) that is tasked to establish collaboration with strategic national and international partners in developing content and capacity of the village centres.

The following highlight the MSSRF experience in Puducherry in the evolution of the VRC and VKC. Details in the setting up and operations of VRCs and VKCs are outlined in next chapter on 'Considerations for a Village Resource Centre'.

2. Local-specific, demand-driven information centres

Situation analysis was the first step done in planning for establishing Puducherry as an information village. The situation assessment focussed on availability of existing technologies/electronic media in the village, information needs and issues of the community members and location identification for setting up the information centres. Several studies and

surveys were carried out from 1993 to 1996 that revealed interesting findings as follow:

- Farmers get most of their information from the local shopkeeper, the market place, and agricultural suppliers.
- The linkages between the farming community and the government departments are poor.
- The reach of electronic media especially television, was reasonably high despite the prevalence of poverty and people used this medium for entertainment and are willing to invest in it.
- There are considerable amounts of information transactions taking place among the rural poor households.

From the findings, it was rationalised that information and communication technologies would play a major role in facilitating rural development. Through access to information and knowledge, the poor would be able to access to resources, facilities, and opportunities that will allow them to achieve food security and improve and protect their livelihoods.

3. Identifying the village to be ICT-enabled

Two decisions had to be made in this stage of piloting the Information Village Project. The first was to identify the particular villages in Puducherry to be ICT-enabled. The second was to select the particular spot in the village where the information centres (the VKCs in particular) would be set up. Since MSSRF was into promotion of 'bio-villages', existing bio-villages became the testing sites.

The first VKCs were initially set up in houses of families who agreed to allow all the villagers access the VKCs without any bias on caste, class, gender and age. Over time, however, one of these VKCs started excluding the socially-oppressed community members to access the VKC. There was also the incidence of misuse of the VKC equipment like the telephone for personal/family use leading to increase in the phone bills. These experiences made the project introduce new processes and instruments in selecting where to set up the VKCs, as well as in selecting the other information villages. These include the following:

- Involving community participation in identifying a common venue for establishing the VKC which would enable social inclusion and community contribution towards electricity and telephone expenditures.
- The introduction of a Memorandum of Understanding (MoU) between VKC and the 'hub' (VRC) to ensure that the arrangements are legally binding and will be accordingly observed.
- Publicising the initiatives of the project through mass media like daily newspapers. This encouraged other villages to come up with their requests for them to be covered by the Information Village Project. This saved time for the project in identifying other potential communities as ICT-enabled information villages.

These new schemes also facilitated involvement of existing community-based organisations/institutions like Temple Trust and women self-help groups.

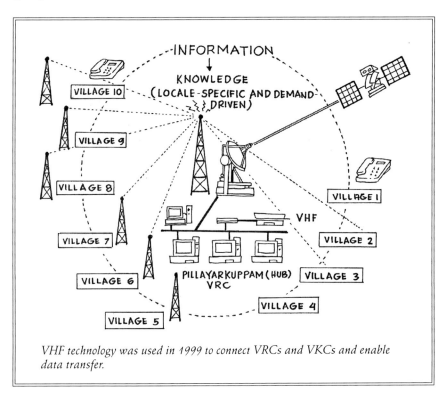

VHF technology was used in 1999 to connect VRCs and VKCs and enable data transfer.

4. Technologies for moving knowledge and information

Planning for the suitable technology to enable information/knowledge (data) transfer to and from the information centres was another major aspect for setting up the VKCs. In Puducherry, there was no phone connection. Through research and testing of different technologies, MSSRF used the VHF (very high frequency) technology to connect the VRC and the VKCs to enable data transfer in 1999. Since VHF is normally used for one-way communication, the project used GM300 and ST869 interface boards to enhance it into a two-way communication.

In 2001, MSSRF installed the Spread Spectrum technology that facilitated evolution of the 'Hub and Spokes' model for information/knowledge transfer. A central information centre plays the role of the 'hub' that processes and responds to the requirements of the villagers by having connectivity with the other VKCs that serve as the spokes.

5. Building local capacity to manage the centres

To further instill village ownership and sustain operations and usefulness of the information centre, MSSRF had to ensure that the project recruited

the right staff and volunteers to manage the day-to-day affairs of the centre. The community was asked to identify two volunteers who were then given training on operation of the equipment, on gender sensitivity, and on packaging information available from the centre in a manner that will attract people to access them and use the centre. The volunteers were also required to send regular feedback and reports about the services of the information centre and how these may be improved.

Two new and important services attached to the information centre that were innovations made by the volunteers were the following:

- Production of a community newspaper which started as news clips or excerpts on farming ideas taken from journals available in the information centre.
- Extending of linkage services to farmers e.g., facilitation of dialogue between sugarcane farmers and the sugarcane factory owner so that the farmers are able to plan/limit their harvests (cutting of sugarcane) to maximum volume that the factory can process in a day.

Staff/volunteers have a critical role to play. They should be there not only to manage the information centre but more importantly, to develop innovations to make them more relevant and responsive to the needs of the rural families and communities.

47

Considerations for a Village Resource Centre

Source:

S. Senthilkumaran and Nancy J. Anabel *M.S. Swaminathan Research Foundation, Tamil Nadu, India, 2010.*

1. Introduction

The M.S. Swaminathan Research Foundation (MSSRF) has been implementing since 1992 the village resource centres (VRCs) and the village knowledge centres (VKC) programmes under its Information Education Communication (IEC) division. In 2003, the VRC and VKC programmes were further strengthened with the creation of the Jamsetji Tata National Virtual Academy (JTNVA) for the purpose of establishing collaboration with several international and national partners for developing the content and capacity building aspects of the programmes for fuller realisation of a sustainable rural development.

The Puducherry experience of the MSSRF in Tamil Nadu, India evolved the 'Hub and Spokes' model for information/knowledge transfer. In its Information Village Project, the VRC is the hub that processes and responds to the community's requirements with connectivity to several VKCs[1] which are its spokes.

This paper particularly talks about the major considerations for the setting up and effective functioning of a VRC based on the Tamil Nadu experience. The steps in establishing the VKC is discussed in a separate paper.

2. Location and other requirements of the VRC

Dividing the Tamil Nadu state into agro-ecological zones such as coastal, delta, dry land and horticulture was one way by which to identify location for a hub centre. Another way is to engage community members in a discussion at the district or block level towards identifying the relevant location for the VRC. Regardless of the method, the main criterion for selecting the area where to set up the hub is to get the block with the most underdeveloped household clusters. As a rule of thumb, the VRC has to be situated that covers a cluster of villages within 30 to 60 kilometres radius for effectiveness and efficiency.

For the building that will house the hub centre (VRC), this should be able to accommodate safely and adequately: (a) the required equipment and

---

1. Also known as the rural knowledge centre.

paraphernalia such as computers, printers, scanners, digital connectivity facilities and others; and (b) as many persons in the covered communities who would like to have access to the different services of the centre.

Setting up a VRC or the hub is costly and a poor community would not be able to finance it on its own. While the MSSRF had to cover for the operation and maintenance of the centre at the onset, adopting a partnership mode for establishing the VRC from the beginning is crucial. This means having a receptive government and other organisations/institutions as strategic partners for the packaging and financing of the VRC services including linkages and capacity building requirements. Community members will be part of this partnership by providing the space for the VRC and bearing the cost of electricity.

3. Connectivity for effective information/knowledge transfer

Speeding up the reach of information/knowledge from the VRC to the VKCs located in the neediest communities is a major consideration and for which digital connectivity is of utmost importance. With this in mind, MSSRF has successfully negotiated with the Indian Space Research Organisation (ISRO) to provide it with satellite-based uplink and downlink facilities to the VRC. This digital connectivity to remote villages through the ISRO has enabled the provision of multiple services such as tele-education, interactive farm and fishery advisories, government schemes and entitlements, weather services and remote sensing applications through a single window.

The ISRO connectivity is a breakthrough in connecting the villagers with the experts. Users located at one node of this network can fully interact with others located at another node through video and audio links. Each node can further be expanded using different technologies such as notice boards, pamphlets, public address system, community newspaper, press releases, cable TV, audio/video conferencing through wireless, telephone, meetings, mobile phone and intranet website for dissemination of the useful and necessary information. Aside from this ISRO connectivity, MSSRF has also demonstrated the effective utilisation of mobile phones and closer user group (CUG) based audio conferences to connect the VKCs to the VRC and provide real time information/knowledge to rural

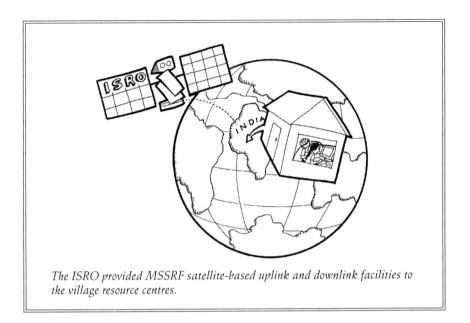

The ISRO provided MSSRF satellite-based uplink and downlink facilities to the village resource centres.

households as bases for making decisions and actions relating to their lives and livelihoods.

With this modern-day digital connectivity, one VRC is able to serve 15-20 VKCs within the radius of 60-100 km that potentially can serve the requirements of over 30,000 people spread across several villages.

4. VRC as one of three management tiers in the information village

The management of the information/knowledge transfer structures and processes as it evolved from the MSSRF experience is essentially three-tiered comprising the VKC, the VRC and the Jamsetji Tata National Virtual Academy (JTNVA) (Figure 47.1).

The JTNVA as the national level hub centre is located at MSSRF in Chennai. It is the first tier which connects several data generators and data providers (universities, experts, financial institutions, corporate sector, technocrats, grassroots academicians, etc.) primarily for connectivity, coordination, content development, capacity building and care and management. This is managed by MSSRF staff.

Empowering people through information access

In Sri Lanka, similar initiatives have been introduced in the tsunami affected villages through Sarvodaya Shramadhana Societies. One such center was the Pathegama Village Information Centre (VIC), located at a coastal village of Southern Provinces. Though started with the challenge of promoting information services in a community where information seeking has not been a practice, the Pathegama VIC has become the conduit through with villagers have become empowered and have better control over their life. In addition to its internet and IT education facilities, the centre has helped create better opportunities for poor families, youth and deserving students by linking them with support systems addressing their education, employment and enterprise development needs. Presently, the VIC offers information services to about 700 members from Pathegama and its neighbouring villages.

FIGURE 47.1

The three-tiered information village

The VRC, or block level hub, is the second tier and located at block level or commune level, which should ideally be at the centre point of cluster of the villages. The national hub is connected with the VRCs through ISRO uplink/downlink satellites for transferring the knowledge/information to VRCs and vice-versa. This is managed by the MSSRF staff (a coordinator and a pool of subject matter specialists) at the VRC level.

The third-tier are the VKCs located at the village or *panchayat* level, disseminating the demand-driven and locale-specific information to the community members. The community manages the centre through the village knowledge centre management committee and knowledge workers.

5. Building strategic partners

A good rapport building with strategic partners produces good services to the community members and this helps in maintaining trust level of community towards VKC. Strategic partners are those who provide resource support in terms of sharing content (data generators) or those who act as resource persons for capacity building and for facilitating access or link to infrastructural facilities for addressing the community needs in the context of VRC. These include officials from government departments, research organisations, academic institution, service organisations like Rotary Club, Lions Club, non-governmental organisations, bankers and individuals with expertise and willingness to share up to national and international levels.

The strategic partners meet regularly every three to six months at the block or commune or *mandal* level in order to address the information/ knowledge requirements of the community. This facilitates, especially for the government departments, to deliver their schemes/training programmes to the community using the VRC as vehicle.

48

Transforming the Value Chains through Business Development Approach

Source:

Ramona Miranda, Chopadithya Edirisinghe, Jayantha Gunasekera and Chaminda Jayantha *Practical Action, Sri Lanka*
Nilantha Atapattu *Rural Enterprise Network, Sri Lanka, 2010.*

1. Introduction

The business development approach used in the project on Strengthening Resilience of the Tsunami-Affected Communities (SRTAC) in Sri Lanka was the Participatory Market Systems Development (PMSD) which draws from the concepts of value chains and participatory market chain analysis.

PMSD is an approach that was used in developing the livelihoods of the communities that the project worked with. This system derives features from the value chain and the participatory market chain mapping approaches and attempts to reduce the vulnerabilities of communities to market pressures.

This approach which focusses on the market system as a whole, has the potential to improve the livelihoods of a significant number of marginalised small-scale rural producers. The PMSD attempts to improve the livelihoods of a significant number of marginalised small-scale producers (MSPs) by: (i) promoting their empowerment so that they can explore and exploit new business opportunities with other market actors and influence on issues that affect them, (ii) promote interaction amongst market actors to improve critical issues and relations that can have a positive impact on farmers, and (iii) promote communication of lessons and business models emerging from the project to inspire others to adopt and adapt to them.

Promoted by Practical Action to develop the livelihoods of underprivileged small producers, the fundamental idea of the PMSD is 'understanding the causes of the root problem collectively and work cooperatively to find solutions' by all players of particular market system. Development agencies are expected to facilitate this process rather than interfering or involving the market system in an unsustainable manner. The key features of PMSD is the participatory and systemic nature of this approach as detailed below (*http://www.itcltd.com/home/pmsd_home*):

Participatory—Understanding market systems, identifying issues, bottlenecks, and opportunities and undertaking corrective actions is better done by sustained participation and collaboration of a wide variety of market actors, civil society organisations, academics, practitioners, researchers and decision-makers.

Systemic—A market system is composed of actors and the relationships amongst them. A successful market improvement requires the transformation of both actors and relationships. And this transformation cannot be in a linear fashion only, and needs to be balanced with flexibility to adapt to the reactions of the market due to these initiatives or to other external events such as economic and political crises, and changes in consumer trends, new technologies, and natural disasters.

Other key features of PMSD is that it is inclusive, sub-sectoral, multi-dimensional and contextual. Inclusive in that it sees market actors as people trying to optimise their outcomes from the role, position and resources they have in the system, and as such to create the conditions for interaction of multiple actors from where initiatives to improve the market emerge. For example, intermediaries are not viewed as exploitative but included in the process. The approach also focusses on a specific product or 'sub-sector', and is multi-dimensional, in that it looks at not just the economic, but also the social, political and cultural factors. Furthermore, the context is seen as fundamental to understand a market system and define strategies to transform it. Market actors do not operate in a void; their decisions and intentions are affected by many issues than can be within or outside of the boundaries of the market system (Figure 48.1).

The PMSD uses the participatory market chain approach (PMCA) as the primary tool to analyse the market in the market mapping process (Figure 48.2). The Market Map has as its main components (Albu and Griffith, 2005), the following two main aspects, seen in the diagram below:

- Market chain actors and their linkages: the chain of actors who own the product as it moves from primary producers to final consumers.
- Enabling business environment factors: i.e., infrastructure, policies, institutions and processes that shape the market environment.
- Business and extension service providers: the inputs and business/extension services that support the chain's operations.

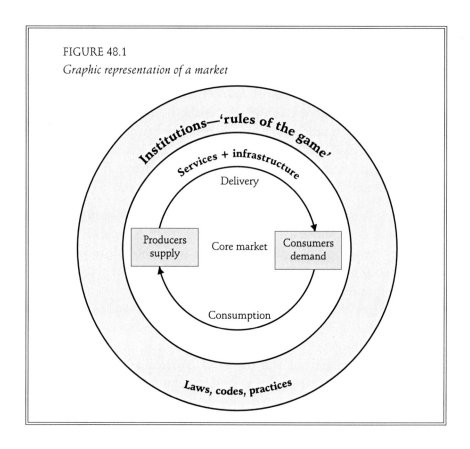

FIGURE 48.1
Graphic representation of a market

These three levels are constantly interacting and influencing one another. The actors in these levels, the relationships amongst them and the forces that influence those relationships constitute a market system.

The process of developing the map, in order to understand the linkages, is better done with the market actors as this will enable them to understand their role in the map and appreciate the value of the others. This it is hoped or believed will encourage their cooperation in improving other linkages in the map other than just those that pertain to them only as this will be seen as beneficial to themselves too. That is, if the whole market system for this product is more stable and sustainable, so are their own transactions and role in the process. Albu and Griffith (2005) argue that participatory approaches to market-chain analysis contribute to

Transforming the Value Chains through Business Development Approach

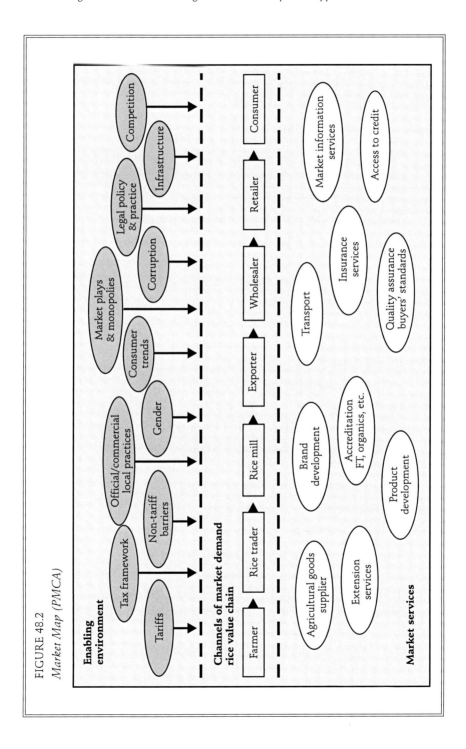

FIGURE 48.2
Market Map (PMCA)

market maps that are more likely to be accurate and to represent a wider range of knowledge.

2. Key PMSD processes and tools
(Alison and Osorio, 2008):

- *Preliminary market mapping*: it helps the project team to capture and organise preliminary information about the market system gathered from secondary sources and key informants.
- *Hooks*: issues that will attract and engage market actors.
- *Market opportunity groups*: groups that represent and empower target producers to explore more and better business opportunities and engage with other market stakeholders.
- *Interest forums*: to convene and engage stakeholders to work together to improve the the market system.
- *Participatory market mapping (PMM) workshops*: to bring together market system actors around joint analysis, relationship, trust-building and negotiation.
- *Moving from analysis to action*: based on the relationships, knowledge and trust generated throughout the process, concrete actions are agreed by market stakeholders to improve the market system (composed by the business environment, the market chain and the providers of inputs and services).

In this process, we also need to be aware of the constraints of PMSD as well as some of the principles that guide our activities.

Among the constraints is the limitation of the sub-sectoral approach where the sub-sector defines who belongs to the system and what issues or opportunities are important. This has both advantages (e.g. better communication amongst actors) and disadvantages (e.g. risk of neglecting external factors to the system but critical to its functioning). Therefore, few sectors have to be selected by the livelihood development programme at the project site.

Another limitation is that some of the market actors are far from the producer and are not interested or willing to participate in the PMSD

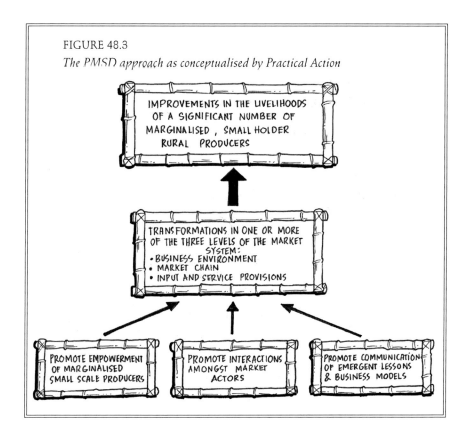

FIGURE 48.3

The PMSD approach as conceptualised by Practical Action

(Figure 48.3). For example, promotions are held in the capital city, in which these suppliers participate, but organisers of such promotions may not be interested in being part of the process even though they may be a key player as the handicraft industry is not so much a demand-led sector.

Examples of activities that development practitioners must avoid in the use of PMSD for poverty reduction through livelihoods improvement of small-scale rural producers is to remember that they are only facilitators of the market transformation processes that should be led by market actors themselves in order to be sustainable. However, the practitioner facilitating the process also needs to have adequate knowledge and expertise about the markets that are to be transformed in order to have credibility to promote successful dialogue and collaboration. A case study of an application of the PMSD is seen in Part 4, 'Applying Market-Systems Approaches in Wanduruppa.'

References

Albu, M. and A. Griffith (2005). *A framework for rural enterprise development policy and practice*. Practical Action.

Alison, A. and L. Osorio (2008). *Lessons and insights in participatory market system development: An international study of PMSD applications in Bangladesh, Sri Lanka, Sudan, Peru and Zimbabwe*. Practical Action. *http://practicalaction.org/?id=marketchain_report*, cited 30 October.

Practical Action (2008). "Participatory market systems development (PMSD)". *http://www.itcltd.com/home/pmsd_home*, cited, 9 September.

Roduner, D. (2005). *LBL, value chains, rural development news*, 2/2005.

Wikipedia (2009). *http://en.wikipedia.org/wiki/Value_chain*, cited 30 September.

49

Planning Coastal Revegetation Programmes

Selection of suitable species and plantation techniques are among the considerations in coastal revegation.

Source:

Bhathiya Kekulandala *Practical Action, Sri Lanka, 2010.*

1. Background

The earthquake and tsunami of 26 December 2004, and the events that followed caused incalculable suffering to millions of people around the Indian Ocean. Communities from Banda Aceh (Indonesia), to the tourist resorts of south Thailand, to the fishing villages of Sri Lanka, and people on African coast were severely affected (UNEP, 2005).

The tsunami impacted nearly two-thirds of the coastline of Sri Lanka affecting about 13 coastal districts in four provinces. The north and east coasts took the brunt of the wave accounting for two-thirds of the deaths and displacements (FAO, 2007).

Sri Lanka has a coastline of approximately 1,660 kilometres. Within this coastal zone[1] is 27 per cent of the total population, 70 per cent of the tourism infrastructure, and 62 per cent of industries. Furthermore, 89 per cent of the fish production comes from the coastal fishery (CCD, 1997; UNEP, 2005; FAO, 2007). The coastal zone is very diverse and contains a variety of habitats such as lagoons and river basin estuaries (many including fringing mangroves). There are also fringing and offshore reefs, mangrove swamps, seagrass beds, salt marshes, beaches, sand spits, rocky shores, and dune systems (CCD, 1997 and FAO, 2007).

Assessments of the tsunami impacts on the coastal ecosystems and especially on the vegetation types revealed a link between the alteration/disturbance to the natural ecosystem and the severity of the impact (IUCN, 2005; CCD, 2005; FAO, 2007). For instance, degraded sand dunes and mangroves had a higher degree of damage than intact sites (IUCN, 2005). These assessments, led to two priorities for the medium to long term: developing defensive ecosystems around the coast and rehabilitating damaged ecosystems (FAO, 2007).

---

1. The area lying within a limit of 300 metres landward of the mean high water line and a limit of 2 kilometres seaward of the mean low water line and in case of rivers, streams, lagoons or any other body of water connected to the seas either permanently or periodically, the landward boundary shall extend to a limit of 2 kilometres measured perpendicular to the straight baseline drawn between the natural entrance points (defined by the mean low water line) thereof and shall include waters of such rivers, streams and lagoons or any other body of water so connected to the sea (CCD, 1997).

Because terrain, natural ecosystems, and human pressures vary throughout the coast, there will be no one solution for all places. Thus, several measures were combined in various ways in a more location-specific and comprehensive approach to strengthening the resilience and productivity of Sri Lanka's coastal zone (UNEP, 2005; FAO, 2007).

2. Coastal revegetation programmes

Five government agencies and six NGOs participated in post-tsunami rehabilitation of coastal ecosystems and revegetation programmes in the coastal zone. Furthermore, the NGOs supported various community-based organisations (CBOs) in carrying out the revegetation programmes.

The issues surrounding coastal revegetation programmes stem from site selection (specifically, land use and land tenure), selection of suitable species, and plantation techniques, maintenance and care. This paper examines the issues on land use, land access and land tenure.

Observations

Initial assessments emphasised proper planning in scientific coastal revegetation. Thus, leading government agencies such as MOENR, DFC, DWLC, CEA and CCD with the help of IUCN, UNEP and UNDP developed a series of guidelines for coastal revegetation programmes (FAO, 2007). The country is yet to commission a countrywide assessment of the post-tsunami coastal revegetation programmes, but some effects are quite observable (Personal Observation).

3. Land use

The National Physical Planning Department categorises the coastal zone as a 'fragile area' and hence recommends proper planning to reduce negative impacts to the zone. Therefore, a clear understanding of the land use of the area and the surrounding landscape is a prerequisite for any revegetation programme. For instance, the coastal zone is highly heterogeneous with anthropogenic or highly modified natural areas predominating. On the other hand, areas of the southern, eastern and northern provinces are more homogeneous with natural land use.

This knowledge is vital to ensuring any coastal plantation's compatibility to the existing land use and to the larger landscape. Since coastal plantations function partly to enhance the aesthetics of the landscape, disregarding the planning process means the plantation would be out of place.

Planning allows the consultation of district and division level planning cells (government) to get information on proposed land use patterns in the area. This information ensures that plantations are compatible with future or proposed land use for the area.

4. Land access and land tenure

Access to land is often based on custom and traditions. Often, community leaders assign customary rights to the community members. These rights of access may develop over a long period.

Land tenure is both a relationship and an institution. Customarily or legally, it is the relationship among individuals or groups with respect to land. (Often 'land' includes other natural resources such as water and trees.). As an institution, it consists of rules invented by societies to define how property rights are allocated within the society. They define how access is granted, including who has the rights to use, control, and transfer land, as well as associated responsibilities and restraints. Simply, land tenure systems determine who can use what resources for how long, and under what conditions (FAO, 2002).

Except for protected areas that were gazetted by various agencies, much of the land in many parts of the country was privately owned before the tsunami. After the tsunami, government declared a setback line of 100 metres in the southern province and a 200 metres setback line in the eastern and northern provinces. This means that the area from the mean high water line to either 100 metres or 200 metres landward is a no-build zone. Consequently, these setback lines were reduced and adjusted based on the recommendations in the revised coastal zone management plan of 1997.

The lands within the setback lines could be private but construction on them as well their use is regulated based on the Coast Conservation Act of 1997. Aside from private owners, district secretaries and divisional secretaries also are vested with ownership of land areas in the coastal belt. There are several extensive protected area systems in the coastal zone and these are governed by the respective line agencies such as the Department of Wildlife Conservation (DWLC) and Department of Forest Conservation (DFC). Hence land tenure and land ownership issues have to be clearly identified and studied for the success of coastal revegetation programmes.

Clearly, coastal revegetation programs have to consider the varying land ownership scenarios in the coastal belt. Many post-tsunami revegetation programmes were unsuccessful in the long term because they ignored the land tenure and ownership status.

In the western and in some areas in the southern and eastern provinces, only narrow strips are available for plantations because most lands are privately owned. Consequently, plantations were placed in exposed beachfronts where they were eventually destroyed by waves.

Experiences from coastal revegetation programmes point to two factors that are vital to their success and sustainability: a clear understanding on land ownership and a consensus among the landowners and implementers of the programmes. The following two scenarios elaborate on approaches that can be adapted to minimise the above-mentioned issues.

4.1 Wanduruppa in Southern Sri Lanka

The area selected for plantation was a sand dune planted to Casuarina. The community-led plantation programme was developed with the consensus and approval of the Divisional Secretary, divisional level coast conservation department. However, the DWLC had gazetted the area as a sanctuary. As such, DWLC owned the management rights to the land. Therefore, the local CBO had to get the approval for plantation from the DWLC and involve them in the decision-making process to ensure the long-term management of the area.

4.2 Karaitivu, Eastern Sri Lanka

No state land was available so the site selected for the plantation was a narrow strip of land between the Karaitivu lagoon and the sea. The lands are privately owned but the owners were resettled in a different area after the tsunami. Although the lands were currently unoccupied, the CBO implementing the revegetation programme got legal consent from the landowners to use the land for plantation. The Divisional Secretary of the area facilitated the acquisition of the consent.

The landowners transferred the user rights of their lands to the CBO to develop the plantation. They agreed to safeguard the plantation and not harvest for five years. Furthermore, economically valuable cashew was introduced to this plantation in addition to other species. Finally, landowners entered into a tri-party agreement with CBO and Divisional Secretary to maintain the plantation beyond the project life.

5. Long-term participation

The above experiences clearly indicate that coastal revegetation planners should consider the dynamics of the local context prevailing in the area in addition to the existing legal and policy frameworks in establishing

plantations. Early engagement with local communities, local level planners and administrators are important to ensure the participation throughout the process. Dynamic or adaptive management mechanisms should be scoped and developed with the participation of all stakeholders with specific responsibilities to ensure maintenance aspects beyond programme/project time frames.

References

CCD (1997). *Revised coastal zone management plan of Sri Lanka*. Sri Lanka: Coast Conservation Department, Ministry of Fisheries and Aquatic Resources Development.

FAO (2007). *Assessment of tsunami damage to coastal ecosystems in Sri Lanka and development of guidelines for integrated coastal area management*. Rome: Food and Agriculture Organization of the United Nations. p. 75.

―――. (2002). "Land tenure and rural development", *FAO Land Tenure Studies* 3. Rome: Food and Agriculture Organization of the United Nations. p.50.

IUCN (2005). "After the tsunami: Restoring terrestrial coastal ecosystems", *Series on best practice guidelines Sri Lanka*. p. 6.

UNEP (2005). *After the tsunami, rapid environmental assessment*. United Nations Environmental Programme. p.140.

50

Role of Coastal Bioshield in Strengthening the Coastal Resilience

Coastal tree vegetation reduces vulnerability of coastal communities to floods, tsunamis and cyclones.

Source:

V. Selvam, R. Ramasubramanian, A. Sivakumar and J.D. Sophia *M.S. Swaminathan Research Foundation, Tamil Nadu, India, 2010.*

The coastal zone of the east coast of India is vulnerable to hazards such as cyclones, floods and tsunami. These natural hazards cause loss of life, injury, other health impacts, property damage, loss of livelihoods and services, social and economic disruption and severe environmental damage. However, the presence of coastal tree vegetation such as mangrove and non-mangrove forest and sand dunes with vegetation reduces the damage.

The coastal tree vegetation along with sand dunes is called as bioshield. Considering the effectiveness of bioshield, the National Disaster Management Authority of India (NDMA) recommended raising mangrove and non-mangrove bioshield as an important measure in cyclone and tsunami management.

The super cyclone of Orissa in October 1999 and tsunami in December 2004 provided opportunities to collect field-based evidences on the role of coastal vegetation in reducing the impact of such natural disasters. On the basis of the inputs obtained, the following indicators were selected: i) loss of lives within the village due to forceful entry of the tsunami waves, ii) number of houses damaged, either fully or partially and type of houses, and iii) height of the water inundating the villages (identified through lines left on the walls of the houses). The project, 'Strengthening Resilience in Tsunami-affected Communities' (SRTAC) brought out following features:

- Local communities in Tamil Nadu were aware of the effectiveness of bioshield and have been calling mangrove forest as aalayaathi kaadu (wave mitigating forests) for thousands of years. Experimental studies in the field and laboratory have strengthened this traditional wisdom.

- Effectiveness of a coastal forest in mitigating the impact of natural hazards depends on the width, density, and structure of the forest and the tree characteristics (height and diameter at breast height).

- Theoretical studies show that 30 trees— either mangrove or non-mangrove—per 100 square metres spread over 100 metres wide may reduce the tsunami flow pressure by more than 90 per cent.

- For a tsunami wave height of 3 metres, the effective forest width should be about 20 metres and for 6 metres high tsunami wave the effective width of forest should be about 100 metres.

- It has also been estimated that trees with 10 to 35 centimetres diameter at breast height will be effective against 4 to 7 metres tsunami waves.
- Loss of life and damage in mangrove protected villages was nil whereas in mangrove-unprotected villages there was 2 per cent death and 85 per cent damage to the houses.
- The fishing community, which is normally reluctant to participate in restoring, conserving, raising mangrove and other coastal vegetation programmes, now shows lot of interest in restoring degraded coastal vegetation as well as raising plantations in new areas. They are now demanding help from government agencies and non-governmental organisations for raising bioshield.
- Administrators and planners are willing to allot large parcels of wastelands including saline-affected areas to the coastal community to raise tree plantations.

M.S. Swaminathan Research Foundation (MSSRF) conducted a study in randomly selected areas covering the entire coast of Tamil Nadu. The study covered: a) the species used to raise coastal shelterbelts, b) the techniques followed, c) the level of community participation in raising shelterbelts, and d) the community initiatives in conserving sand dunes and associated coastal vegetations. Major suggestions of this study are:

- It is essential that local community and Panchayati Raj Institutions (elected local government), which has a constitutional mandate of disaster management at village level, be taken as lead partners in planning, implementing and monitoring bioshield programmes.
- Due to the lack of proper ownership, participation of the community and long-term sustainability of these plantations cannot be guaranteed.
- Multispecies non-mangrove bioshield with a mixture of species, which are native to the region, part of local landscape, and economically important species should be raised along the coastal areas.

- Participatory conservation of sand dunes with suitable sand binding vegetation should be given priority. Sand dunes act as barriers against cyclones and tsunami and play a dominant role in water conservation in coastal areas where water scarcity is a recurrent feature during the summer season.
- Prevention of planting of casuarinas very close to the tide line as it affects the nesting ground of turtles and biodiversity of the beaches. Furthermore, the monocultures of Casuarina equisetifolia do not address the issue of linking livelihood security of the coastal communities with ecological security of the coastal areas.
- Part of the beach from the high tide line to landward portion should be left barren because the sand binding property of the plantation will reduce the supply of sand to the littoral current. To compensate for this, current and waves will remove large chunk of sand in other areas, leading to sea erosion in those areas.

Conclusion

Both mangrove and non-mangrove vegetations can be part of any programme that tries to enhance resilience of the coastal community to natural disasters. However, in order to develop and sustain effective coastal bioshields, active participation of the local community and local self-government is necessary. The sustainability of the bioshield can be achieved only if the ecological security of the coastal zone and livelihood security of the coastal community is given concurrent attention. There is a need to develop and demonstrate models of mangrove bioshield integrated with fisheries and multispecies non-mangrove bioshield that has a mixture of ecologically and economically important species.

The workshop by Food and Agriculture Organization (FAO) in Bangkok in 2006 also brought out the following findings:

- Although forests and trees can act as bioshields, the degree of their effectiveness depend on many variables such as characteristics of the hazards themselves, the features of the site and characteristics of the bioshield.

- It should be established that bioshields cannot alone protect against hazards.
- The use of bioshields should be considered within the framework of disaster management strategies, which also include effective early warning systems and evacuation plans.

51

Non-Mangrove Bioshields in Coastal Areas

Mangrove and non-mangrove vegetations reduce the impact of coastal hazards.

Source:

V. Selvam, R. Ramasubramanian and A. Sivakumar *M.S. Swaminathan Research Foundation, Tamil Nadu, India, 2010.*

1. Introduction

The coastal zone is endowed with a variety of features and systems such as estuaries, lagoons, tidal wetlands, coral reefs, seagrass beds, sandy beaches, sand dunes etc. In most places, tidal wetlands are characterised by the presence of a group of saline tolerant trees and shrubs called mangroves whereas sandy beaches and sand dunes are characterised by the presence of natural terrestrial vegetations and man-made plantations. These mangrove and non-mangrove vegetations play significant role in reducing the impact of coastal hazards such as cyclones, storm surges and tsunami.

Dr. M.S. Swaminathan, renowned agriculture scientist and proponent of Evergreen Revolution wrote an article entitled 'Mangrove can act as shield against tsunami' on 27th December 2004, a day after the 2004 Asian Tsunami, and since then the word bioshield has been used to refer to mangrove and non-mangrove vegetations of coastal areas in the context of disaster risk reduction (DRR) and preparedness not only in India but globally. Previously, shelterbelt was the most popular term used to indicate man-made non-mangrove plantations raised along the coastal areas including in India for protecting lives and properties of coastal communities and the coastal resources from hydrometeorological hazards such as cyclones, salt sprays and storm surges.

2. Shelterbelt or bioshield

The practice of shelterbelt was introduced initially in inland areas of European countries long time back mainly to save agriculture crops and livestock from unfavourable weather conditions. Shelterbelts, which are linear planting of trees and shrubs, are designed to alter the flow of wind and snow, thereby altering the microclimate in an immediate area to make it more habitable for crops and livestock. It was started in late 19th century in the Ulyanovsk Oblast in Russia and it was a multiple species shelterbelt with oak, maple, pine, birch, linden, elm trees and popularly called Genko's Forest Belt (Genko was the Forester who started this practice). It is still existing and being protected as a nature reserve. Considering effectiveness of shelterbelt in reducing wind speed and soil erosion, they were also introduced in non-saline sandy areas along the coastline.

In India, raising coastal shelterbelt as a protective measure against cyclone was started in late 1960s, if not earlier. In 1964, a severe cyclone wiped out a coastal village namely, Dhanushkodi of Ramanathapuram district in Tamil Nadu, which triggered Forest Department of Tamil Nadu to plant casuarina as shelterbelts, in large scale along the coastal areas. The Mitra Committee commission in 1970 on Disaster Mitigation also strongly recommended coastal shelterbelt as the only substantial solution for breaking intensity of cyclones and reducing damages from such natural disasters. Backed by this policy recommendation and lessons learnt during 1977 super cyclone in Andhra Pradesh both Central and state governments promoted coastal shelterbelts in large scale with the financial support of World Bank. Records of the Forest Department indicate that in Andhra Pradesh alone about 25,800 hectares has been brought under coastal shelterbelts since 1978. Likewise, 2004 tsunami also triggered raising of coastal shelterbelts in large scale and in Tamil Nadu alone shelterbelt has been raised in about 4,700 hectares between the year 2005 and 2007. Most of these shelterbelts were established only after a disaster and it is not considered as a mitigative effort in strengthening people's lives in the absence of a disaster. Besides which, one of the more serious problems both in the past and present is nearly 90 per cent of the coastal shelterbelts area is covered by single species namely casuarina. Hence, many presume that coastal shelterbelts are nothing but monoculture plantation.

Shelterbelt model can either be monospecies or multitier species. This article shares both benefits and disadvantages over monoculture and multiple species model, based on the experiences from east coast of India. There is no specific thumb rule in implementing the shelterbelt; however, there is always a need to consult the community to improve the model based on local needs.

3. Issues in casuarina shelterbelts

Casuarina is an exotic species and growing them in large scale as a monoculture plantation, occupying an entire beach area, may have serious ecological impacts. Monoculture casuarina plantation completely changes

> *Casuarina equisetifolia*
>
> Common names: casuarina, ironwood, coast she-oak, horse tail, Australian pine. Casuarina is a deciduous tree with a soft, pine-like appearance. It is a heat-loving plant, grows well in sandy soil. It is tolerant to saline condition and salt laden wind. It has ability to fix atmospheric nitrogen and thus able to grow even in nutrient poor soil and thrive well in low moisture condition. It is not prone to any serious pests and diseases. Because of the above characters, survival rate is always high. Its growth is rapid during the first seven years, about 1.5 to 2.5 metre/year. The wood of casuarina is hard and has high calorific value. Because of these, casuarina is the most preferred species by the farmers for commercial plantation in sandy areas and by the Forest Department to raise shelterbelts in the coastal areas. However, raising monoculture plantation of casuarina as shelterbelt has many disadvantages.

the landscape of the coast, wherever it is planted in large scale, it affects local biodiversity. It produces dense shade and a thick blanket of leaves and small but hard, pointed fruits, which completely cover the ground beneath and make the ground ecologically sterile. Once established, it radically alters light, temperature and soil chemistry of beach habitat as it out-competes and displaces native plant species and destroys habitat for native beach insects and wildlife. Secondly, raising of casuarina plantation right from high tide line interferes with turtle nesting. Different species of crabs lives in different vertical zones near the high tide line and planting of casuarina close to high tide line affects ecological niches of these crabs. Economically, there is also not much benefit from casuarina shelterbelt to local community, unlike an exception in case of Andhra Pradesh. Casuarina plantations fetch huge economic returns. For example, every four years, almost Rs.60,000 is earned from an acre as reported in coastal villages of Andhra Pradesh, where casuriana is grown as commercial plantation by farmers. Some portions of the earning is allocated for village development and cultural events, while the remaining is used to replant. However, the community has no access and rights to use this resource. It is simply argued by the Forest Department that casuarina trees are planted to protect villages from cyclone so it cannot be felled.

Such lack of clarity in access, rights and economic benefits from casuarina shelterbelt is one of the main reasons for lack of genuine participation of local community in developing and managing these shelterbelts. Above all, raising monoculture plantation of casuarina defies scientific guidelines in establishing effective shelterbelts.

4. Model of non-mangrove bioshield

In order to provide an alternative model to single species shelterbelt and to link bioshield and livelihood issues, MSSRF developed a multispecies non-mangrove bioshield model, which primarily consisted of only two components namely, barren beach followed by multispecies bioshield. While sharing this concept, the affected community also contributed to the suggested model. The final model is shown below and the four components of this model are:

4.1 Component 1: Barren beach

Enough space is given between the high tide line and the bioshield. This gap takes care of the coastal ecological process such as supplying sediment to littoral drift and also providing enough space for nesting of marine

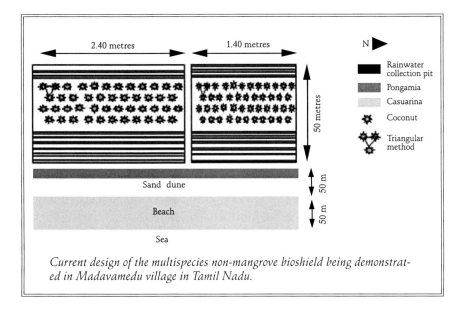

Current design of the multispecies non-mangrove bioshield being demonstrated in Madavamedu village in Tamil Nadu.

turtle. In addition, barren beach provides space for boat landing and net-mending hence not affecting the needs of fishermen. The breadth of the barren beach depends on the total breadth of the beach exciting in the identified area. It is normally advised to keep at least 30 metres between the high tide line and bioshield.

4.2 Component 2: Small sand dune

Next to barren beach, the model consists of a small sand dune. According to local community, continuous sand dunes are more effective in preventing sea water from entering into the villages during storm surges and tsunami. The height of the sand dune may be around 2 metres and breadth could be around 10 metres. This sand dune can also be established artificially and its height can be increased by following traditional method of using small machineries. The dune should be stabilised with sand binding local creepers such as Ipomea pes-caprae and spinifex sp in order to sustain them for long time. However, planting of these creepers are not recommended on the seaward side, as this side is usually utilised by sea turtle for their nesting.

4.3 Component 3: Rainwater harvesting structure

This is another component introduced into the model by the local community. The shallow ditches can be dug on the landward side of the dune and parallel to sand dune to harvest and store rainwater. This also helps in recharging ground water and takes care of water requirement of the bioshield.

4.4 Component 4: Multispecies non-mangrove bioshield

Multispecies bioshield is present next to water harvesting structure. It consists of shrub or medium height species in the front rows, economically important tall tree species such as coconut in the middle rows, which is followed by tall evergreen tree species on the landward side. The number of rows for planting the vegetation is usually determined based on the breadth of the beach and its availability (Figure 51.1).

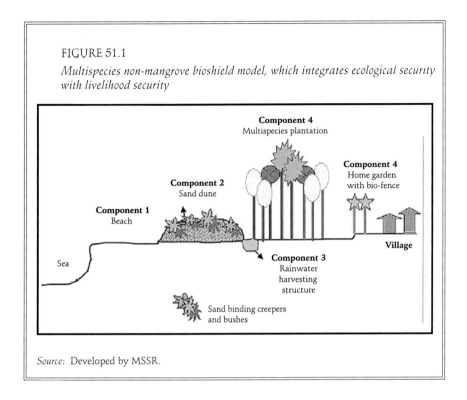

FIGURE 51.1
Multispecies non-mangrove bioshield model, which integrates ecological security with livelihood security

Source: Developed by MSSR.

Conclusion

Shelterbelt or non-mangrove bioshield should be multispecies and established with native trees and shrubs. Single species shelterbelt with uniform height will create wind turbulence on the landward side and thus, may cause severe damage to the area during cyclone. Effectiveness of shelterbelts in mitigating the impact of natural hazards depends on the width, density, structure of the forest and the tree characteristics (height and diameter at breast height). The height of shelterbelt determines what area of land is to be protected. The area of land protected by the shelterbelts is approximately 20 times the height of the trees in the shelterbelt. Generally, longer shelterbelts, covering entire exposed area in front of a village, are more desirable than shorter ones since the later tend to channel wind sideways around them. Shelterbelts should be linked to other corridors of vegetation that naturally exists for greater reduction of wind speed. Shelterbelts should be semi-permeable or intermediary in density because very porous shelterbelts have very modest capacity to reduce the wind speed. Though high density shelterbelts are very effective in reducing wind speed they allow winds speed to recover more quickly and thereby limiting the area protected by them in the landward side. For agriculture purposes, shelterbelts of intermediate width are preferred but for reducing the risk and impacts caused by disasters, wider shelterbelts are more effective. In the case of mangrove bioshield, indicative width for different coastal hazards such as cyclone (100-300 metres), storm surges (200 metres) and tsunami (500-2,000 metres) has been developed but for non-mangrove bioshield it is yet to be defined. Regarding design, shrubs should be planted in the seaward side in order to reduce velocity of seawater flooding during storm surges. This can be followed by medium height trees and finally tallest trees can be planted. Above all, non-mangrove bioshield should have species that can safeguard ecological integrity of the coastal areas, livelihood security of the coastal communities and protect the community from intense forms of natural calamities and unpredictable changing seasons.

Part 4

Risk Reduction Experiences from the Field

The fourth part is probably the most important part of this compilation, because it deals with practical experiences from the field, capturing diverse interventions and limitations faced by the communities and implementing institutions. These cases were written by practitioners in Sri Lanka and India, who have demonstrated different approaches for vulnerability and poverty. Almost all cases and case-stories are derived from the project Strengthening Resilience in Tsunami-affected Communities of India and Sri Lanka (the post-tsunami project), co-funded by International Development Research Centre (IDRC) and Canadian International Development Agency (CIDA). Three main recipients—M.S. Swaminathan Research Foundation, Sarvodaya and Practical Action were involved directly in the project implementation. The earlier parts of the source book set the context and frameworks for understanding community-driven efforts to reduce vulnerabilities of coastal communities.

52

An Introduction to the SRTAC Project

1. Introduction

In 2004, the intensity and unimaginable magnitude of destruction caused by the Indian Ocean tsunami across 14 coastal countries shocked the whole world. Indonesia, Sri Lanka and India suffered the most, as mortality rate reached 73.5 per cent, 15.4 per cent, and 7.1 per cent, respectively.[1] More than 228,000 people lost their lives. The economic loss was estimated at 10 billion[2] as 1.7 million coastal populations and their livelihoods were displaced. The losses and the destruction were along the coast line, about 2 to 4 kms inland. With altered weather patterns, it is expected that many such intense and unpredictable disasters will impact more people along the coast.

The number of natural disasters has grown from an average of 150 a year in 1980 to over 450 a year.[3] Globally, 1.7 billion farmers are highly vulnerable to climate change impacts.[4] The majority of the world's 200 million full and part-time fisher folk (fishers, fish processors, traders and ancillary workers) and their dependents live in areas vulnerable to the impacts of climate change.

Source:

Priyanka Mohan *IDRC Programme Manager SRTAC Project.*

---

1. Cosgrave (2007).
2. Losses and Damage from the tsunami: (BRR *et al.*, 2005; IRIN, 2005a, 2005b).
3. World Economic Forum (2008).
4. Oxfam International (2009).

2. Vulnerability of coastal communities to disasters

More than a billion communities live within 100 kms of the coastal line globally. They are exposed to different types of natural calamities that occur every year. These include—unpredictable rainfall, sea water intrusion, change in monsoon cycles, natural erosion, cyclones, floods and droughts. Hence, empowering coastal communities helps minimise the risks, and allows them to cope with natural disasters. Yet, resilience building on a large scale is a process that requires time to ensure that effective systems are working successfully and are in place for recovery and are working successfully.

3. Fighting vulnerability with resilience

Part evidences has shown that losses can be minimised and vulnerability decreased if measures were taken to build the communities' resilience, the most basic of which are:

- Putting in place natural mitigating measures. A community from Muthupet with whom MS Swaminathan Research Foundation (MSSRF) had worked with in the past discovered that mangroves vegetation can provide a natural barrier to tsunami waves. Their experience showed that mangrove trees along the shoreline dissipated the energy of tsunami waves. Tidal creeks and canals associated with the mangrove wetlands also absorbed large quantities of tsunami water and hence reduce the amount of water reaching inland.

- Building the capacity of communities to deal with natural disasters. Prior to the tsunami, some organisations worked towards building community capacities that could tackle disasters. Villagers from Samiyar Pettai, Cuddalore had benefitted from the disaster recovery and preparedness training received from an UNDP—Government of India funded project. The loss of lives was much lesser in comparison to nearby village that had not taken the training.[5]

---

5. UNDP (2005).

- Providing Access to information and communication technology. Appropriate early warnings and easy-to-use technology, saves lives. The village knowledge centers (VKCs) established by the MSSRFand Sarvodaya provided the community with timely information before and after the tsunami struck. For example, when the VKC of the Navallu village received information from Singapore regarding an on-coming tsunami, it spread the information right away through the public address system, giving people ample time to move out and thereby preventing the loss of lives in the village. Communities utilised the VKC for mobilising relief materials.

- Using Indigenous Knowledge. Evidence also shows that traditional and indigenous knowledge can save people's lives. Besides the signs of agitated animals, some of the tribal communities in Andaman and Nicobar, Kerala, Indonesia, and Thailand, recognised the disturbance in the sea, who evacuated to a nearby hill.

4. The Strengthening Resilience in Tsunami-affected Communities of India and Sri Lanka (SRTAC) project

More popularly called the post-tsunami project, co-funded by International Development Research Centre (IDRC) and the Canadian International Development Agency (CIDA) was implemented in India and Sri Lanka. Recipients from India, MSSRF and from Sri Lanka, Sarvodaya and Practical Action, worked with the affected communities since April 2006 to help reduce vulnerability of coastal communities to natural calamities.

The project introduced resilience building elements for strengthening the capacities of the coastal fishermen and farming communities to cope with any coastal hazards. The rehabilitation effort for the target community was implemented through three main strategies:

- By introducing some of the tested and replicable models such as habitat creation and restoration (mangrove and non-mangrove green belts) and access to village information centres (VICs).

- By identifying the needs and priorities of the communities and integrating these needs into such disaster planning and livelihood rehabilitation.

- By identifying cross-cutting issues such as gender (youth, women and men), developing partnerships, and holistic community development for better governance, while building the capacities of local groups.

5. Overview of the project

The specific objectives of the project are:

1. Identify suitable village-based sites for mangrove and non-mangrove bioshield development in the states of Tamil Nadu and Andhra Pradesh; and in the districts of Hambantota, Kalmunai, Batticaloa and Trincomalee in Sri Lanka.
2. Develop and demonstrate new techniques, methods and management strategies to rehabilitate, establish and sustain mangrove and non-mangrove bioshields.
3. Increase awareness and benefits for women and woman's groups, men and youth in technical, social and economic terms from the management of coastal forest bioshields and related resources.
4. Document and disseminate best practices of bioshield development and management through multimedia toolkit and distance learning among various stakeholders and VKCs/VICs.
5. Enhance the capacity of managers and government organisations, local NGOs, local elected officials, young scholars and scientists, administrators and policymakers in ITC- enabled coastal development and bioshield management through training, seminars, education and awareness programmes.
6. Strengthen the livelihoods of women and the poor in tsunami-affected village sites through improved access to natural resources and to information services.
7. Increase the quality, quantity and local relevance of VKC/VIC information content to improve conditions for both men and women.
8. Build the capacity of local organisations to prepare for and mitigate natural disasters using local knowledge and external expertise accessible through VKCs/VICs.

9. Analyse current policies and programmes relating to bioshield development and management and village information centres with reference to coastal disaster preparedness and livelihood development and advocate guidelines for better policy framework.
10. Analyse links between gender, livelihoods and natural disasters in tsunami-affected coastal villages, particularly in Sri Lanka.

6. The project's approach

The overall approach of the project was science-based, people-centred and process-oriented. The key elements of the process were expected to involve the following steps:

- Situation analysis and site identification;
- Participatory methods;
- Community-based institution building;
- Micro-planning; and
- Monitoring and evaluation.

7. Implementation

Sri Lanka activities were implemented by Sarvodaya and its local partners in a total of 12 tsunami-affected villages selected from four different districts: Kalmunai (Ampara), Batticaloa, Trincomalee and Hambantota. Two sites in Sri Lanka were managed by Practical Action formally known as ITDG. Practical Action played a key role in demonstrating and transferring disaster-resistant livelihood frameworks through pilot programmes and advisory support to the other Sarvodaya-managed project sites. Linkages were established with the Rural Enterprise Network (REN) for market access.

Indian activities were implemented by the MSSRF in two southern states of Tamil Nadu and Andhra Pradesh. In Tamil Nadu Vembar, Manamelkudi were covered. In Andhra Pradesh sites from Krishna and Kakinada districts were included. MSSRF implemented its activities through boundry NGOs: People's Action for Development India (PADI), Society for Participatory Research and Integrated Training (SPRIT) and Praja Pragathi Seva Sangham (PPSS).

8. Facing challenges

The initial years of the project, proved to be challenging as there were many practical difficulties that had to be assessed and understood prior to implementation. The difficulties mostly revolved around coastal land acquisition and ownership for both bioshield and information centres. In addition to these, relevance of meeting all project objectives in each of the identified villages and the risks of working in conflict zones (in Sri Lanka) proved to be challenging as well. But once the target groups were identified, the partners were able to identify suitable sites, resolve issues around land ownership, get access to land entitlements in order to implement planned activities and provide access to information centres. In the meantime, partners also liaised with local stakeholders, grassroot NGOs and government officials to network and build effective partnerships. The effectiveness of such partnerships was very evident in Sri Lanka, especially in the area of coastal resource management (CRM) and disaster preparedness planning.

9. Involving the communities

From the very beginning, target communities were involved in planning, implementing and monitoring all identified tasks. Various processes and practical approaches were adopted for building their capacities. Institutional building ensured transparency and provided a forum for participation and decision-making process for both men and women. Through local institutions, various sub-groups were formalised and were involved in the planning and implementation of different components of the project.

10. Meeting the needs of the communities

10.1 Providing information needs

Today, more than 23,000 coastal villagers have benefited from the project. More than 350 ha of bioshield and 16 VKCs were established in India and Sri Lanka. A network of different ICT tools such as public address systems, Internet radios, mobile phones, and VKCs were established across 22 villages of India and Sri Lanka, with an emphasis on those located in the remotest of the locations. During the project period, there were many warnings for tsunami and cyclonic storms that was disseminated to the locals through these facilities. On a day to day basis, these centres provide computer literacy programmes for the youth and children. They also provide a space for community discussions. Besides, information on many government schemes, service, employment and health related information, is also disseminated through notice boards and local community newspapers.

10.2 Rebuilding and providing alternative livelihoods

Efforts were taken by the partners to enhance fishing and farming practices of the lives of the poor and marginalised, including women and men. Disaster risk reduction (DRR) and resilience building is the core of this project. It was realised, however, that appropriate linkages are required while implementing alternative livelihoods. As a part of the project activity, focal points were recognised between disaster, livelihood, market linkages and coastal resources that helped to approach the livelihood intervention in an integrated manner, notable especially in Sri Lanka. Local specific detailed disaster preparedness and CRM plans were developed for each of the villages across Sri Lanka. These are now available at both the village and district offices. An important strategy adopted across the countries was to recognise the value of bottom-up approach for rebuilding existing occupations.

10.3 Flood mitigating agricultural measures

Farmers at project sites are now mitigating floods and improving irrigation systems throughout the year. Three key outcomes of the project is the community-based approach to reviving fallow land to agriculture, introduction of eco-friendly (organic) prawn farming and mitigation measures adopted for floods.

In a village of Andhra Pradesh, the local community is now cultivating two crop cycles of paddy and three crop cycles of shrimp over 300 acres of land, since 2008. These lands were once left fallow for more than a decade due to increased salinity, and lack of technical knowledge on shrimp farming. Likewise, villagers in Hambantota (Sri Lanka), that face more than a dozen floods annually, have now built drains to clear excess water into the sea, on time. Both these sites needed facilitation, negotiations and appropriate linkages with local stakeholders. This approach is now recognised by the Sri Lanka government which is planning to adopt this community based flood mitigation in all flood-prone regions.

10.4 Financing

Besides improving income generation of farmers, partners have also explored potential niche markets for linking producers. Small capital support was provided to the women groups which now ventured into other business such as fish vending, lime making, grocery shopping, coir making etc. Minimal interest rate is charged and funds are managed by the community as revolving credit. Repayment is affordable in comparison to the interest imposed by money lenders. This relief from money lenders has demonstrated to the communities that they can be less dependent on outside lending systems. Women have begun to question the funding system in their respective villages and have started negotiating with the traditional panchayat to bring down interest rates to affordable limits.

11. Monitoring and evaluating the project's impacts and achievements

The project initially conducted exchange visits to encourage cross-learning, joint internal meetings to review project status. Yearly advisory committee meeting were organised at which issues were discussed and solutions were sought.

The impact of the project and its achievements were subject to extensive monitoring; ongoing meetings with constant follow-up on the activities. A three-tier monitoring system was prepared with the participation of partners. The partners utilised two methods namely, logical framework analysis (LFA) and outcome mapping, which were consolidated into one reporting system. While the result-based LFA helped measure results, the outcome mapping probed to understand behavioural changes to greater depths. While this system was executed, the partners were given opportunities to discuss the field challenges and appropriate solutions were provided for effectively measuring the designed indicators. This rigorous task took more than a year.

The final project evaluation focussed on the resilience that the community has built during the stipulated time. Again, a participatory and utilisation focussed method was adopted so that the learning element can be built in. The assessment flagged many achievements and potential improvement areas for further strengthening. This exercise, although short, was useful to those who participated as they were able to use the findings in strengthening and wrapping up the final project activities. This evaluation also played a key role in identifying the need to document the achievements, good practices, and challenges from the project for wider sharing.

In the past four years, the project has looked into various aspects of resilience building to assist the community in building back their lives beyond tsunami. The various interventions were mainly to enhance their adaptive capacity to climate related disturbances. Although disasters bring about massive destruction, they also widen the opportunity to create, innovate, test and demonstrate new methods and approaches in reviving the communities. While the poor are vulnerable to handle disasters, the Project has broadened their views and showed them long-term mitigation efforts that they can execute themselves in the event of future disasters. Still, the process of building community resilience on a large scale requires time to put in place all the working systems ready for disaster recovery and risk reduction.

References

Cosgrave, J. (2007). "Joint evaluation of the international response to the Indian Ocean tsunami", *Synthesis Report: Expanded Summary*. London: Tsunami Evaluation Coalition.

Oxfam International (2009). "People-centred resilience: Working with vulnerable farmers towards climate change adaptation and food security", *Oxfam Briefing Paper* 135. 16 November www.oxfam.org(pdf file, 43 pages)

UNDP (2005). *The post-tsunami recovery in the Indian Ocean: Lessons learned, successes, challenges and future action*. (April), United Nations Development Programme Bureau for Crisis Prevention and Recovery.

World Economic Forum (2008). *Building resilience to natural disasters: A framework for private sector engagement*. The World Bank, United Nations International Strategy for Disaster Risk Reduction.

53

Networking and Partnerships in Rural Sri Lanka

Source:

Vishaka Hidellage *Practical Action, Sri-Lanka, 2010.*

1. Introduction

Partnerships and networking are important strategies being adopted widely in participatory approaches to rural development. These strategies are defined in a variety of ways depending on the context in which these are operated.

Although working together to help each other is an important aspect of traditional Sri Lanka, this is gradually disappearing with only remnants of the culture remaining. Social capital was built through local networks and partnerships and played crucial roles in the day-to-day life of the community. Today, these roles are diminishing. In some villages, however, people still adhere to traditional norms and practices in specific situations, e.g. village tank management.

In many villages, people have banded together to develop mechanisms for cooperation such as funeral, village temple societies, and cultural events. These popular societies are very active and function quite well in delivering their objectives. Their existence and success also indicates that the practice of working together is still respected in rural Sri Lanka.

There are many other networks and partnerships that also exist locally, although most are facilitated by the government or other external agencies for particular purposes. These mechanisms have limited coverage and are generally confined to helping members get access to services like government grants or inputs such as fertiliser subsidy or seeds.

2. Project experience

The post-tsunami project was designed primarily to identify and build on these local networks and partnerships and secondarily, to strengthen some of those already existing societies. Also, building suitable alternative mechanisms with the aim of improving local social capital was also considered.

Extensive awareness creation and community mobilisation paved the way for the community to identify the appropriate local networks and partnerships. Livelihood based networks existed in the villages, while networks on environment and information either did not exist or were defunct. The project focussed on either creating or serving these networks and partnerships to address their priority problems. When needed, agreements were forged to activate some of the defunct networks or partnerships so that they could take on similar responsibilities.

3. Building on existing networks and linkages

Paddy farming is one of the main livelihoods in Wanduruppa and surrounding villages. A key problem for the paddy farmers was inadequate infrastructure such as road to paddy field and lack of irrigation water. Existing roads and paths were in bad shape and were submerged during frequent flash flooding. While frequent flash flooding is also a key problem, the various stakeholders could not agree on a universally accepted solution. Through the project's facilitation and extensive consultation, the stakeholders eventually arrived at various options.

Building the foundation

Networking is the coming together of individuals and institutions or organisations of diverse interests, to achieve common objectives based on common agenda.

Partnership is defined as coming together of two or more parties who have different expertise, to work jointly on an agreed programme, with each party undertaking specific aspects and responsibilities of the work.

Each village has a farmer organisation that operates under the supervision of Agrarian Service and Irrigation Departments. The village organisation was mobilised to identify ways to address the problem. Farmers' organisations from 10 affected *grama niladhari* (GN) divisions formed a network for developing a coordinated approach to flood mitigation.

The authorities and officials were mobilised to improve existing relationships with the community for mutually beneficial partnership. The community established stronger links with the Divisional Disaster Management Coordination Committee (DMCC) too, hence, active participation by all local stakeholders.

The improved relationships through facilitation helped both parties understand each other's view of the different aspects of the problem and the solutions. This ultimately led to the government's endorsement of the practical and viable suggestion proposed by the community—building on a traditional approach to flood mitigation. The community in turn, agreed to involve government officials in supervisory and regulatory capacity to implement flood mitigation plan.

> Building up and strengthening the collective capacity of the individuals in the committee resulted in:
> 1. Sustainable local institutional capacity to maintain the appropriate infrastructure.
> 2. Active and capacitated local network with knowledge and experience to collaborate on other similar work in the interest of the village.

In the project implementation, a more active and task-specific Flood Mitigation Committee was formed with a team selected amongst farmers. This committee was involved in research, discussions, debates and decision-making which finally led to redesigning of roads to withstand floods. The committee led and coordinated local inputs to the road construction.

The road was constructed through partnerships that the committee built with local and external actors. This resulted in transfer of knowledge and experience to the members of the committee. Community and the local officials had specific responsibilities, wherein community level responsibilities were also divided among the different members of the community. These included monitoring water levels for early warning of floods, coordinating with the office of Divisional Secretariat (DS), and the related agencies and implementing mitigation activities with the supervision of DMC committee.

4. New partnerships

Creating new partnerships is necessary to bridge the information and build new and beneficial partnerships beyond village level. To do this, the Participatory Market System Development (PMSD) approach (discussed earlier in the source book) was adopted to connect local products with external stakeholders for sustainable links with dynamic markets.

PMSD facilitated the entry of external inputs that helped in improving quality and diversifying local products. In addition, the approach also provided opportunity for producers to market their produce outside the village, further strengthening links with external markets. In some instances, this meant gradually reshaping any existing external links that community already had with markets. They gained knowledge, experience in business development, and bargaining power, hence, transforming their ability to develop relationships that provide more equitable market linkages.

Another example is the solution to the flooding of Karaitivu (post-conflict area in East Sri Lanka). In comparison to Wanduruppa, flooding in Karaitivu was less serious. In the absence of a local network, a village disaster management committee was created. This community-based organisation (CBO) committee partnered with the local authorities to create awareness among the community and get inputs to facilitate a solution to their flooding problem.

Against the background of a relatively inactive community networking and the strength and capacity of the newly formed CBO, the Divisional Secretariat and Pradeshiya Sabha (PS) played crucial roles. Local authorities led the implementation and their leadership effectively increased participation of the community than the CBO could.

This community behaviour in post-conflict Karaitivu may be attributed to the erosion of social capital caused by the long-standing civil unrest. Discussions with CBO and local authorities led to the preparation of flood mitigation plans. These included rebuilding of the flood designing floodgates and operating them. Unlike in Wanduruppa, a local NGO District Economic Social Mobilisation Organisation (DESMIO) supported

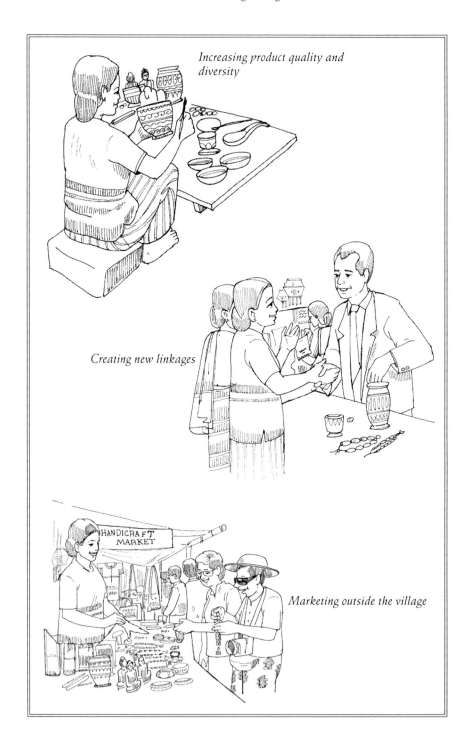

the CBO in increasing community involvement and mobilisation, and constructing the flood mitigation structures.

Technical support from the PS was key to ensuring quality and proper construction. The DS helped to resolve some of the community problems such as conflict due to higher demand for hired labour and land issues among community members. The initiative ended positively and the flood mitigation plan was eventually implemented. On the community's behalf, the CBO monitors flood level, informs GN and DS early and gets their endorsement, and coordinates local inputs for operating the flood management system. It also assists local officials in maintenance.

5. Strengthening existing village level hierarchies

The voice of marginalised and vulnerable groups within traditional and non-traditional village networks is rarely heard. In relation, there are few opportunities for partnerships based on their specific needs. Through the Post-Tsunami Project, these marginalised groups were identified and brought together. Their networks were also strengthened giving vent to their individual and/or collective voices without antagonising the established power groups.

The women in the coastal village Wanduruppa showed the positive social and economic changes that their voicing out brought about. These women have no other livelihood options and have been producing handicraft based on raw material from coastal vegetation. The traditional designs and products were limited and confined to local markets. Moreover, women worked separately and earned individually at cottage industry level.

The tsunami's impacts on coastal vegetation affected livelihoods of poorer women in the village who were dependent on agriculture. It affected the

---

Rural Enterprise Network (REN)

A facilitated network of small producers at national level, who have come together for competitive marketing of their produce.

supply/input in the business as well as the markets due to the disrupted socioeconomic setup. Compared with fishing and agriculture, handicraft-based livelihoods were not considered important. Thus during rebuilding, the community ranked handicraft revival very low along with coastal green belt development.

In the earlier example, a PMSD approach was adopted and a range of actors within the total market chain came together to contribute towards sectoral improvements. Aside from women, the actors included raw material suppliers, buyers of products, consumer groups from local areas, and service providers such as National Craft Council (NCC), Agromart Foundation and Rural Enterprise Network (REN).

Through this initiative, the following consequently took place:

- The stakeholders understood the value of cooperation over competition. Each player became interested in networking for increased overall productivity and profitability because it helped to reap maximum individual gains, too.
- The handicraft producers worked together to revive and strengthen the defunct handicraft society; developed their business skills. During business development planning, they identified priorities and key gaps in asset building; and due to awareness, strengthened the traditional women's network on collective purchasing and selling.
- New partnerships emerged or were formalised. Some women took on specific roles (trainers, market liaison, quality assurance, etc.) but all shared the benefits. They helped build a critical mass to respond to market demand by, in turn, building capacity in neighbouring villages. Moreover, they shared their customers with the entrepreneurs engaged in ecotourism and built on each other's experiences for improving strategies for mutual benefit.
- Women networked with other handicraft societies in other villages and with REN; thus becoming more aware of problems and opportunities and responding better to markets. The DS through the National Craft Council (NCC) has recognised their capabilities and they now act as trainers for the disabled who are interested in

producing handicrafts for livelihoods. This recognition served as promotion of the community's sales centre as well.

6. Building on the community's priorities

It is difficult to mobilise people to work together when the issue is a low priority or when they believe that the issue can be resolved through a different agenda. Particularly challenging was the coastal green belt development component of this project that included home gardens as a secondary bioshield.

Community based strategy envisaged through the project did not work well. Partnerships or networks on related issues prevailed at the local level, but were not adequately committed to the issue. For example, a biodiversity task force (ecological conservation society) formed through a donor-funded government initiative, did not have the recognition, capacity or attitude needed to assume leadership of a relatively long-term green belt initiative. Most importantly, the community did not prioritise this issue although they understood the links between coastal environment and the bioshield.

To get better interest and participation, the project introduced perennial crops such as coconut, cashew and timber species with food and income potential. In Karaitivu, the positive outcomes resulted from a high input from the project rather than motivation of the community *per se*. This resulted from pending ownership issues with relocation of communities away from the coast. Partnership was developed with a local network formed with the leadership of the Divisional Secretariat and include local actors such as PS, CBO, forest, cashew and agriculture departments for supervision and maintenance until the green belt could be safely left alone.

The home gardening programme introduced as secondary bioshield was not adopted as expected. It was initially promoted as an additional livelihood option but the organisers failed to consider that the fishing community might not be interested to take up agriculture. This was especially so in the East where livelihood demarcations are much stronger.

The farmer women society at Wanduruppa, organised by the Department of Agriculture to promote agriculture related work, was used to promote home gardening. The initiative did not result in development of expected processes and common agenda at the local level for sustainable home garden development. Certain individuals, however, mostly from poorer section of the society have developed successful and sustainable home gardens from which they get food and income.

There are number of reasons for relative low success rates of bioshield development initiatives. First, communities were never convinced of the need for intensive resource inputs for development and maintenance of bioshields as against allowing natural process to take their course. Second, complications arising from land availability and ownership in green belt development locations were barriers to getting required commitment from the community. There are, however, some good examples particularly from Andaragasyaya where perennial home gardening was promoted in private lands with water issues addressed through the introduction of rainwater harvesting and pot irrigation.

7. Recommendations

- Experience shows that community and local resources can be mobilised to seriously rally around and find ways to address crucial common issues.
- The networks that exist locally usually reflect traditional local hierarchy and have equity issues, and projects should be conscious of this and have adequate measures to address this in an empowering and harmonious manner.
- Local authorities and officials too can be persuaded to work with community and share some of their responsibilities with community if they are sufficiently convinced of community capacity and commitment.
- Projects can become a useful facilitator to forge these links and facilitate partnerships and networks that will help continue local development initiated through project interventions in a sustainable empowering manner post-project completion.

54

Village Development Planning Processes in Sri Lanka

Community-based approaches tap into local knowledge and experience to arrive at relevant development solution.

Source:

Vishaka Hidellage, Vajira Hettige and Buddika Hapuarachchi *Practical Action, Sri Lanka, 2010.*

The concept and application of participatory village development planning is discussed as an approach with the potential to empower communities, strengthen their social networking and capacity to negotiate with government officials to pay adequate attention to village level priorities, and to be able to better bargain with actors in external market/service value chains. The article provides details of piloting the concept in post-tsunami rebuilding context in Sri Lanka and lessons learned.

1. Context

Community-based approaches are considered an important tool in bottom-up participatory development. Its value with respect to tapping into local knowledge and experience to arrive at relevant solutions, local capacity strengthening and building local ownership and commitment are well known.

Devolution of certain Central government functions and responsibilities to the local governments is considered bottom-up development in the government sector. The Government of Sri Lanka adopted devolution in 1987 as a solution to the armed conflict that prevailed for almost 20 years then in the country. The devolved structures and systems at provincial Provincial Councils (PC) to local Pradeshiya Sabha (PS) levels were put in place. After devolution, the decentralised government systems of Sri Lanka that prevailed before continued to function more or less in the same manner. The office of the Divisional Secretariat (DS) that is line managed by the Central government continues to carry more power than the PS. Similar to the experience of other South Asian neighbours, the local level authorities are confined to certain service delivery and maintenance functions and occasionally get engaged in implementing projects designed at national or provincial levels.

Local government authorities (LGAs) need to play a more proactive role and engage in designing local development to push a locality, and in turn, a country towards sustainable development. LGAs are well positioned to engage in participatory development than other levels of government planning. Viable models of village development that LGAs can effectively engage in would be essential in making a decentralised development work. It is hoped that lessons learned from piloting participatory development in two tsunami-affected villages in Sri Lanka; Wanduruppa in the South, and Karaitivu in the East, will contribute to achieve this goal.

2. Introduction to village development planning process

After the tsunami in 2004, the government encouraged the formation of village development committees (VDCs) through which government assistance may be accessed in addressing community grievances. The project saw this as an opportunity to extensively involve the government officials—to expose them to working with the communities and community institutions and thereby influence them to engage in participatory development.

The project adopted a three-steps methodology to engage stakeholders in participatory village development planning:

1. Initial discussions to form a VDC. The VDC should consist of village representatives and village level government officers as per government directive.
2. Planning process with community groups and local government officers to prepare the village development plan (VDP).

3. Local capacity development, coordination, implementation of agreed actions and monitoring. The DS, the *grama niladhari* (GN), PS members, local officials of government line ministries and other key stakeholder groups participated in the planning process. Participation of the DS was particularly crucial because the DS has powers to authorise or sanction village level initiatives.

The planning process included holding series of consultations with divisional and village level government representatives, leaders and members of CBOs, with various livelihood groups, low income families, women groups, senior citizens, youth groups, etc. The process captured priority needs for village development as perceived by each stakeholder. Methods such as focus group discussions (FGDs), daily routines capturing, resources and land use mapping, transect walks, interviews, etc., were used in this process.

The project further adapted the VDC concept introduced by the government, to meet village development needs. As a result, the VDCs functioned as an institution recognised by local officials representing community in the village and their diverse needs. VDC was the key institutional mechanism which facilitated village level planning process in each village.

3. Experience of formation of VDC and developing VDP

The tsunami devastation caused the pouring in of huge humanitarian support to Sri Lanka. This has resulted in high expectations and negative attitudes among affected communities. Simply waiting for assistance to be delivered to them and even demanding and complaining rather than working together with officials and being party to the solutions were common among the affected. In this context, the project initially sought participation of the GN office and key village leaders to identify main village development priorities. In a community meeting, these priorities were presented and members of the VDC were appointed. The VDC included representatives of all CBOs in the village (e.g., farmers'/fishers' associations) including women's representation. The GN officer, agriculture research and production assistants (ARPAs), Samurdhi officer of the division, etc., were invited to the community meeting to obtain advice, endorsement and support.

Village Development Planning Processes in Sri Lanka

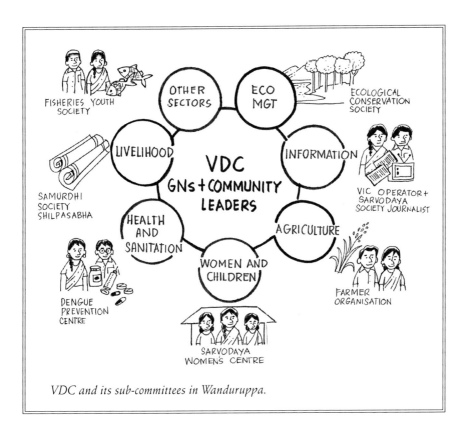

VDC and its sub-committees in Wanduruppa.

The VDC members worked in sub-committees based on livelihood, common infrastructure and environment sectors. Each sub-committee had extensive consultation with communities preparing detailed suggestions for development of their respective area. The needs and suggestions identified by the sub-committees were presented to all stakeholders including the DS, the PS and the other development agencies. Through this process of interaction, discussions and negotiations, the VDPs were formulated.

Discussions with wider stakeholders included resource implications and limitations leading considering alternative options. Cultivation of traditional paddy to tackle resource issues was a suggestion offered in Wanduruppa. The multi-stakeholder planning process also led to getting support of government and other agencies to implement the

VDPs. In Wanduruppa, the Disaster Management Centre (DMC) took responsibility to lead the disaster risk reduction (DRR) planning process, the Divisional Secretary and Irrigation Department took the lead for preparing flood mitigation mechanisms, and the Agriculture and Agrarian Services Departments agreed to include Wanduruppa in home gardening programmes. The GN officers and political leaders agreed to repair village roads, and the National Crafts Council committed to provide technical and administrative assistance to strengthen the handicraft sector. Some of the INGOs too pledged to support VDP. For example, in the planning process in Wanduruppa, Helpage Sri Lanka took the responsibility of developing the village disaster preparedness plan for Ambalantota South GN division with DMC, Mercy Corps agreed to develop eco-tourism sector in and around the village, and Caritas agreed to strengthen flood preparedness activities.

The interactions during the village development planning led to its strengthening, increase in confidence, capacity and bargaining power of community and VDC as well as VDC gaining recognition among other

stakeholders. The process gave special emphasis to include voices of the marginalised (women, disabled, youth and elderly) through separate discussions within each such group. Interestingly, the youth and the elderly stated that this was the first time their views were considered as important, as the elite/dominant groups in the village usually made decisions on their behalf.

4. VDC in a post-conflict situation

The establishment of VDC in Karaitivu took longer than in Wanduruppa. The village situated in the conflict affected East has been dependent on relief long before tsunami. The conflict situation had a negative effect on getting community cooperation which was compounded by post-tsunami expectations. Mobilising community and LAs towards participatory development was a challenge. The caste systems and social recognition patterns in Karaitivu added to this challenge; engaging women and low income groups such as fishing communities was a slow process. It required extensive and committed community mobilisation with longer term view focussing on trust and social capital building. This was done through a local NGO based in Karaitivu. Capacity of the NGO was built to carry out social mobilisation for development rather than for the distribution of relief which they were used to. The project had to invest more time to work with the NGO, community and LAs in Karaitivu compared to Wanduruppa.

Certain strategies were used to earn trust, confidence, and commitment of people and officials. Construction of flood control mechanisms, and access roads for evacuation during the flood season was an effective incentive proposed during the planning process and made it less difficult to get the involvement of the local authority to support the VDC and subsequent community proposals.

5. A few things to keep in mind about VDCs and planning process

- The VDP process is a viable option to facilitate links between local authorities and communities and engage in participatory local level development planning.

- More attention and resource allocation is essential to implement the same process effectively in conflict affected areas; extensive and sensitive mobilisation of community and other actors, with a long-term view is essential in development work in post-conflict areas.
- The VDC is not a standard mechanism or social structure but is a process that will vary from village to village depending on the local context and operation of government mechanisms.
- The VDC is a suitable mechanism for collective capacity building and liaising with external agencies. However, it should have adequate precautions to prevent manipulation of forces who want to prioritise other agendas over overall community well-being.
- Capacity to influence, bargain (negotiate) and understand local policies and policymaking are essential to the VDCs for them to effectively carry out their functions.
- Special strategies are needed to ensure that the VDC does understand and take equity and inclusion as part of its responsibility.

55

Local Participation in Mangrove Management

A conventional village level institution versus a contemporary village level institution.

Source:

J.D. Sophia, R. Ramasubramanian and A. Sivakumar *M.S. Swaminathan Research Foundation, Tamil Nadu, India, 2010.*

1. Introduction

The MS Swaminathan Research Foundation (MSSRF) promotes sustainable rural development and greater community participation across gender and economic classes using a people-centric and bottom-up approach in project interventions. In 1997, it piloted the use of village level institutions (VLIs) in the Joint Mangrove Management (JMM) initiative to enhance the capacity of the local community, the Forest Department, and other agencies in restoring, conserving, and sustaining mangrove wetlands through participatory analysis and action. It eventually put in place 45 VLIs in Tamil Nadu, Andhra Pradesh and Orissa. Fourteen of these VLIs, eight in Tamil Nadu and six in Andhra Pradesh, were formed under the auspices of this project.

2. Conventional *versus* contemporary VLIs

The MSSRF studied existing VLIs in the coastal villages and found that although the vast majority of the rural population is still not organised into groups, the coastal villages of Tamil Nadu and Andhra Pradesh have formal internal governance structures such as the *panchayat*. The *panchayat* structure and processes evolved because of the need to prevent and resolve intra- and inter-community conflicts. It also fulfilled other functions like maintaining community rituals and structures, conferring membership and belonging, resources distribution, redressing grievance, and dispensing justice.

The village level institution (VLI) is a formal body that ensures sustainable development in the villages. VLIs empower and enable villagers to take charge of their own well-being rather than rely on external agencies.

The MSSRF found the conventional VLIs to be very active in carrying out their functions and decided to revive them for the project activities if they change their policy and include women as members. However, the idea was unacceptable to the *panchayat* members. For them, women should have no role in community management initiatives.

The MSSRF decided not to impose its preference but drew an important lesson—that although the traditional VLIs had strong controlling mechanism, it has weaknesses that necessitated the formation of new VLIs. Active participation of rural people can only be brought about by organising VLIs that are democratically governed, autonomous and are capable of overcoming weaknesses.

3. Process of VLI formation

The objectives of VLIs organised under the project are: a) to provide a platform for people to participate in project planning, implementation, and monitoring; b) to provide an opportunity for women and marginalised communities to participate in decision-making; and c) to create ownership of all project activities implemented in the village.

Weakness in traditional panchayat

1. No space for women in the Governing Board and the Executive Committee.
2. Limited opportunity for external people to interact with General Body.
3. Traditional institution interested to play a token role in natural resource management including bioshield development.

Organising new VLIs was process-intensive (Figure 55.1). To educate people and motivate them to organise, the implementers employed the following tools: one-to-one interactions, group discussions with various stakeholders, village meetings, cultural events, and exposure visits to villages where VLIs are functioning. Exposure visits as cross-learning exercises to successful locations, i.e., JMM sites, were eye openers and made the community agree to have a VLI organised in their place. In the case of Andhra Pradesh and one of the sites in Tamil Nadu, the government and NGOs had already formed village development committees (VDC). In Manamelkudi, Tamil Nadu, the entire process started from scratch.

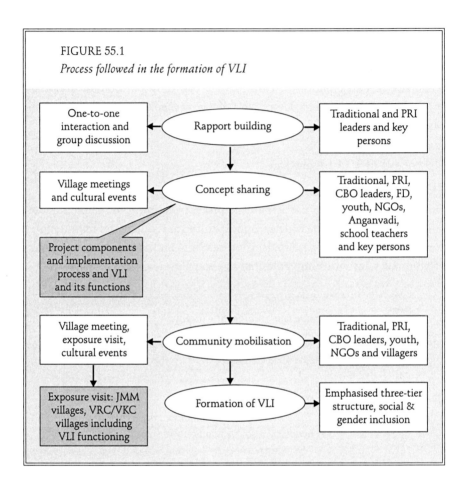

FIGURE 55.1
Process followed in the formation of VLI

4. Structure and functions of VLIs

Figure 55.2 shows the three-tier structure of the VLIs adopted in all the project villages. The bottom tier, the General Body (GB), is comprised by adult male and female representatives of each household in the village. The GB acted as the decision-making body for micro plans, budget and expenditure. The middle tier is the Executive Committee (EC) that consisted of selected representatives from the GB who are well informed about the village and are committed to village development.

From among the EC, the top tier consists of the office bearers (OB), i.e., president, secretary, were selected to lead the VLIs with the consensus

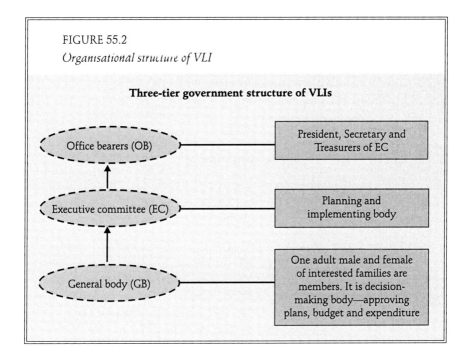

FIGURE 55.2
Organisational structure of VLI

of the GB. The community agreed on the need for gender equality and gave 50 per cent representation in the EC to the women. In some areas in Andhra Pradesh and Tamil Nadu, women members of the VLI outnumbered the men and at least one of the OB was a woman. The GB met once in six months while the EC met on fixed dates every month to review their progress and to make future plans. Apart from the monthly meetings, special meetings were convened as and when required. Decisions taken and resolutions passed in the GB and EC were recorded in the minutes of meetings.

5. Roles and responsibilities

The VLIs played a vital role in planning and implementing the project components, namely, community mobilisation, bioshield development, management of the village knowledge centre (VKC), and livelihood interventions. They facilitated the community needs assessment and livelihood analysis. The VLI ensured the proper identification of potential members—the poorest of the poor and women-led households—and prioritised the groups that were to receive livelihood strengthening services. The funds provided by the project to implement activities and the revolving funds were meticulously managed by the EC. The EC also distributed the revolving funds to new targets and ensured that repayments were made.

6. Impact of the VLIs on the community

Yet another significant change that the VLI concept brought into the community was narrated by Mr.Sundaramanickam, a leader of the Thangamalpuram VLI:

> People in our village used to be self-centered and never showed interest in community development. This situation was aggravated by the heterogeneity of the people owing to the caste system. When the People's Action for Development (PAD) introduced the concept of VLI, it was formed with only the men as members. It was restructured with the advent of this project by mainstreaming gender and the poor into the system.
>
> Initially, people did not participate but did not take long to convince them especially when we started reaping the benefits from the VKC, the interventions that strengthened our livelihoods, and the new income generating activities. Those who were only concerned about their own progress realized the fruits of collectiveness and became concerned about community development. The VLI also encouraged networking with external agencies. For example, we expressed concern on the safety of our drinking water after the VKC helped educate and inform us about it. Now, our VLI is networking with a line department to establish a low cost water treatment plant in the village to supply safe drinking water to our community.

7. Creating space for women

VLI members of other villages in Tamil Nadu also related their experience on how the entry of women into public spaces, especially in traditional institutions, was once restricted. They said that it was often asked, *'Aangal pesum podhu idathil pengalukku enna velai?'* (What is the role of women in public forum where only men should interact?) Now, the men are allowing women to participate in public forum after demonstration.

Twenty years of enmity vanished and unity flourished

Two decades back, internal clashes divided the Kattumavadi villagers into two traditional institutions that were nevertheless, performing parallel functions. The enmity severely compromised group dynamics because group members ceased to mingle with each other. When the MSSRF and the Society for Participatory Research and Integrated Training (SPRIT) approached the villagers to talk about the project, the members of the two groups did not immediately respond. But the MSSRF and the SPRIT did not give up. Instead, they continued to talk about the project concept to the two groups separately. The villagers did realise the protective functions of mangrove forest especially after the 2004 tsunami, appreciated the role of information centres in disseminating early warning, and were impressed by the project components but still they could not seem to reconcile their differences.

This changed after an exposure visit to the Pichavaram mangrove forest and the MGR Nagar village. The cross-learning exercise gave the villagers insights on JMM, VLI formation and functions, and the chance to interact with the people who encouraged them to organise a VLI in the hamlet. They finally decided to set aside differences and organise the VLI on the condition that both groups are equally represented in the organisation. The VLI had a three-tier structure. At the first level was the General Body which was composed of one male and one female from each of the 110 households, the Executive Committee with 15 men and 15 women as members, and the Office Bearers comprised three men and two women. The women now play very dynamic roles in community management. The VLI brought sweeping changes to the hamlet and caused 20 years of enmity to vanish and unity to flourish.

— Mr. Kandasamy, Mr.Sivasami Pillai and
  Mr. K.T.Gandhi—VLI community members

In Kattumavadi, women participated in the monthly meetings of the *panchayat* where their ideas were heard by the traditional leaders. They share their grievances which was normally represented only by the male members of their families. These changes in the practice of the traditional *panchayat* attributed to the leadership demonstrated by women in the VLI.

8. The way forward

The participatory approach used in planning, implementing, and monitoring project interventions is the key to greater involvement and response of the community to the VLIs. Demonstrating social and gender inclusion in the VLIs through project implementation enhanced equality

Cultural events were one of the tools used to educate villagers on the role and function of VLI for the poor and women.

The VLI EC meet once a month while the GB meet once in six months on fixed dates. Decisions made were recorded in the minutes of the meetings.

and equity in the village. Institutionalising the monthly VLI meetings and quarterly monitoring meetings increased project performance and improved community participation.

The VLIs became highly effective in building the livelihood assets in their villages. The villagers of Tamil Nadu and Andhra Pradesh said that the physical infrastructure, i.e., computers, enabled the villagers to access the Internet while a public address system helped disseminate real time information, i.e., the weather forecast to fishermen.

In the long term, the community aims to establish a mangrove and non-mangrove bioshield as a natural barrier to protect lives and the livelihoods of the community. They plan to integrate the lessons and similar activities to enhance fish resources in the community. The project investments will strengthen the livelihood assets and help earn additional income especially for the poorest of the poor and the women-led households in the years to come.

The VLIs set up village knowledge centres equipped with computers that have access to the Internet and public address system used to give real time information to villagers.

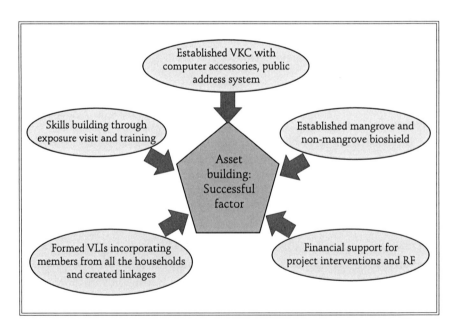

The VLIs have become a social capital showcasing democratic functioning through enhanced transparency, accountability, gender mainstreaming, and social inclusion. The villagers acknowledged the effectiveness of this 'social capital' and how it has empowered people because of its efficient project design, bottom-up and demand-driven approaches, and its services. Similarly, the linkages created with the line departments empowered the VLIs to negotiate and access services and entitlements.

Community mobilisation, project monitoring, exposure visits, and skills building programmes ensured the people's participation and decision-making in the project. This in turn enhanced their human capabilities after they learned new knowledge and skills in community-led efforts.

Today, there is a momentum building up in the villages and the people are primed to take action for their own development. A slow change can be observed as the VLIs are efficient in managing the resources, maintaining transparency by sharing information, taking appropriate decisions and delivery mechanisms. It is evident that people-centered approaches that are systematically implemented, emphasise social and gender inclusion, employ relevant interventions that meet community expectations and needs and will encourage high level of people's participation and greater sustainability in the long run.

56

Unleashing Women Power in Tamil Nadu

Women's participation and mobilisation efforts during disasters helped increase resilience of their community.

Source:

Asha Sitaram *Swayam Shikshan Prayog, Mumbai, India, 2010.*

1. Background

The 2004 Indian Ocean Tsunami devastated thousands of coastal communities and resulted in the death of more than 225,000 people, over 18,000 of whom were from India. Tamil Nadu was one of the most affected with about 8,000 deaths. In its Nagapattinam and Cuddalore districts, up to 196,184 and 99,704 of the population, respectively, suffered direct and indirect damages.

The number of deaths among women (25-300 per cent more than men) in the two districts underscored a gender differentiated impact of the tsunami. Women were more vulnerable to the killer waves because of their familial roles and the nature of their livelihoods. As the family's caregivers, many of them stayed behind after the first wave to look for their children and other family members. Moreover, while many men were out at sea fishing, a large proportion of the village women were at the beach conducting their daily livelihood activities of purchasing and cleaning fish in order to sell at local markets. On the other hand, men at sea were relatively safe from the waves that only passed under their fishing boats. These same waves hit the shore with full force, swept away and killed many of the women.

While the Tsunami brought horrific destruction and suffering, ironically, it drastically changed gender roles and the local power structure in the two Tamil Nadu districts, unleashing the potent and critical role that women have since played in the reconstruction and development of affected communities. A wealth of lessons can be drawn from their story by people and groups who are in relief work, post-disaster reconstruction, and community development. More importantly, this case study shows how important it is to understand how women can and should be included in the process of recovery and resilience building.

2. How women organised after the tsunami

Several months following the Tsunami, women slowly began to play a larger role within their community. One of the biggest outcomes of the Tsunami in these communities was the impressive expansion in abundance and strength of women's collectives. Women, on a large scale,

Unleashing Women Power in Tamil Nadu

Livelihood-based activity groups enabled women to collectively demand assistance from relief organisations to continue their livelihoods.

The Keelamoovarkarai village disaster Task Force

An exposure visit by Arogya Sakhis for Health Awareness and Action (ASHAA) members to Vellakoil village led to the creation of the disaster Task Force of Keelamoovarkarai village in July 2007. The women mobilised the community and raised awareness about the importance of a village based disaster Task Force. Of the willing participants, they selected 15 female and 15 male members for the Task Force. The task force members were divided into the five subgroups (early warning, rescue, first aid, relief and monitoring). Then, the Keelamoovarkarai Task Force trained two other villages in disaster Task Force organising. It also provided information to all village community members on the critical aspects of the Task Force, re-enacted the tsunami, and showed how the disaster Task Force could save lives. Their involvement also gave them the confidence that they did not have in the past. They now speak in public and at trainings.

began to organise into collectives and to mobilise around their needs and the needs of their communities.

The women formed groups after incentives were made available only to those who shared the same income generating activity and who grouped together to voice their collective needs. In other cases, the formation of women's livelihoods-based activity groups was triggered by the negligence of relief organisations to meet the needs of those who lost their livelihoods during the Tsunami. In some fisher community villages, for example, the focus of relief organisations was on providing for the loss of fishing boats and nets. However, this solved only the problems of the men and neglected the needs of women who had also lost the equipment required to continue their livelihoods such as fish vending. Consequently, women fish vendors gathered into activity groups in order to collectively demand for compensation and livelihood related supplies to restart their businesses. Activity groups provided an opportunity for women to collectively solve livelihood related problems and to also collectively save money and take loans for productive and personal purposes.

2.1 Self-help groups (SHGs)

After the immediate relief phase following the tsunami, NGOs began to gather women into SHGs focussed on savings and credit activities, and to encourage solidarity and collective problem solving. SHGs were already in existence then, although in significantly smaller numbers. Following the expansion of the SHG movement, many if not most women are now saving money and have greater knowledge on money management. They are also able to take loans for their personal and livelihood activities and are more financially independent.

2.2 Disaster Task Force

Following the tsunami, disaster task forces were created in many villages specialising in any of these concerns: early warning, rescue, first aid, relief, and monitoring. Women and men were divided equally into these subgroups based on their strengths and interests.

3. ASHAA groups

Arogya Sakhis for Health Awareness and Action (ASHAA) groups were formed starting in May 2005. The Swayam Shikshan Prayog (SSP) began working with communities and women soon after the tsunami and together they identified unaddressed needs in the health and sanitation sector. Some women were identified and trained as health guides who provided training and awareness on health issues to their peers and this eventually lead to the official formation of ASHAA collectives in August 2005 in Cuddalore and Nagapattinam.

Changes reported by women grassroots members

	More	No change	Less
Confidence	97.6%	2.4%	0.0%
Speaking in public	69.5%	20.7%	9.8%
Influence over personal and family decisions	68.3%	24.4%	7.3%
Participation in village activities	69.5%	15.9%	13.4%
Social status/respect	90.2%	7.3%	1.2%
Environmental awareness	92.7%	7.3%	0.0%
Awareness of health issues	95.1%	4.9%	0.0%
Awareness of water and sanitation issues	97.6%	2.4%	0.0%
Awareness of women's issues	97.6%	2.4%	0.0%
Knowledge of money management	70.7%	20.7%	8.5%

Reason for increased confidence	Respondents (%)
Participation in meetings	78.0
Knowledge and skill gained	64.6
Improved speaking skill	58.5
Health, water and sanitation knowledge gained	17.1
Helping others through ASHAA	14.6
Knowledge on government schemes	12.2
ASHAA group activity	11.0
Meeting government officials	11.0

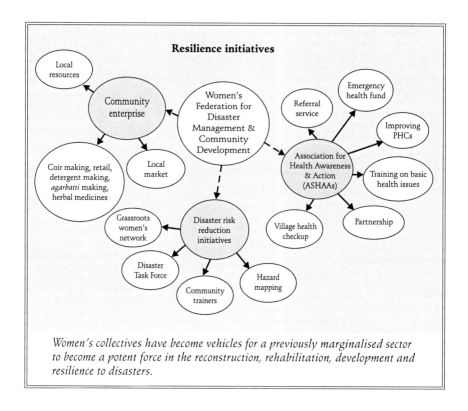

Women's collectives have become vehicles for a previously marginalised sector to become a potent force in the reconstruction, rehabilitation, development and resilience to disasters.

SSP linked the women to primary health centres (PHCs), government offices and other organisations while building their capacity to sustain these relationships and to lobby the government and hold its institutions accountable. SSP also facilitated and provided training for the organisation of health and sanitation related activities, such as health checkup camps, food *'melas'* (festivals), health awareness talks, and vermin compost projects. ASHAA members have also been trained on new livelihood activities and have begun to earn income from *agarbatti* production, coir making, vermin compost production, sale of sanitary products and herbal medicines, and production of seashell products.

As ASHAA members are for the most part SHG leaders, the two groups are very closely linked. The creation of ASHAA has had a significant impact on the activities of SHG members. Prior to ASHAA, members were focussed mainly on savings and credit activities but now many SHG members are involved in health, sanitation and general village development activities.

ASHAA groups later unified under the Women's Federation for Disaster Management and Community Development, with each district having one federation. The federations coordinate, strengthen and standardise ASHAA group activities across each district.

4. Elements for increased resilience to disasters

Organising into collectives has not only allowed the women to save and take loans from one another for both productive and personal purposes. It also provided the vehicle to involve themselves in the reconstruction, rehabilitation and development of their communities. More importantly, it has increased their communities' resilience to disasters. The women leaders and women's groups in the Tamil Nadu villages see the link between the role that they have played in their village after the Tsunami and their communities' increased resilience to disasters. What changes took place such that organising into collectives enhanced communities' resilience to disasters?

The women have included health awareness and action among the areas in which they initiated activities.

4.1 Stronger networks

Women's networks in affected communities were strengthened after the Tsunami through increased involvement and participation in women's collectives that were organised into federations. The federations served to reinforce and strengthen the women's collectives. Stronger networks have equipped the women to make collective demands to relief organisations and government agencies and to provide a coordinated response to their communities such that resources and aid are used most efficiently and effectively.

The women's groups have also strengthened informal community-wide networks, even with groups that have men or mixed members. In particular, groups that are involved in community development activities have greatly contributed to the strengthening of both inter- and intra-community networks. Men now recognise that these women's groups are beneficial to their communities. These groups are very active in the village, therefore they are well known, trusted, and have strong links to all community members. Stronger community links are critical to successfully mobilising people pre- and post-disaster.

Stronger networks also improve access to information such as government programmes. This is critical to disaster resilience as there are now more concrete and effective channels of information flow on disaster warnings as well as governmental and non-governmental relief schemes and programmes following a disaster.

4.2 Alternative livelihoods

Another measure of disaster resilience is the capacity of community members to return quickly to income generating activities following a disaster. Traditional livelihoods of fishing and farming are very reliant on environmental conditions and are vulnerable to natural disasters. Therefore, the creation of alternative sources of income is an important component of disaster resilience. After the tsunami, many organisations trained the women on alternative livelihoods such as *agarbatti* production, coir production, pickle production, and seashell product selling. They have taken these up as livelihood activities either as individuals or as groups.

4.3 Activity groups and federations

Should another disaster strike the tsunami-affected communities, livelihoods-based activity groups and their federations will also provide a support network through which affected people sharing the same livelihood activity can collectively voice their needs and make demands. This will enable affected communities to 'bounce back' more easily in terms of restarting their livelihoods and accessing necessary resources and compensation.

4.4 Women's ability to make demands and lobby government

After the tsunami, women developed the capacity to make demands and hold the government and other institutions accountable to their needs. This is one of the most important outcomes of the tsunami and will inevitably be critical to women's and communities' resilience to future disasters. While relief organisations and government bodies are

The most important outcome of the tsunami is that the women found their voice and developed the ability to make demands from government and other institutions and lobby for what they believed is best for their community.

continually refining their approaches to disaster response and recovery based on past experience and knowledge, community voices need to be heard in order to ensure that disaster response is conducted effectively. For example, women in certain villages noticed that there was an unnecessary duplication of aid and informed the NGOs about same.

4.5 Women's community involvement

Women's roles have changed from before to after the tsunami. They have, in many cases, become caretakers of their communities and not only their families. In the future, should a disaster strike, women will be in a better position to immediately respond to the needs of their communities. They will be likely to mobilise quickly and efficiently for relief if a disaster strikes. In the future, women's mobilisation and participation in the immediate relief of their communities will likely occur on a larger scale and more quickly than previously.

5. Conclusion

The years following the tsunami have seen a rise in the organisational capacity of women. In many villages, a majority of women have joined SHGs. Additionally, women have formed collectives for various purposes: 1) to improve the health and sanitation of their communities; 2) to solve livelihood related problems; and 3) to collectively overcome caste-based barriers. The activities of women's collectives and individual women within these communities resulted to a change in their societal roles and an improvement in their socioeconomic status. Women have taken on multitude responsibilities in their communities following the tsunami and have become empowered to go beyond their traditional roles in society.

Women's participation post-disaster and the initiatives that they have undertaken within their communities have contributed to increasing the level of disaster resilience of their communities. Stronger networks, linkages with service providers, increased health and sanitation awareness, the presence of village disaster task forces, creation of alternative livelihoods, and growth in women's collectives all contribute to increased resilience of communities to disasters.

With the apparent rise in frequency and intensity of disasters throughout the globe, innovative ways of looking at disaster response and recovery are needed. This case study provided examples of how including women in response and recovery processes following disasters not only served to empower them and improve their socioeconomic status but also helped communities take ownership of the recovery process, increased their resilience and ability to find appropriate and sustainable solutions in the aftermath of a disaster.

References

Byrne, B. and S. Baden (1995). "Gender, emergencies and humanitarian assistance", *Bridge Report* 33 (November). *http://www.ids.ac.uk/bridge/reports/re33c.pdf*. Accessed May 2008.

Narayan, D. (ed.) (2002). *Empowerment and poverty reduction: A Sourcebook*. Washington, DC.: The World Bank.

Oxfam (2005). "The tsunami's impact on women", *Oxfam Briefing Note*. (March). *http://www.oxfam.org.uk/what_we_do/issues/conflict_disasters/downloads/bn_tsunami_women.pdf*. Accessed May 2008.

Pan American Health Organization (2002). "Gender and health fact sheets: Gender and natural disasters", *Women, Health and Development Program*. *http://www.paho.org/English/DPM/GPP/GH/genderdisasters.Pdf*. Accessed May 2008.

Tamil Nadu Tsunami Resource Centre. *http://www.tntrc.org/index_01.php?ref1=menu_gt&ref2=gt_tngov*). Accessed April 2008.

UNDP. "UNDP & tsunami recovery—India," *http://www.undp.org/tsunami/india.html*. Accessed May 2008.

UNICEF Press Centre. *http://www.unicef.org/media/media_42236.html*. Accessed May 2008.

57

Exposure Visits: A Tool for Gender Mainstreaming

Exposure visits gave participants the opportunity to interact with members of the SHG about mangrove restoration and management.

Source:

J.D. Sophia, A. Sivakumar, T. Nedumaran, K.G. Mani and V. Sivanesan *M.S. Swaminathan Research Foundation, Tamil Nadu, India.*

A. Raja, A. Oliver Freeman and A. Amaravati *Society for Participatory Research and Integrated Training, Tamil Nadu, India, 2010.*

1. Introduction

In a tsunami resilience project, exposure visits were identified as a possible way to enhance community participation specifically for project interventions and gender inclusion. Lessons were derived from a research conducted in the coastal village of Kattumavadi, India where fishing is the primary source of income. Although weather events such as cyclones, storms and tsunamis are rare, the possibility of natural disasters is still threatening to the lives and livelihoods of the Kattumavadi people. This project was implemented by the Society for Participatory Research and Integrated Training (SPRIT) and MS Swaminathan Research Foundation (MSSRF).

Participatory approaches were applied during planning and implementing project interventions such as the mangrove conservation and management, establishing and sustaining village knowledge centre (VKC) and strengthening local livelihoods. Although, initially the villagers' lack of

MSSRF implemented a Joint Mangrove Management (JMM) Project in all the major mangrove wetlands located along the east coast of India from 1996 to 2003. The implementation was done with stakeholder participation including the mangrove user communities and the state forest departments of Tamil Nadu, Andhra Pradesh and Orissa. The rigorous participatory process of JMM resulted in formation of 33 village level institutions (VLIs) comprising approximately 5,240 mangrove user families as members. About 1,500 hectares of degraded mangrove wetlands were restored through these VLIs. A number of socioeconomic developmental programmes were also implemented in the project villages to enhance community participation. Evaluation revealed that these VLIs continue to protect the restored and nearby healthy mangroves with the support of the Forest Department. Rigorous participatory processes in tandem with technical knowledge sharing, social and economic empowerment of the local community are some of the major success factors of JMM project.

These JMM project villages not only restored and continue to manage mangrove wetlands but are now also deemed as model villages for other villages seeking cross-training in related community development programmes.

Exposure Visits: A Tool for Gender Mainstreaming

understanding of the participatory processes proved to be a limiting factor when carrying out community mobilisation processes despite the people's growing interest.

Observations showed that women were generally not included in the local institution particularly in traditional governance structures. Such power imbalance put women in a disadvantaged position. Although there are some projects aimed to empower women, they were mostly limited to microcredit activities.

An exposure visit for women and men of Kattumavadi were brought to successful JMM places. They observed villages that were steadily making progress in terms of:

1. mangrove development and management,
2. VKC establishment and management, and
3. formation of VLIs with an emphasis on the gender inclusion at every level.

2. Cross-learning: Seeing is believing

In May 2007, the exposure visit was organised to villages practising the JMM. The villages handpicked for the exposure were Pichavaram of the Cuddalore district, Tamil Nadu, and the VKC in Puducherry.

At first, 15 men and 15 women were selected to join the exposure visit. Many of the men, who also performed leadership roles in the village, could not participate because of a prior commitment to a local temple festival. Some women took advantage of situation and volunteered to join the exposure visit to Pichavaram.

A total of 17 women and 4 men participated in the two day exposure visit to Pichavaram. The visit encompassed classroom sessions and a field visit. On the first day, the participants learned the process of JMM from the MSSRF exoerts on how mangrove and non-mangrove coastal vegetation played a vital role in reducing the damages of 2004 tsunami. Establishing mangrove and non-mangrove bioshields were also demonstrated. Methods on how to link these techniques to real life situations were shown meticulously through field exercises. Other topics discussed were

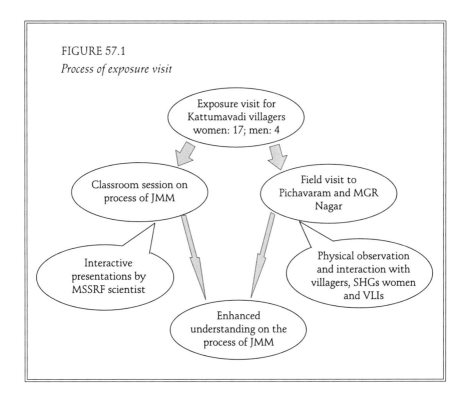

FIGURE 57.1
Process of exposure visit

community mobilisation and participation, processes relating to PRA, formation of village level institutions, the preparation and execution of village level plans.

Activities on the second day focussed on observing the Pichavaram mangrove forest and the nearby village of MGR Nagar. Participants rode in boats to fully visualise and experience the canal method of mangrove plantation. An example of this system was the fishbone method. Upon seeing this systematic approach, the participants realised that their disorganised ways of growing mangroves at Kattumavadi can be improved. Some were eager to adopt this method upon their return to their village at Kattumavadi. Two people from MGR Nagar, who were fishing in the restored area, informed the group that crab culture was also being done in the sub-canals of mangrove plantations. These practices alerted the women to alternative ways of livelihood by rehabilitating their mangrove forests.

Exposure Visits: A Tool for Gender Mainstreaming

The participants were given opportunities to interact with members of SHGs involved in nursery raising and members of the village mangrove council (VMC) about mangrove restoration and management and institution building.

Traditionally, women who share views and ideas in the public forum are taboo in coastal villages. But this situation was an enabler for the women of Kattumavadi since they comprised 81 per cent of the delegation. The women utilised this opportunity to conduct productive sessions with the local community, including the VLI whose members, comprised both women and men, on mangrove development. Questions raised included clarification on what tangible benefits can be garnered from adopting JMM as a practice to how the VLI formed its current structure and functions. The women further explored Pichavaram's practices on how villagers divide labour and clarified the role of women. MSSRF facilitated only when the discussion was inclined to technical aspects.

Since MGR Nagar was largely an egalitarian society, women were given more opportunities to respond to the queries. This approach was quite impressive and further stimulated the Kattumavadi women's interest

towards project intervention. They were surprised that women could take on a substantial role in JMM and VLI. MGR Nagar's tendency to practice democracy made it easier for the community to assimilate to new practices and work together as a community to achieve the project components and the common goal of development.

Technical knowhow and commitment coupled with the financial support and continuous monitoring of MSSRF was identified as key to success of the MGR Nagar JMM effort. The formal handing over of the project to MGR Nagar will be done once the village reaches a self-sustaining level when conserving and managing mangrove plantation. In MGR Nagar, the villagers' shift from being silent observers to actual doers was a gradual and steady process guided by MSSRF. MSSRF's dedication to the MGR Nagar people impressed the Kattumavadi women who decided to try to incorporate JMM in their village.

3. Learning into action

Back at home, women shared their learning experience on mangrove development and management and VKC during the first VLI meeting. If replicated in Kattumavadi, the women believed that there will be significant societal benefits. The women also suggested restructuring Kattumavadi's VLI to mirror some successful components of the JMM villages in Pichavaram. After sharing their views, they then identified place to begin a mangrove bioshield. With the insights drawn from Puducherry, they also influenced the VLI to invite MSSRF and SPRIT to set up a VKC in Kattumavadi. Once in place, SPRIT and MSSRF started planning and implementing the interventions together with the VLI. Before the exposure visit, the community members thought that developing and managing mangrove is simple, but after observing the mangroves in MGR Nagar and interacting firsthand with the villagers, they realised that the communal cooperation and commitment is essential. The participants realised that mangrove restoration requires technical skills and could be achieved only with the support of MSSRF. In response, MSSRF contributed technical expertise and other support to SPRIT and the community to accomplish the project components.

4. Impact of exposure visit

A few months after the beginning of the project, Kattumavadi village was evaluated to determine whether the role of women in decision-making grew. This evaluation is part of a quarterly monitoring process and was conducted largely through questionnaires.

It was determined that many women who were at first silent spectators' are now playing dominant roles in deciding many of the planned activities. As shown in Table 57.1, Kattumavadi's decision-making process, nearly 46 per cent of the activities were either led by women or were jointly coordinated by men. While the decision for the VKC establishment was done by women, the location was decided by men, which constituted 9 per cent. Women who were responsible for independent decisions were also assuming the role as keeper of the mangrove plantation including identifying locations. The need for new VLI was decided by women, but building the structure was decided jointly by both. A revolving fund for women's groups was provided to strengthen the existing livelihood activities and/or to begin additional income generating activities. These activities were solely decided by women. Decisions like finalising signatories of bank accounts, fixing and collecting interest on revolving

TABLE 57.1
Kattumavadi's decision-making process

Decisions	Who made the decision?		
	Women	Men	Both
Mangrove bioshield—Idea and place	✓		
VKC			
Idea	✓		
Place		✓	
Need for a village level institution			
Idea	✓		
Structure			✓
Signature to bank account			✓
Revolving fund for women	✓		
Interest on revolving fund			✓
Selection of knowledge workers			✓
Selection of animators			✓

fund, the selection of knowledge workers and animators were jointly undertaken by women and men. It was a large change that was, at first, surprising to the community. The VLI members who once refused space for women, now recognise the role of women as decision-makers in community development.

5. Conclusion

Since few men participated in the exposure visit, women were given an opportunity to interact without restrictions during the orientation and field visit. Since social inclusion was a priority, the projects' sensitivity to gender issues expanded the roles of women by mainstreaming them in existing activities.

This project determined that organising exposure visits exclusively for women prior to the visit of men would help facilitate the active participation of women in decision-making at village levels. Both women and men benefited equally from project interventions.

Mainstreaming JMMs must be gender-specific, particularly when women are known to be in disadvantageous positions in the community. Although the field visit was not exclusively planned for women, it was found that this circumstance was and will be highly successful to directly and indirectly address discrimination. Similar exercises can be replicated on extensive scales for further science based interventions where the male dominance is high in a specific society.

By acknowledging that experiences, knowledge and the interests of both women and men are integral to project planning, implementing and monitoring, gender equality can be gradually achieved.

58

Community Disaster Resilience Fund: Learning from a Pilot Initiative

The Resilience Fund will give us money, we will take it, multiply it, give it to someone else to multiply, then someone else..

— Member of a women's collective in Ranipur, Bihar

Source:

Asha Sitaram *Community Disaster Resilience Fund, Learnings from the Pilot Initiative in India, National Alliance for Disaster Risk Reduction, October, 2009.*

1. Introduction

To implement the Hyogo Framework for Action (HFA), the key instrument for implementing disaster risk reduction (DRR) that was adopted by the member states of the United Nations, substantial funds are being allocated for developing national level institutional capacities to manage disaster risk, prepare for response, and to cope with emergencies brought on by natural disasters. Largely, the focus has been on preparedness for response and only a small proportion of funds reached hazard-prone communities, that are treated more as victims of calamities, rather than actors who can actively contribute to DRR and risk reduction activities.

The process of DRR programme design, implementation, and monitoring leaves little scope for at-risk communities to define priorities, agenda, and undertake or contribute to DRR efforts. Numerous community driven DRR projects are being undertaken by community organisations and women's groups living in disaster-prone areas and disconnected from local and national processes.

Against this backdrop, the government of India piloted the Community Disaster Resilience Fund (CDRF) in line with the country's commitment to the HFA. The CDRF is a mechanism to direct resources for DRR to risk and vulnerable communities in the context of the HFA. The initiative was endorsed by the National Disaster Management Authority (NDMA) at the Second Asian Ministerial Conference on DRR at New Delhi in November 2007 and coordination by ProVention Consortium and the National Alliance for Disaster Risk Reduction (NADRR), a network of over 150 NGOs.

The CDRF aims to demonstrate how a funding mechanism can promote a decentralised, pro-poor community driven approach to DRR; develop the capacity of local communities to identify vulnerabilities and reduce risks by creating sustainable livelihoods through linkages with development programmes; enhance the general understanding of community resources and resilience initiatives by engaging in research, monitoring, impact studies and cooperation with the grassroots; and upstream lessons and leverage resources and partnerships for community-led disaster resilience priorities.

Community Disaster Resilience Fund: Learning from a Pilot Initiative

CDRF is broadly used in this document to refer to the project and processes to enhance community resilience. In particular, it refers to the community-owned and managed funding mechanism piloted by grassroots women's groups and community based organisations (CBO) to reduce vulnerability and disaster risk in poor communities.

Its key actors and stakeholders are multi-hazard prone communities, women's self-help groups (SHGs), local, state and national level governments, and non-governmental and community organisations. NADRR constituted a Project Advisory Group (PAG) in October 2008 to advocate for and publicise the concept of CDRF at the local, national and global levels. The PAG acted as a feedback mechanism to converge existing local CDRF processes with key development programmes.

2. Approaches and key principles

2.1 Promoting locally-led DRR

Community-based resilience efforts began with the promotion of the concept of DRR, grassroots community ownership, and the communities themselves establishing baselines and mapping vulnerabilities during calamities. Most commonly identified vulnerabilities are the lack of safe access to drinking water and sanitation facilities, the depletion of agro-based livelihoods, the lack of proper shelter, and damage to schools, health clinics, and community centres during calamities.

Community-based organisations (CBOs) were selected as managers for the CDRF pilot areas across a cluster of villages. CBOs formed local CDRF committees consisting of Panchayati Raj Institution (PRI) members and women leaders, with the facilitation of NGOs. They conducted baseline surveys to identify vulnerabilities and hazards faced by the communities. The communities held meetings and came up with options and proposals for DRR that were then submitted to the local CDRF committees. The CDRF were responsible for disbursal, guidance and monitoring of projects.

2.2 Establishing baseline indicators, mapping vulnerabilities and capacities

The NADRR held a workshop in March 2009 for community groups and NGOs to share methods, jointly evolve resilience baseline indicators, map vulnerabilities, and identify DRR priorities. Groups of women and men trained on resilience concepts created baseline information on water and sanitation facilities, health care access, safe/unsafe places, assets, livelihoods, and vulnerable groups to identify factors that place the community at risk. The grassroots communities identified the following vulnerabilities during the mapping exercises:

1) Poor selection of construction sites and limited access to facilities;
2) Potentially unsafe drinking water with water pumps and sanitation facilities constructed near each other in low-lying areas that are flooded during rains or cyclones;

Women's self-help groups are some of the key actors and stakeholders of the CDRF.

3) Schools and heath centres constructed in poorly selected locations;

4) Depletion of agro-based livelihoods (resultant high migration rates); and

5) Lack of safe housing.

Agriculture-based livelihoods are particularly vulnerable, as they are severely and continuously affected by calamities and climate change, and are characterised by oscillating incomes and slow development. Preparedness measures and effective community interventions are urgently needed to reduce rural vulnerabilities and strengthen livelihoods and food security cycles. The community listed the following set of priorities:

1) Better preparedness strategies;

2) Early warning and emergency response measures; and

3) Stronger community resilience practices focussed on the improvement of livelihoods and access to basic services, particularly health services and sanitation and drinking water facilities. Mapping allowed the establishment of a baseline against which development and results of the CDRF pilot could be assessed.

2.3 Capacity building for women's groups

Women's groups were encouraged to present findings and learning during village meetings and to adopt safer and replicable practices. Groups visited other communities in order to share their practices and raise awareness about the concepts of DRR and the CDRF. Many of the priorities and projects that emerged were focussed on livelihoods such as women-managed weekly markets, seed and fodder banks, while others such as the creation of task forces focussed on using traditional and women's innate coping practices. Others evaluated ongoing development programmes and infrastructure projects to make them more resilient.

Facilitating organisations provided capacity building that emphasised the need to change mindsets from response and relief to building resilience by playing an active role in managing and implementing DRR initiatives, forming CDRF committees, monitoring and evaluation, and fund management.

2.4 Role of women's groups

Women's groups planned, implemented and managed the CDRF initiative. They have played key roles in managing the funds and in decision-making, and as representatives of the committee in monitoring the process. They are actively involved in the development of the village resilience plan and in the CDRF committee, having a voice on the setting of DRR priorities. Women's groups in these villages were earlier limited to savings and credit and their participation in decision-making was largely informal. This initiative has enabled women's groups to formalise their participation and voice out their opinions, and influence decisions in disaster management. Women's collectives in Bihar address health issues during floods, livelihoods pre- and post-floods and build community-based preparedness, and early warning and emergency response mechanisms.

> If we have the means to generate decent incomes, our efforts to combat disasters will be sustainable. It will fulfill all that I want.
>
> — Nagina Katham, Kharatia, Bihar
> CDRF Committee member

3. Results: CDRF—making victors out of victims

The CDRF has changed the way people are treated and how they see themselves especially in times of calamities: from victims, they have become victors who vanquished weaknesses—their own and that of institutions around them—by being actors who actively contributed to DRR.

3.1 Taking the lead in handling local DRR funds

CDRF committee members are mostly women leaders (70%) who are experienced in running SHGs and federations. The first village level CDRF committees were formed in May 2009 with 12-15 democratically elected members from SHGs, local government, NGOs, youth clubs, farming groups, and disaster task forces. The CDRF committee was designated as

Coir making is one of the livelihood activities that the CDRF is supporting.

a community monitor to handle accounting of funds, procure materials for DRR initiatives, dialogue with local government to leverage funds, provide training and awareness to communities, and scale up and sustain activities. The fund is kept in a nationalised bank with controls in place so that only two members can transact and sign documents on behalf of the members.

By experiencing DRR fund management, women's groups have strengthened their capacity for understanding development priorities and linking these to DRR initiatives. Members have volunteered their services to visit and monitor progress in the pilot areas. In Bihar, for example, a supra committee manages funds and initiatives in all five CDRF villages. The committee of 37 members holds meetings on a rotational basis in all the pilot villages, in a way exerting some amount of pressure to ensure that funds are well managed.

Committees have worked to strengthen disaster task forces. Women in particular see their work as resilience building and have become involved in decision-making on DRR priorities. Women's groups have used their networks and the local media to propagate messages on CDRF as a means to enable leaders to share practices and solutions, and thus transfer their knowledge and skills to other communities.

3.2 Conceptualising community-led innovations

Community groups have identified some DRR strengthening initiatives. These are livelihood promotion through management and marketing of dairy products, palm leaves, sea shell, coir, and first aid boxes; grain and seeds banking; cashew production; planting of drought-resistant Sevan grass (*Lasiurus Sindicus*); addressing health, water and sanitation issues in flood-prone areas; and creation of an emergency and social insurance funds.

3.3 Livelihood initiatives to enhance DRR in Tamil Nadu

Women's groups in Kanyakumari district in Tamil Nadu have strengthened traditional DRR strategies by minimising disaster vulnerability and by creating more secure livelihoods. Following the tsunami that hit the eastern Indian coast in December 2004, communities pointed out their lack of knowledge and awareness of traditional DRR practices. An example is the Kanyakumari CDRF that aims to:

1) strengthen livelihood activities linked to local resources such as palm leaf, coir making, and sea shell handicrafts;

2) promote women leaders as resource persons in DRR, linking livelihoods and sustainable development;

3) exchange such practices and enhance their contemporary value through the validation of findings and learnings among communities in order to have better products and services; and

4) identify linkage possibilities to ensure self-reliance in resource management.

Diversification and strengthening of livelihoods under CDRF has offered an opportunity to reduce community vulnerabilities and overcome impediments to growth such as household income and cash flow management. Resilience building, by way of the livelihood approach and learning exchanges, calls for a long-term action by the doers (designers, artisans and producers) and the facilitators. Balancing immediate, short-term income earning opportunities and long-term economic and social development gains to the artisan communities has to be attended to.

> Planning together, sharing problems and finding solution to the problems together was a very positive and a strength of the community.
>
> - Krishna, KGBKK, Assam - Committee member
>
> Women's empowerment was a striking feature in the village, the money taken as loan from the SHG is used in livelihood generating activities, there is unity among the women.
>
> - Bharti Doley, KGBKK, Assam - Committee member

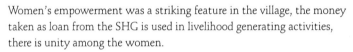

3.4 Women's weekly markets to strengthen livelihoods

With the help of the rural volunteers centre (RVC), women's groups in 14 villages of the Lakhimpur and Dhemaji districts of Assam have created a women-owned weekly market to address heavy economic and livelihood losses caused by the floods. Amar Bazaar is now recognised as a federation and is registered under 1968 societies act, as Matri Amar Bazaar Kendriya Samiti. It has formed linkages with industries where it provides raw materials and handloom manufactured by women. In disaster and flood situations, Amar Bazaar Samiti takes over and helps communities in arranging emergency relief shelters, and evacuation and medical services.

Floods have a devastating impact on Assam's economy as majority of the population is dependent on agriculture and livestock rearing for their livelihood. To address the flood hazard, RVC has formed a village level CDRF committee, strengthened women-owned weekly markets through leadership, DRR and marketing training, and supported all the project villages in showing the link between livelihood/ income generation with disaster resilience. Forty members (28 members from 14 members

Women also manage milk production and marketing as one disaster resiliency initiative supported by the CDRF.

of CBDRR committees and 12 members from Amar Bazaar) are part of the CDRF committee. The CBOs have used funds to improve market infrastructure and acquire small equipment that can prevent loss of livelihoods.

RVC has also initiated orientation programmes for CDRF committees and the Amar Bazaar on the concept of DRR. Participatory workshops focus on identifying existing disaster related vulnerabilities and risks, traditional DRR practices and gaps, and the possible scaling up of practices to reduce risks of disaster. The initiative has a positive outcome as communities now better understand the risks of disaster in a local context and have developed knowledge and understanding to reduce the risks.

3.5 Encouraging the planting of drought resistant crops and setting up of a fodder bank

The Bikaner district in Rajasthan is a drought prone area and the sparse rain cannot meet the agricultural, domestic, and drinking water needs of the local communities. Livelihoods in the area have traditionally depended on agriculture. The government has sold fodder at a controlled rate during drought periods. In order to cope with recurrent droughts and water scarcity, and the resultant unavailability of fodder and marketable crops, communities in Bikaner district have cultivated a drought resistant crop, Sevan grass, and have set up a fodder bank.

Approximately 20 most vulnerable families in two pilot villages have received the Sevan grass seed along with traditional crops such as *bajra*, *jawar* and *kejri beej*. The planting of a drought-resistant crop has increased fodder availability in the area. Surplus produce are sold at the market to generate income, and thereby reduce economic vulnerabilities. Women understand drought to be a perpetual occurrence and are preparing to grow fodder locally and stock it for use in water-scarce times. The creation of water storing systems to alleviate drought has also been considered.

4. Achievements

The CDRF developed the capacity of communities to identify vulnerabilities and reduce risks, to enhance the understanding of grassroots community resources and to upstream lessons in community-driven DRR. Overall, it has developed community level demand for reducing the impact of disasters and developing livelihood activities; connected communities to resource persons by developing their understanding of vulnerability and disasters; created local platforms for engagement on DRR between communities and government and other institutional actors; enhanced the capacities of grassroots and women's groups in DRR; and created a local network of community based organisations, NGOs, government bodies, and other actors that together craft the national DRR agenda.

One of the key learnings in this initiative is the change of approach by facilitating partner NGOs. In the past, they were more focussed on implementing DRR for short periods. Now, they have realised their role in improving the capacity and skills of grassroots women in building community resilience. This also brings an opportunity for them to understand other NGOs/communities' work and to learn from each other. So far NADRR has conducted three orientation and capacity building workshops for NGOs and community leaders to understand the DRR in their local context and how it is linked to development and their roles as NGOs and communities.

The CDRF initiative was an opportunity for grassroots groups and NGOs to use the alliance to share a common platform, share the learning, and lobby with policymakers including the National Disaster Management Authority (NDMA) and state/district level authorities. Learning exchanges proved that communities across states were keen to learn, implement, and transfer the knowledge. The workshops that were held provided community leaders and NGOs a first opportunity to perceive their work in the light of DRR and view affected communities as a resource.

A slow but steady process is going on to bring the grassroots network of community women groups and link them to policymakers. Already, there are networks in a few implementing states like Tamil Nadu, Maharashtra, Bihar and Orissa that are seeking to highlight community initiatives in DRR. Challenges remain in scaling up and improving communication between grassroots communities and local institutions. Communities already involved in the initiative are beginning to serve as local champions for neighbouring villages and district administration. Increased support is needed from district and block administration to enlarge the CDRF network to contribute to the voice that the grassroots groups have in influencing the national and international DRR agenda.

59

Building Bonds, Breaking Bondages in the Gulf of Mannar

Freedom from debt from the moneylenders/fish merchants meant that fishers became free to directly market their catch and determine the best price for it.

Source:

P. Sekar *People's Action for Development, Tamil Nadu, India, 2010.*

1. Introduction

Many development organisations implemented rehabilitation activities in India, one of the countries that were hit the hardest by the Indian Ocean tsunami in 2004. Among these was the People's Action for Development (PAD) that organised the Financial Initiative for Sustainable Human Economic Resource Regeneration (FISHERR) in coastal villages facing the Gulf of Mannar on the southeast border of India's southernmost maritime state of Tamil Nadu.

Through participatory rapid appraisal, focus group discussions, and validation procedures, PAD found that the damage the tsunami wrought on the area was prefaced by a situation that had, through several decades, served to deeply entrench a dependency of the poor and marginalised sectors (i.e., women and the landless) to a mercantile caste that wielded power and influence on all aspects of life in the coastal society.

The PAD implemented FISHERR in the coastal areas of Vilathikulam and Tuticorin blocks of Thoothukudi district and the Kadaladi, Thirupullani and Mandapam blocks of Ramanathapuram District where it identified fisherfolk, Palmyra tappers, the Dalit, the coastal poor, small and marginal farmers, the landless, and children and women as its target clientele. The ultimate aim was to break the bondages of cyclical indebtedness and poverty that the well-entrenched dependency had caused among the poor sectors in the area.

In Thoothukudi and Ramanathapuram districts live 125,000 fisherfolk of whom about 57,000 are active fishermen. The local fishermen practice collective fishing in *vallams* or *vathais*, wooden crafts that are restricted to the shallow waters around the coral reefs and within the lagoon. Other active fishermen are labourers on mechanised trawlers, fishing outside the lagoon in the Bay of Bengal while the poorest among them do individual fishing close to the shore with boyars, a self-made contraption of expandable polystyrene (Styropor). Women who fish for a living engage in reef-based and allied activities such as marketing dry fish and net mending. The major problem for those dependent on fishing is the seasonality of adequate catches and the general decrease of the catch due to an overexploited marine ecosystem.

Building Bonds, Breaking Bondages in the Gulf of Mannar

Palmyra tappers from the Nadar caste.

In other villages where most of the families subsist on Palmyra tapping, work is even more seasonal because tapping is possible for only five months of the year. During the rest of the year, some families resort to fishing, though not as intensive as the fishing communities, while some work as agricultural or other labour. As agriculture is not practiced much in these villages, many of the men migrate to the towns in search of employment. A third sector is comprised by a few Dalit villages where the people work as agricultural labour, or as labour with contractors engaged in charcoal making.

2. FISHERR: Its economic, social and political context

The highly seasonal practice of artisan fishing, the overexploitation of the fragile marine ecosystem of the Gulf, and intense competition from within and outside the fishing communities have marginalised small-scale boat operators, pushing them to resort to temporary migration as a coping strategy. The second major, also seasonal, income source in the region is provided by the Palmyra tree from where men of the specialised Nadar caste tap nectar manually that the women process into molasses. Women also weave baskets from palm leaves while the men regularly migrate to cities in search of petty jobs. The incomes of more than 60 per cent of these families are well below poverty line and starvation during the lean season is common. In a survey by PAD in 1998, the average household income among coastal villages of the Gulf of Mannar was at or below the nationally recognised poverty line of Rs. 24,000 per annum, a figure that has not changed according to more recent diagnostics.

In order to survive during the off-season, almost all families living in these villages take loans from moneylenders. The moneylenders who are also usually the fish merchants/mechanised boat owners do not charge formal interest on the loans. Instead, they buy all the catch, invariably for a price that is considerably lower than the local market price, the difference representing interest on the loans. The equivalent rate of interest for the loans is estimated at between 5-10 per cent per month.

The families themselves do not see the moneylenders as exploiters but as essential support. Because of the seasonality of their produce, a sustainable cycle, in terms of being able to fully repay loans and be eligible to take a fresh loan, is 3-5 years. This means that at least once or twice in 3-5 years the family becomes a 'defaulter' and take high risk for further loans. Unlike the moneylenders who understand the multi-year cycle as a feature of the business, take the additional risk by giving more loans and just charge a higher rate of interest as the fee for higher risk, the regular finance system does not accommodate 'defaulters'. It requires the full repayment of the existing loan before a family can become eligible for another loan.

This is the fundamental reason for the success of the moneylender. Moreover, the moneylender is also a trader who earns his income from selling the produce of the fisher folk and the Palmyra tappers. Therefore, he also provides another important service—a regular market channel for the fisher and Palmyra tapper families—ensuring that whatever they produce is bought, albeit at lower prices. However, the high cost (in terms of price difference) means that often the families cannot repay the moneylenders fully and practically become slaves or bonded labour to the moneylenders for generations. The control wielded by the mercantile castes on the economy of the area serves to perpetuate the inequity.

Gender roles are also a source of inequity—women are discriminated against although their unpaid labour turns the raw material into marketable, priced products. Among the Palmyra tappers, despite women's crucial contribution to household incomes, decision-making and actual spending of money is controlled by men. During the male's temporary migration, women and children are often left in the villages without any means to secure their basic needs. An indicator for the negative impact of the traditional migration pattern is the literacy rate of rural women in Ramanathapuram district which is only 43.42 per cent (compared to 64.55 per cent state level average) and in Thoothukudi district, 47.08 per cent (men: 52.91 per cent; state: 82.33 per cent).

Both major occupations in the project area bear high risk of accident. A reasonable number of men also desert their first families, thus 18-36 per cent of the households in the villages are headed by women. Another aspect of the seasonal male migration is the prevalence and high risk to HIV/AIDS.

Moreover, political power rests with the fish merchants/boat owners who also dominate the local elected bodies (Panchayati Raj Institutions). Before PAD's intervention at the local level, they successfully prevented women's political participation under reserved quotas. The area is also prone to infiltration from militants by sea from across the Gulf of Mannar. In turn, this has implications for the fishermen of the area who are frequently attacked by Sri Lankan Navy boats on suspicion of illegal activity. Regular infiltration of Sri Lankan refugees is also very common.

Moneylenders collect the fisher's catch as payment for debt.

3. Fishers' chain level cooperative markets breaking the bondage of indebtedness

A tsunami rehabilitation initiative implemented by PAD and supported by Welthungerhilfe-Germany, Terre des Hommes Suisse and Christian Children's Fund of Canada has demonstrated a means to break the bondage created by the vicious debt cycle. The PAD conducted village level analyses with the full participation of men, women, youths, and children who came to a consensus that the continued indebtedness to moneylenders was the principal problem facing community members that has prevented them from becoming self-sufficient.

The PAD facilitated the formation of 34 village development committees (VDC) as an advisory body of the village. The members are from the livelihood groups (LG), Panchayati Raj Institution (PRI)-elected members, traditional leaders, and primary school teachers. PAD's base in the intervention areas are LGs with 15-20 members each. From 2002 to 2005, PAD had already developed 342 LGs with 6,138 members (2,874 male and 3,264 female) in 45 villages.

To liberate the fisher families from debt, the project purchased 57 new boats and distributed this on a joint ownership basis to 342 fishing families. These families, organised as an informal cooperative, appointed one person per village to take care of direct marketing of their catch. Though the boats were given for free, the groups repaid the cost to help other poor fishers in the community to free themselves from debt. The VDC and the beneficiaries worked out the mechanism, not forgetting the role played by the moneylenders of providing the safety nets to the poor fishers and Palmyra tappers by setting up a common saving that they could access during emergencies. Through the scheme, FISHERR has rescued 62 more boats from the clutches of moneylenders/merchants after clearing their informal loans using the repayment, common savings and through formal loan linkage with banks.

The repayment norms from every catch are as follows: 15 per cent towards the repayment, 5 per cent towards common savings, and 5 per cent towards the administration. An administrative fee of 2 per cent is deducted from the sales of each *vallam's* catch as a way to sustain itself (i.e., pay for the administrative costs of running the cooperative), 5,000 artisanal fisher families have become significantly more economically independent by using the cooperative as a preferred alternative to the merchants/moneylenders. They have established an emergency fund of Rs. 1,500,000, have contributed Rs. 1,000,000 to the administration of the cooperative and the poor community members have collected Rs. 17,888,000.

4. Conclusion

This village is characterised by significant social homogeneity: the fishing hamlets tend to be inhabited by two traditional fishing castes, the Muthirayar and Fernando, while the Palmyra tappers belong to the Nadars. This has encouraged higher social cohesion than in most Indian villages and allowed PAD an effective social mobilisation and integration of all villagers in their development efforts.

Moneylenders collect the fisher's catch as payment for debt.

The project has directly and indirectly benefited more than 5000 families. The system has turned into a revolving loan mechanism, with the FISHERR Cooperatives and VDCs administering the repayments and keeping proper accounts at the village level. Sustainability is ensured in the project: the debt payments are repaid to the cooperative and revolved among its members. More *vallams* are involved based on need. Every quarter, two new *vallams* are rescued and included in the FISHERR cooperative.

Sustainable change of the general equity and power structure of the fishing economy in the intervention villages has begun: the FISHERR-cooperative supported fisher families to reclaim the ownership over their traditional boats by settling their debts and freeing them from bondage of local merchants/moneylenders. A significant number of women-headed households were able to participate in fishing activities by being part-owners of boats. More than 20 per cent of debt from outside sources has already been replaced by the FISHERR Cooperatives members.

According to Mr. Muthukumar of Vembar village,

> *Fisherr varuvathurkku Nangalam parambaraiya kadanaliyakthan irunthom. Ithuthan engalai kappatrritu. Fisherrkku munadi nan Rs. 3,000 sambarithen anal ippoluthu nan Rs. 7,000 mathathirukku sambarikiran.*
>
> (Before forming FISHERR, we were all indebted. FISHERR saved us. Before FISHERR, I earned Rs. 3,000 per month but now I earn Rs. 7,000 per month.)

FISHERR has proven to the people that by building bonds through the LG/cooperatives, they will be able to help each other break free from the bondage of cyclical indebtedness.

60

Insurance: Investment towards Security and Resilience in Palakayatippa

Microinsurance increased communities' resilience to disasters and strengthened their capacities to restore normalcy quickly.

Source:

P. Venkateswara Rao *Praja Pragathi Seva Sangham, Andhra Pradesh, India*

J.D. Sophia and R. Ramasubramanian *M.S. Swaminathan Research Foundation, Tamil Nadu, India.*

1. Background

Palakayatippa is a coastal village in Hamsaladeevi *panchayat* of Krishna district, Andhra Pradesh, in the southeastern coast of India where people earn a living from fishing (60 per cent) and agriculture (40 per cent). Located to the west of the Bay of Bengal, Palakayatippa village is extremely vulnerable to hydrometeorological hazards such as cyclones, storms and floods.

In 1977, a severe cyclonic storm killed 800 in Palakayatippa. Storms of similar intensity in 1990, 1996 and 2006, and the 2004 tsunami wrought heavy damage to property and livelihood and caused death to thousands, highlighting the people's poor resilience to such hazards. These disasters also emphasised the fact that the people were invariably dependent on the government and funding agencies for relief, recovery and rehabilitation.

This situation made the local NGO, Praja Pragathi Seva Sangam (PPSS), see the importance of microinsurance as a tool for managing disaster risks especially for low-income families. PPSS also found that the Palakayatippa villagers did not have access to insurance coverage because they were not aware of its importance. In 2005, the PPSS facilitated the availability of microinsurance in Palakayatippa through the Royal Sundaram Alliance Insurance Company Ltd.

2. Context and existing scheme

The PPSS first conducted a study to assess the impact of the 2004 tsunami on Palakayatippa village. The people lost food grains, household articles and other livelihood assets related to fishing and agriculture. The worst part, according to the villagers, was losing the family's breadwinner, and as a result, experiencing severe financial crisis that oftentimes meant sacrificing the female children's education. After every calamity, the villagers awaited relief support from external agencies, including the government. The disasters proved that the coping capacity of the community to safeguard their lives and assets is very poor. The most shocking reality shared by the community was that not even 20 per cent of the households have insurance policies because they are not aware about these despite the availability of insurance schemes in the village. The people, however, were aware of Adarana or helping hand,

The insurance policies benefited the women more than the men.

an insurance scheme of the Andhra Pradesh government that covers accidental deaths for below poverty line families.

The Adarana scheme for fishermen is applicable for accidental deaths even at sea and entitles families Rs.50,000 (approximately 1,076 USD). The nodal agency that provides the entitlement for accidental deaths while at sea is the Fisheries Department. The department requires that claims should be supported by first-hand information and post mortem reports. Under the scheme, the district allocates a maximum of Rs.10 lakh to cover about 20 accidental deaths. In case the number of deaths exceeds 20, the process to compensate the affected families will take more than a year. Under these circumstances, PPSS explored an alternative insurance scheme that would cover both life and non-life to strengthen the coping capacity of the poor. It was also at this juncture in 2005 when the Royal Sundaram Alliance Insurance Company registered with the Government of India (GoI), approached PPSS and other coastal NGOs to promote and sell the microinsurance scheme in the disaster-prone villages of coastal Andhra Pradesh.

> Sequence and topics discussed in the meetings—
>
> Meeting 1: Although micro insurance is likely to result in losses for Royal Sundaram, it will implement the scheme under its Corporate Social Responsibility programme.
>
> Meeting 2: NGOs explained the real problems of rural people, and identified the vulnerability risk factors and their implications on rural lives and livelihoods.
>
> Meeting 3: The company proposed premium amount and coverage. NGOs suggested modifications based on community concerns.
>
> Meeting 4: The company incorporated the request of NGOs and developed Janashakthi and Janashakthi Plus insurance policies and submitted these to the Insurance Regulation Authority (IRA) for approval.
>
> Meeting 5: After approval by the IRA, the NGOs were given insurance agency accreditation status and the MOU between the NGOs and the Royal Sundaram was signed.

3. Process and products

Based on their working experiences in the coastal villages, the local NGOs advised the Royal Sundaram that unless the microinsurance scheme helps the coastal villages in safeguarding their lives and livelihood assets, it will not be patronised by the people. Taking this information into consideration, the insurance company, through the sponsorship of Care India, convened a series of meetings with representatives of 15 NGOs. Using the input from the meetings, Royal Sundaram developed two insurance packages, namely, Janashakthi for women's life and assets, and Janashakthi Plus to include the women's spouses. The policy prioritises women as they are the most vulnerable to disasters. Moreover, giving women access to and control of such benefits helps the family in the absence of the breadwinner.

4. Insurance package

The insurance policies do not require medical checkup and are mainly for women aged 18-70. The annual premium is Rs.60 for Janashakthi and

Rs.90 for Janashakthi Plus. The insurance package covers eight types of coverage (Figure 60.1). Except for household articles and education grant, both the insurance holder and the spouse get benefits as per amount assured. Household articles covered by the insurance package are food grains, utensils, home appliances, and fishing and agriculture implements.

PPSS convened a village level meeting to orient representatives of 204 families on the new insurance product in July 2006. Since the annual premium was very low, around 180 families availed themselves of the insurance while the remaining 24 families did not because they already had existing general life insurance policies. At a premium amount of Rs.1,000 per annum, enrolling in these other life insurance policies was quite prohibitive. Out of 180 families, 165 insured under Janashakthi Plus while 15 families enrolled in Janashakthi. Total premium paid by 180 families under both policies was Rs.15,750. To avail themselves of the policy benefits for accidental death, the insurance agency only required a certificate from a local registered medical practitioner and the facilitating NGO, making for a simpler claim process compared to other insurance schemes.

FIGURE 60.1
Types of insurance coverage

Insurance package of Janashakthi and Janashakthi Plus

Coverage	Amount (Rs.)
Five days compensation @ Rs. 100 per day	500
Medical expenses of accidents	1,000
Cremation expenditures	2,000
Education grant	5,000
Household articles	5,000
Partial disability	12,500
Permanent disability	25,000
Accidental death	25,000

5. Benefits from the insurance

In October 2006, Cyclone Ogni hit the coast, bringing with it 900 mm of rainfall in two days. Heavy run-off damaged irrigation and drainage channels flooded the entire Palakayatippa. In the village, water stood at three feet for nearly seven days, causing extensive damage to household articles and food grains. Immediately PPSS informed the insurance company. The insurance team acted with dispatch, however, it could only complete the assessment after 15 days since they had to cover a large area affected by the cyclone in Krishna district.

The insurance policies signify freedom from worry of loss of assets, crops and lives.

Assessment results showed that to recover food grain losses, around 148 families received coverage ranging from Rs.150 to Rs.750. Two families received Rs.1,500 each to cover medical expenses incurred due to accidents and Rs.500 each for wage losses for five days. Similarly, for one accidental death, Rs.25,000 and Rs.2,000 for funeral expenses was provided. In addition a deposit of Rs.5,000 was made towards the education of the female child of the deceased family. A victim of partial disability due to accident was compensated with Rs.12,500 while two other fire accident victims received Rs.4,500 (Figure 60.2).

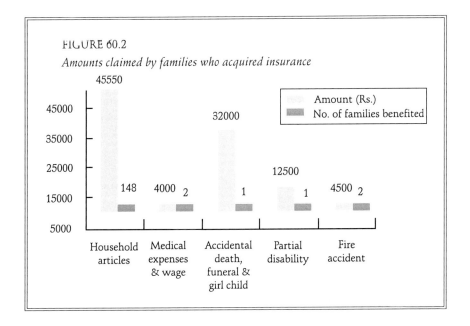

FIGURE 60.2
Amounts claimed by families who acquired insurance

The community expressed satisfaction at the insurance package because it has a very low premium and it provides good benefits. On the whole, 154 families received Rs.98,550 from insurance claims within a year of paying the first premium. However, the people felt that the 45 days it took for the NGOs to settle claims was very long, and suggested that the insurance company should provide emergency funds for the facilitating NGOs to use for immediate relief that could then be paid back after claims are released. The NGOs are in the process of negotiating with the insurance company to make this possible.

The community was also of the view that the insurance coverage starts on the same day that the premium was paid. The NGOs quickly addressed this concern and successfully negotiated with the insurance company to make the policy effective on the day it was purchased. It was also a challenge for the NGOs to educate the villagers on the process and procedures of microinsurance but they have increasingly become aware of the security that it gives them during vulnerable situations as they have availed themselves of its benefits every time disaster strikes. Because of this, the PPSS was able to convince 5,500 to subscribe to the insurance in 25 villages, in 10 mandals of Krishna district.

6. Conclusion

Normally, government and other NGOs concentrate more on the vulnerability and the immediate needs of the disaster-affected communities. The microinsurance initiative was used as a tool to increase resilience of the community to disasters by minimising risks and strengthening their capacity to restore normalcy as quickly as possible. Instead of just being reactive, the villagers have become proactive by taking out insurance policies.

The key success factor that enabled the NGOs to build community resilience was engaging the community and partnering with the insurance institutions. The NGOs were also able to negotiate with the insurance institutions to explore ways to consider community needs in conceptualising the insurance policy.

The need for an emergency fund in order to shorten the claim period from 45 days is something that can be addressed by the government by providing disaster mitigation funds to the local governing body. The fund could be incorporated into the community-based disaster preparedness initiative and used for immediate disaster relief.

61

Community-Managed Microfinance: The Case of Danavaipeta, Andhra Pradesh

The village revolving fund helped strengthen the livelihood activities of women in Danavaipeta, Andhra Pradesh.

Source:

J.D. Sophia, S.V.V. Prasad, R. Ramasubramanian, P. Suvanna Raju and N. Chitti Babu *M.S. Swaminathan Research Foundation, Tamil Nadu, India, 2010.*

Fishing and agriculture are the primary occupation in the Danavaipeta village of Andhra Pradesh. Few minor businesses like petty shops, *paan* shops, tailoring, vegetable vending, saree business, and goat rearing exist also to supplement the primary income.

1. Gender roles

Both men and women equally share the responsibility of family earning. Men are engaged in fishing while women, in addition to the domestic chores, are predominantly involved in fresh and dry fish vending in the nearby town and villages. A total of 98 women are engaged in fish vending; of which 26 and 72 are fresh and dry fish vendors respectively.

Danavaipeta village	
Total households	: 474
Total families	: 789
Poorest of the poor	: 158
Poor	: 473
Middle	: 158
Women headed HH	: 76
Identified through PRA - Wealth Ranking	

2. Existing microcredit and related issues

Villagers are constantly dependent on external sources as none of them have investments. There are different credit sources prevailing in the village and these are:

2.1 Self-help group

The self-help group (SHG) was established by the government of Andhra Pradesh across state and promoted savings and internal lending since 1996. It also provided Revolving Fund (RF) of Rs.600,000/- to Danavaipeta village through the District Rural Development Agency (DRDA) for income generating activities.

2.2 NGO-promoted microcredit system

In NGO's promoted credit system, there is no incident of defaulters. However, women expressed that the rate of interest was 18 per cent which was comparatively higher than the government. The operational procedures were likewise rigid and more inclined towards penalising the members for late repayment and not attending the meetings as well. It affected their self-esteem but still continued to avail credit support because there were no other options available. Gradually, this created aversion among women and few decided to withdraw from the NGOs network.

2.3 Local moneylenders

Local moneylenders collect payments on a daily, weekly, and monthly basis with interest rates varying from 5 to 10 per cent. With such interest rates beyond the paying capacities of the fish vendors, the become trapped in constant debt cycle.

3. Establishing a revolving fund (RF)

Realising the struggle of fish vending women, the MS Swaminathan Research Foundation (MSSRF) introduced RF to the community to provide credit support to these women. The purposes of which are: (i) to reduce the dependency on moneylenders and other microfinance institutions, (ii) to inculcate the habit of saving, internal lending, and enhance financial management, and (iii) to revive the linkage with DRDA so that the government schemes including microcredit could be accessed by these villagers in the long run.

4. Micro planning and RF distribution

The RF was taken to the Village Development and Management Committee (VDMC) for discussion. The members of VDMC decided to provide at least Rs.300,000 as RF based on the existing number (613) of poorest of the poor and poor families. Before fund distribution, norms were developed in a participatory manner by a set criteria for members.

> Norms for revolving fund:
> 1. Upper age limit for accessing: Revolving fund is 55 years.
> 2. Credit seeking women should be involved in fish vending or other enterprises.
> 3. Currently should not have pending microcredit.
> 4. Should attend monthly meetings regularly.
> 5. Priority is for women headed households.
> 6. Credit takers should accept the interest fixed by the VDMC.
> 7. Credit takers should abide to pay stipulated monthly instalments fixed by the VDMC.
> 8. Should form small groups among themselves (10 members) to channelise repayments.

5. Target group identification

Based on the set norms, the MSSRF along with the VDMC members identified the target group by first classifying the total families into poorest of the poor, poor and middle through wealth ranking exercise. Women headed households were identified from the first two categories and included as poorest of the poor. Determining the amount that each woman would get and the repayment strategy was done in consensus with the VDMC and the community.

6. Institutionalisation of the RF

The RF system consisted a three-tier structure (Figure 61.1). Since the RF has been adopted for socioeconomic development of the village by strengthening the ongoing livelihood activities and supplementing through additional income generating activities, the overall management of the RF should be under VDMC. Hence, there is Executive Committee (EC) of VDMC. As far as office bearers (OB) are concerned, one woman each from the EC and General Body (GB) were selected as president and secretary while a staff from MSSRF is selected as treasurer. These three are the signatories of the bank account of RF. All transactions related

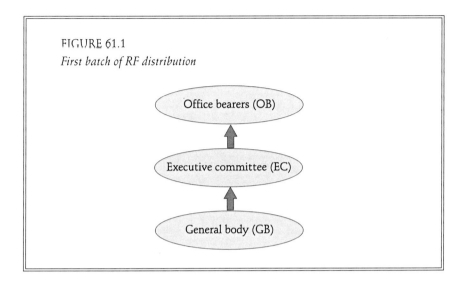

FIGURE 61.1
First batch of RF distribution

to RF are done only through bank. All the identified target women are formed into smaller groups comprising 5 to 12 members. Each group elects one woman as a leader to monitor the transactions and collecting the repayments. The VDMC provided the loan to the group leader through bank cheque. Members pay the monthly instalments along with the interest to the group leader who collects and deposits in bank joint account. The deposited money along with the interest is given as a loan to the other women as per the priority list provided by the VDMC.

7. Benefits of the RF

Community women use RF to strengthen livelihood activities like dry and fresh fish vending, goat rearing, and putting up and maintaining a petty shop business, among others. The following are the perceived benefits of the RF:

1) There is an assured source of microfinance within the village under the control of their own people (VDMC).
2) Funding can be availed hassle-free, i.e., there is very less paper work involved.
3) The norms and approaches are user-friendly.

Community perception

Ms. Arjilli Bangari, a fishing labour working for large dry fish maker said that she started a petty shop to supplement the income generated from the fishing labour activity. With the Rs.2,000 she got from the RF, she was able to purchase commodities for the petty shop. Although the business was very small and majority of the customers were children, she was able to pay the amount within the stipulated time and avail Rs.4,000 for the second time to expand the business. With the latter amount, she expanded the business by adding new commodities especially vegetables and other provisions, keeping in view the requirement of local families.

4) When compared to the moneylenders and other microfinance institutions, the rate of interest is low.

5) The monthly payment procedure with fixed instalment amount and interest is more practical for the borrowers. Unlike in other lending systems, the interest collected from the RF is ploughed back to the principal amount, increasing the financial asset of the community.

The realisation that RF is a community asset made women of the community strive to pay their loans regularly.

6) The community and women play significant roles in decision-making. The use of the RF, i.e., amount to be distributed on loan, repayment procedures, and rate of interest are democratically made.

7) There is reduced dependency on moneylenders and the other microfinance institutions.

8) There is a sense of ownership among women i.e., they started realising that RF is a community asset and therefore payments should be regular as to continuously build the asset.

9) The initiative of RF has given women confidence towards a better future.

Community managed revolving fund is one of the vital tools that ameliorate the conditions of community people while releasing them from moneylenders' exploitation and constant debt cycle. In addition to access to financial resources, managing RF as common fund helps address women's immediate or practical needs and enhance their self-confidence. It likewise increases participation of women in community level activities, thereby paving the way towards their achieving of their strategic needs.

62

Community Participation: The Case of the Andaragasyaya Canal

Community consultation and participation were key success factors in the successful restoration of the Dorawabasna Ela Canal.

Source:

Chintha Munasinghe *Laymen's Den (Pvt) Ltd, Sri Lanka*
Indika Dilhan *Somarathne Sarvodaya, Sri Lanka, 2010.*

1. Introduction

The Dorawabasna Ela canal irrigates 200 acres of farmland and supports the livelihood of farmer families in Andaragasyaya, Sri Lanka. Unique to this topography, the canal—'Dorawabasna Ela', that brings down spillage water from a cascade of five irrigation tanks (Tissa tank, Weerawila tank, Pattegamuwa tank, Debarawewa tank, and Yogawewa tank) across the villages and provides water to the paddy fields, falls into the sea from Andaragasyaya district. As described by a farmer, this is the only place where spillage water crosses a cultivation land. However, once a year for a decade, the same life source threatened the livelihoods of more than 60 farmer families living and farming in half of this vast fertile agricultural land. But tsunami worsen the situation, adding salinity to the water and also further damaging the canal system.

Every year, torrential rains and monsoons caused the canal to overflow, flooding 100 acres of paddylands in Karijjawelyaya and Odipolayaya villages. As a result, the cultivable land was left barren for many years. Production loss was estimated at 270,000 kg of rice per season or an income of about Sri Lankan Rs. 2,800,000 per season, a loss of revenue not only to the people of Karijjawelyaya and Odipolayaya but to the whole country.

2. Andaragasyaya canal restoration project

2.1 Community needs assessment

It was after the 2004 tsunami that Sarvodaya's Strengthening Resilience in Tsunami Affected Communities (SRTAC) Project arrived in Andaragasyaya. Along with Practical Action, through assistance from International Development Research Centre (IDRC) and Canadian International Development Agency (CIDA), Sarvodaya approached the villages to understand the situation and work out interventions addressing livelihood issues of tsunami-affected communities.

2.2 Consulting the community to identify the problem

Sarvodaya organised a needs assessment meeting to discuss the problems the community was facing. The main issue they raised was floods that

Community consultation and participation were key success factors in the successful restoration of the Dorawabasna Ela Canal.

affect their land during rainy seasons every year. In consultation with, and by involving the community, the project team did further exploration of the problem. While taking the community recommendations, the team also consulted the Technical Officer of the Irrigation Department. These actions built community confidence in the project team.

2.3 The problem

Through consultation with farmers, village elders, affected families in the village Sarvodaya officers were able to map the key reasons that have contributed to the problem.

The canal was 1.7 kilometres long from Andaragasyaya to Modara-Moya. Due to lack of management and proper maintenance system, it was narrow (5 feet) at many places while the depth of the canal was also found to be not more than 1.5 feet. In some places, the soil was washed away from the land during floods which got deposited inside the canal, making it less deep. Having not managed properly for more than 20 years, the canal has been encroached by the banks with high vegetation of plants narrowing its width. As a result, the capacity of the canal to bear the volume of spillage water from five tanks was considerably low.

2.4 Consultative plotting of interventions

The community was going through a process of problem tree analysis, resource mapping, etc., during the designing period. Their recommendation to increase the depth of the canal was well taken by the project team and technical officer, who led the surveying, calculations and technical plan for restoration of the canal. During this calculation, it was decided that the width of the canal should be 30 feet and the depth to remain at two feet above sea level.

The farmer organisation with the project team were able to work with the technical officer of the irrigation department who were involved at every stage of designing and planning the restoration work.

2.5 Participatory planning of action

The designing process was highly participatory with most of the affected families and farmer organisation officials included in the process. During the consultative meetings, they, in small groups, first worked out the causes of the issue; discussed among themselves, and with the help of the officials, identified the possible interventions, worked out plans and schedules and assigned a committee and sub-committees to support the field coordinator in the restoration work.

3. Facing the challenge with community participation

Since the canal was only 5-6 feet wide, the challenge was to make it 30 feet wide throughout its length of 1.7 kilometres. The elevation of Andaragasyaya village is just above the sea level. Digging the canal deeper may cause sea water entering the land. Therefore, they calculated the possibility of widening the canal to increase its capacity to hold the water spillage from five tanks without causing floods.

Community Participation: The Case of the Andaragasyaya Canal

Sarvodaya Shramadana Society (SSS) and village farmer organisations (VFO) have a long standing history in the community. During its 15 years of existence in the area. SSS has introduced many important development ventures to the community, such as pre-school, nutrition and health programmes, savings and credit programme, business linkages, agriculture extension programmes with the involvement of many families in the community. It also has a strong link with Sarvodaya Shramadana Movement and its affiliated companies such as Sarvodaya Economic Enterprise Development Services (Guarantee) Ltd.

With 20 days of continuous work, the community were able to complete the restoration, successfully. Although the muddy and hazardous condition of the canal restricted community participation in actual restoration work, the farmers were able to help in the clearing of the canal banks and in supplying meals for the workers.

4. Reaping the results of cooperation

After the restoration work, this 1.7 kilometres canal was made 30 feet wide with strong banks. Culverts and spillage bridges were also developed to avoid heavy flooding during rains. When it rained heavily for two days after completion of the restoration work, the farmers observed the free flow of water through the 30 feet canal without flooding any their paddy fields.

Today, as a result of this intervention, there are about 60-65 families peacefully cultivating paddy and other vegetables, with little or no loss to their harvest. Community consultation and knowledge is critical in project planning and development efforts. Involving affected community members at the stages of analysing the causes of the issues, and allowing community to decide on technical planning contributes to successful implementation of the intervention.

63

Revival of Agriculture in Sorlagondi Village, Andhra Pradesh

Source:

R. Ramasubramaniam, J.D. Sophia and V. Selvam *M.S. Swaminathan Research Foundation, Tamil Nadu, India*
Venkateshwara Rao *Praja Pragathi Seva Sangam, Andhra Pradesh, India, 2010.*

Sorlagondi village, located in Nagayalanka mandal of Krishna district, Andhra Pradesh is a delta village with flat topography prone to natural hazards such as the cyclone and flood. Diviseema cyclone in 1977 killed 714 persons and devastated the livelihood assets of the village. Fishing, agriculture and shrimp farming are the primary sources of income for the villagers. The villagers get two paddy crops per year, one during *kharif* season (June to October) and the other during *rabi* season (November to May). Irrigation water is available only from July to April from Prakasham Barrage through canal system.

1. Occupational transitions and its implications

Sorlagondi village has clayey soil in which paddy was cultivated both during *kharif* and *rabi* seasons till 1990. All 430 acres of village land was under paddy cultivation. In the year 1992, 230 acres of land was diverted to shrimp farming. For the next three years people enjoyed the profit. Annual profit was Rs.20,000/- per acre which was 100 per cent higher than the profit from agriculture. However, the sudden outbreak of viral disease in 1995 caused huge loss in shrimp farming. Shrimp farms were abandoned and left fallow for a decade and farmers faced huge debts. The fallow land was affected with salinity. As a consequence of

Demographic details	
Total no. of household	: 434
Households of Vadabalija community	: 424
Households of Yenadi community	: 10
Total no. of families	: 512
Total population	: 2,052
Total male	: 972
Total female	: 1,080
Agriculture	
Total extent of cultivation land:	430 acres
Total no. of farmers	: 352
No. of large farmers (> 5 ac)	: 3
No. of small farmers (2 to 5 ac)	: 50
No. of marginal farmers (< 2.5 ac)	: 299

this, out of 65 households engaged in shrimp farming, 50 households migrated to the nearby towns as agriculture labourers for two months i.e., during November and December while remaining 10 months they worked as agriculture labourers within Sorlagondi and nearby villages. The incident of tsunami aggravated the issue of salinity and therefore the land continued to remain unproductive. Not only Sorlagondi, but also the adjacent villages faced the same problems due to failure of aquaculture farms.

2. Process of land reclamation

Immediately after tsunami, the partner NGO Praja Pragathi Seva Sangam (PPSS) intervened to reclaim 150 acres of fallow shrimp farm and transform it into a cultivable land. Meanwhile the same village was also identified for implementing the components of the post-tsunami project. In the process of Participatory Rural Appraisal (PRA) and livelihood analysis, 48 farmers with 80 acres of land prioritised the need for reclaiming their land. Debt-ridden farmers depended on the project due to lack of financial resources for reclaiming the land. The project was successful because of two reasons; the land had access to water from irrigation canal and clayey soil was suitable for paddy cultivation. The aim was accomplished by adopting a step by step process with the participation of community. The activity was initiated by preparing

a micro plan with the guidance provided by the Assistant Director, Agriculture Department of Avanigadda mandal. Soil test at 15 locations revealed a high salinity (15 parts per thousand or ppt) which is very high for agriculture. The soil test after reclamation process revealed a drastic reduction in salinity to 5 ppt. Agriculture department certified the suitability of land for paddy cultivation based on the soil test result after land reclamation. Land reclamation involved levelling, ploughing and leaching. Levelling and ploughing was done using machines. Leaching involved a long process of flooding the land for 15 days with fresh water lifted from irrigation canal and draining the lands. This was repeated three times. After first flooding and draining, one tonne of gypsum was administered per acre. Application of gypsum promoted water infiltration and also removed sodium—the element that caused increase in salinity. Defunct and damaged canals were repaired using time, money and local human resources.

3. Community contribution

Apart from 10 per cent contribution of the total land reclamation cost, the community also extended manual labour support. Forty-two men and six women monitored the whole process of reclamation in their respective lands.

Wherever machineries could not reach the land, levelling was done manually. Community also participated in works such as carrying gypsum from main road to their fields, applying gypsum, releasing and draining fresh water during the leaching process.

Cost of reclamation of land

Levelling @ Rs. 3,500/acre for 80 acres	= Rs.28,000
Ploughing	= Rs.40,000
Addition of gypsum	= Rs.20,000
Total cost	= Rs.400,000
Community's contribution	= Rs.40,000

The overall supervision was done by the village level institution (VLI). Mr. Radha Krishnamurthy, a retired agricultural officer and currently the staff of PPSS visited the field twice a week during the first crop and provided guidance to the farmers. He also prepared agro-advisories pamphlets in the local language and distributed them to farmers.

4. Impacts of intervention

From August 2008 to April 2009, two crops have been harvested from the 80 acres of reclaimed land. During the *kharif* season (August to December) around 50 acres was cultivated of which only 32 acres yielded better. Though similar process of land reclamation was adopted for the remaining 18 acres, the yield was less due to high salinity. The intervention had following impacts:

- Harvesting of two crops per year restored.
- Labour opportunities for landless families increased.
- Migration decreased.
- Yield was comparable to other fields that were not affected by salinity.

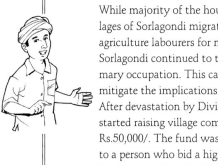

Sorlagondi adaptation strategies

While majority of the households from neighbouring villages of Sorlagondi migrated to the other coastal districts as agriculture labourers for nine months in a year, the People of Sorlagondi continued to treat fishing and agriculture as primary occupation. This capacity enabled them to adapt and mitigate the implications of downfall of shrimp farming. After devastation by Diviseema cyclone in 1977, Sorlagondi started raising village common fund. Initially they generated Rs.50,000/. The fund was auctioned every year in January to a person who bid a highest rate of interest. Thus the fund increased to Rs.1,500,000. When the shrimp farming was abandoned and the subsequent drought added to the vulnerability of Sorlagondi, the fund was distributed equally among 512 families.

- Food grain availability increased.
- Fodder was available for milch animals.
- Land value increased from Rs.20,000 to Rs.100,000.
- Third crop production is in the anvil.
- Visit to field centre of N.G. Ranga Agriculture University exposed farmers to System of Rice Intensification (SRI) that requires less water.
- Adaptive mechanism in place for the community.

5. Conclusion

The post-tsunami project helped the village to understand the importance of adaptive mechanisms and need of community resilience. Villagers have requested for hands-on training for the SRI cultivation and the PPSS will address it in collaboration with the agriculture department. Reclamation process has empowered the community to plan the village development. The project has not only helped the present generation but also has laid a strong foundation for future generations. Interventions in the field need not always be driven by a natural calamity. This is a classic example of anthropogenic unsustainable practice led to degraded environment and unexpected loss. Villagers were able to recognise their wrong-doing, helped NGOs address important livelihood and migration issue.

64

Transfering Ownership to Communities: Flood Mitigation in Ambalantota

The WRMIDMC introduced a new flood management methodology and successfully prevented flooding.

Source:

Dilhani Thiruchelvarajah, Buddika Haporarachchi and Menaka Liyanage *Practical Action, Sri Lanka*
Chintha J. Munasinghe *Laymen's Den (PvT) Ltd., Sri Lanka, 2010.*

1. Background

This is a story of a flood mitigation programme which clearly illustrates that innovations to old methods can indeed produce remarkable outcomes.

Wanduruppa, a village with high population density, is located in Ambalantota (District Secretariat Division) in the Hambantota district of southern Sri Lanka. It lies on the borders of the Walawe river estuary and the Indian Ocean. Flooding has been a problem even before the British Sevara/Dikets period. In the past, a state-appointed leader, designated as 'Modara Arachchi' and subsequently 'Patabendiarachchi', was responsible for the management of floods with the support of the community members. They used to drain out flood water at the river mouth by manually removing sand barriers between the river and the sea. It was a job which requires high level of skill as it must be done methodically especially in monitoring the river flow and the seashore level. Overtime, this system collapsed with new governments, policies, and administation leaving people with no choice but to bear with the consequences.

Every now and then, the villages along the Walawe river experienced flooding especially when the river overflows due to continuous rain. The tsunami of 2004 worsened the situation as it caused the formation of a sand bund at the river mouth everytime the river level goes up due to heavy rainfalls. The tsunami also diverted the river from its natural course as a result of heavy sand deposit along the river mouth. Unless action is taken to rid the river mouth of heavy sand barriers and restore its natural flow, the Wanduruppa village and nearby villages was bound to suffer from frequent flooding. Tsunami gave opportunity for the local institution to visit the village, trace the issue and history of flooding and lost community ownership.

Floods, which occur about 10-12 times on the average per year, risked nearly 2,940 acres of paddy lands in 10 nearby Grama Niladari (GN) divisions (village administrative units) and 462 households. The sand barriers also caused the backing up of water heading to the upstream along the banks, giving rise to varying salinity concentration in the river water. Salt water intrusion affected the rice fields and destroyed the crops. River water interruptions at the sand barrier also prevented fishermen to carry out their trade. All these necessitated urgent and collective action from villagers who primarily rely on agriculture and fisheries for their subsistence and livelihood.

2. Laying the groundworks

In December 2006, Practical Action, in collaboration with Sarvodaya, brought together community leaders in the village to form a village development committee (VDC). The VDC took charge of the development of an integrated plan and set of activities that dwell on disaster risk reduction (DRR), livelihood development, bioshield propagation, infrastructure development, and village information service.

Since flooding has been a priority issue in the village, efforts were concentrated on implementing mitigation activities in cooperation with relevant community groups. A series of multi-sectoral meetings were held involving the District Secretariat (DS) of Hambantota, Disaster Management Centre (DMC), Irrigation Department, Pradeshiya Sabha (a local authority under the provincial council), University of Peradeniya, National Aquatic Resources Research and Development Agency (NARA) and the Coast Conservation Department (CCD).

At least three alternative options were identified to address the flooding problem: (1) infrastructure development in the Walawe estuary; (2) procurement of an excavator; and (3) improvement of the existing conventional flood management methodology. The idea of developing infrastructures was strongly opposed as it would require a comprehensive impact assessment (e.g. sea current patterns, sea current variations), it bears impacts on the proposed Hambantota harbour, and due to unavailability of resources. Purchasing of an excavator was likewise ruled out as they foresee complications as regards the ownership, maintenance, and costs (fuel). Hence, the third alternative was considered as the most applicable and practical solution to mitigate flood problem in Wanduruppa.

A committee, the Walawe River Mouth Integrated Disaster Management Committee (WRMIDMC), was then established to take on the responsibility of carrying out innovations to the prevailing flood management methodology, aside from flood monitoring and reporting to the VDC. The 30-member committee comprises representatives from the affected GN Divisions, government, and local communities. A constitution was formulated detailing the responsibilities, financial management scheme, and protocols of the committee. Training on the technical aspects of removing the sand bar, proper way of monitoring the project were offered to the committee and were encouraged to air their thoughts and needs to the local governing bodies.

A Memorandum of Understanding between Practical Action and the DS of Hambantota was then drawn providing for the endowment and transfer of revolving fund for the project from the Practical Action. The DS Hambantota Office apportioned approximately 92 per cent of the fund to a fixed deposit and the remaining 8 per cent as savings for urgent financial needs. Interests generated from the fixed deposit were used to pay for the excavator rental. Other expenses were made in to the bank to generate sufficient interest to meet the machinery. The committee treasurer and chairman or secretary were the only ones authorised to withdraw funds under the supervision of the Irrigation Engineers' Office, with the approval of the DS.

3. Results

For the first time in Sri Lanka, a flood DRR mechanism was introduced and was proven effective and successful. Records and statistics yielded the following encouraging results:

- Seven floods had been mitigated from only two withdrawals from the savings account for a period of six months.
- Profits have increased from 9.67 million LKR in 2007 to 78.3 million LKR in 2008.
- Only 0.1 million LKR had been lost in 2008 as opposed to 47.7 million LKR in 2007.
- The mitigation measures had successfully prevented the potential flooding of about 807 acres of land.

This is based on a comparative study data (Table 64.1) obtained from farmer organisations and divisional agricultural department.

TABLE 64.1
Comparative analysis of the before and after intervention scenario

	2007 Prior to the intervention (LKR million)	2008 After the intervention (LKR million)	% Increase/ decrease
Loss of paddy	47.7	0.1	(99.79)
Paddy income	49.59	120.22	142
Paddy profit	9.68	77.99	706

Note: Some paddy lands were abandoned due to heavy losses to farmers. Convincing DRR interventions motivated the farmers in 2008 to use 59 acres of abandoned paddy lands to cultivate paddy successfully.

4. Conclusion

The positive impacts and benefits stemming from the project confirm once more the value of teamwork and institutional collaboration among village people, government agencies and non-government organisations in carrying out development projects such as flood mitigation. A project with favourable outcomes has great potential for replication, particularly the approach and methodology applied. The DS of Hambantota invited WRMIDMC to share their experiences in district development coordination meetings and positive results made the village proud of their achievements and value the support provided by Practical Action and other agencies.

65

The Importance of Providing Solutions that Really Work

Source:

Dilhani Thiruchelvarajah and Bhathiya Kekulandala *Practical Action, Sri Lanka.*
Chintha Munasinghe *Laymen's Den (PvT) Ltd., Sri Lanka, 2010.*

Karaitivu in the Ampara district of the eastern Sri Lanka was affected by the 2004 tsunami and the decade-long conflict. Karaitivu is a highly populated area of the province with a Tamil speaking population. Being a conflict-ridden area for decades, the area was characterised by a number of socioeconomic problems such as devastated infrastructure, broken families, ruined income avenues, collapsed institutions, disconnectivity to service agencies. This situation worsened after 2004 due to impact of devastating tsunami. At the time when post-tsunami project began work in Karaitivu, the area was in a highly degraded condition in social, economic, infrastructural and institutional terms.

1. Persistent flooding problem in Karaitivu

A five-kilometre natural canal flows through Karaitivu in the eastern part of Sri Lanka and meets up with the sea from both ends in Sainthamaruthu and Nintavur. The canal primarily drains waste water from the communities living along it and the flood water during the rainy season. It was also used to support agricultural activities and fishing before the tsunami in 2004.

The tsunami dumped debris in the canal that the local government, aided by an NGO, later on cleared. However, the debris left along the canal banks created earth bunds that the people used as roads or pathways. Because the earth bunds were higher than the settlement areas, water could not drain easily and caused persistent flooding that affected 40-50 families in Karaitivu. Flooding worsened with each rainy season as more and more houses were built near the canal. To prevent flooding, the people often dug small drainage canals across the earth bunds into the canal, further damaging and making the road hazardous to vehicles. The earth bunds that the villagers used as roads also frequently caused accidents to happen because these were narrow and easily eroded during heavy rains.

2. Participatory consultation and community action

The project saw the opportunity for community action to help solve the flooding problem in Karaitivu. Practical Action initially held consultation meetings and other activities which the villagers participated in, to work out broader solutions to address the problem. Among these activities was

The Importance of Providing Solutions that Really Work

a transect walk through the area with government officials, community leaders and technical experts to identify issues and arrive at possible solutions to the flooding problem.

At the end of the community consultations, a support group called the Village Development Committee (VDC) was organised to assist Practical Action in coordinating the work that had to be done. Training and capacity building on leadership, communication and collective work were assigned to members of the VDC to empower them to build ownership of the project. The VDC also took the responsibility of materials purchasing, technology issue identification, recommending working steps and solutions, and in working with the community in the canal restoration process. The VDC monitored village level operations while Karaitivu officials monitored overall project performance. The project identified the immediate construction of a well-planned flood control system as a solution to the persistent flooding in Karaitivu.

Communities along a natural canal flowing through Karaitivu were often inundated in flood waters that caused public health and environmental problems.

Practical Action and a Karaitivu technical officer designed the flood control system which was submitted for approval by appropriate bodies of the Karaitivu Division. They partnered with the District Economic Social Mobilization Organization (DESMIO) for the latter to construct the flood control system's gabion walls. Aside from hired labourers, village men and women volunteered to take charge of packing the gabion mesh boxes with granite blocks. While monitoring the project, Practical Action observed that by the time 75 per cent of the construction project had been completed, the gabions were not packed and laid out correctly such that some had already detached from the walls. The community were assembled by Practical Action to make observations and analysis on the cause of the defects in the construction. After the exercise, the community decided to repeat the whole process and correct the mistakes made in constructing the gabion walls. Correcting their mistakes meant an additional expense (of LKR 75,000) and delay in completing the flood control system but upon consideration of the implications of a defective infrastructure on community life, Practical Action discussed this with IDRC who accommodated the request to undo the mistake and ensure a well-constructed flood control system in Karaitivu.

3. A solution that really works

The flood control system was rebuilt. A strong base was used to hold the gabions that were topped off correctly, completed with a strip of grass turf on both sides. A regulatory mechanism and a system were put in place in the event of floods. The Village Disaster Management Committee, which manages the flood control system, reported no flood occurrences in the lowland area since it was completed. Villagers living on either side of the canal are very grateful that the flood control system worked, solving the flooding situation, and easing the environmental and public health problems caused by flooding. People were able to travel on better roads and transport their products smoothly. They look forward for the rest of the canal to be constructed against floods and erosion. After experiencing the process in which the solution to the flooding was arrived at and construction was undertaken, the Karaitivu Division officials incorporated into their future plans the construction of gabion walls along the entire length of the lagoon and have started to approach donor

The Importance of Providing Solutions that Really Work

Solutions to development problems are best arrived at when the common good is given top priority.

Vigilance in monitoring infrastructure projects and in making sure that they are structurally sound ensure that resources do not go to waste.

agencies to secure funds for this. Meanwhile, the community attitude towards protecting the environment around the lagoon has been observed to have changed as indicated by more community programmes organised to clean the area and to implement a waste management system to reduce the waste disposal to the canal. This shows that the process of identifying and implementing the flood control system paved the way for openness to change and more receptiveness to collective action in solving problems and concerns.

Key lessons

- Human capital of conflict (and disaster) affected areas have to be assessed carefully before the participatory development interventions. Whenever the community/local institutional capacity is not adequate for a development activity, additional external assistance/resources are definitely required.
- Quality parameters of particular activity are defined by the local people living in the area. It will depend on the local capacities, resources and needs of the community.
- Short-term development actions implemented after disastrous events can lead to long-term impacts and some can end up as development induced secondary disasters.

Thus disaster reconstruction programmes should be designed within the broader development framework.

66

Collective Management of Resources in Village Development

The CDC and SSS organised the community to clean along the two-kilometre canal.

Source:

Chintha J. Munasinghe *Laymen's Den (PvT) Ltd., Sri Lanka.*
Indika Dilhan Somarathne *Sarvodaya, Sri Lanka, 2010.*

1. Introduction

Thalalla is a coastal village located south of the Matara district in the Dondra Dimensional Secretariat Division in Sri Lanka. Aside from white coir production, which the village is well-known for, agriculture also functions as the main livelihood to about 35 per cent of the total number of families in the village. For irrigation water, the Thalalla farmers depend on a natural upstream tank called 'Ela Kanda,' which serves 114 acres of paddy fields through a 15-kilometre long irrigation canal system. A two-kilometre water tributary that crosses Thalalla connects the canal to a lagoon that opens to the sea.

In a span of two decades, the irrigation canal has deteriorated. Gradually, the canal system was damaged resulting in upstream leakage. From then on, upstream villages put up with the problem of flooded rice fields, while downstream villages experience water scarcity. Majority of the farmers who rely on the canal system was forced to abandon their rice fields. This situation lingered for years.

The 2004 tsunami affected the lives and infrastructure in this coastal community even more. It completely blocked the already ruined canal system. As if the impacts were not enough, the tsunami even destroyed the entire canal doors and saline gates causing saline water to creep inland. Additionally, thick weed and reed species grew in the uncultivated agricultural lands. The livelihood of the people was on the slump.

After the tsunami, a concerned NGO called GOAL, came and helped the community clean up and restore the canal. The work, not as easy as it seemed, warranted technical expertise, hence, the first community programme failed. In 2006, Sarvodaya and Practical Action revisited the canal restoration to revive the uncultivated land to agricultural land.

2. Success factors on the restoration of the irrigation canal

Several factors led to the success of this Participatory Canal System Restoration Project.

- *Multistakeholder consultation*: At the start of the planning process with the post-tsunami project staff, the Sarvodaya District

Centre (SDC) consulted with the irrigation department of Matara District for the possible reconstruction of the irrigation system. Several meetings with the village agriculture officer, Sarvodaya Shramadana Society (SSS), and the Thalalla Farmers' Organization ensued to formulate a canal development plan integrating community participation. Engineers and resource persons from Practical Action also participated in these meetings with their technical and development expertise in community-based planning processes. The series of meetings led to the formation of the Canal Development Committee (CDC) composed of representatives from SSS and Practical Action. The village temple priest was even involved as an adviser to CDC. CDC was mainly responsible for organising volunteer labour, procurement of materials and services, and financial monitoring.

- *Community participation*: CDC and SSS joined forces and organised a two-day Shramadana to clean up and get rid of the weeds and reeds along the two-kilometre stretch of the canal. A total of 200 community people participated in this clean up drive comprising about 100 from Thalalla and the rest from nearby villages. Even youth groups from Matara Sarvodaya Societies came to support. Only the men worked for the restoration of the canal system.

  Although villagers worked on voluntary basis, the project compensated to cover their daily basic requirements. On top of the average working hours of about 10 hours (7.00 am to 5.30 pm), the community people offered to work for extra hours (one hour in the morning and half an hour in the evening) on a voluntary basis. Majority of the work was done through manual labour, except in cases when the canal was too deep and filthy that it was but logical to use machinery.

- *Facilitation of the process*: The draft plan and proposed budget was approved by the Irrigation Department only after two months of submission. The budget allotted for the project was 3.2 million Sri Lankan rupees and the Irrigation Department also mobilised government funds for the reconstruction of saline gates.

> *Capacity-building initiative*
>
> The purchasing sub-committee was responsible for all the procurement work which includes issuance of materials and replenishing stocks. The Technical Officer in-charge of the canal restoration gives me the list of materials required for scheduled construction work. I then bring this list to the purchasing committee who will then hold a meeting to approve the purchasing of items as well as to call for quotations where necessary. After deliberations, we then decide where to purchase the items. It was a tough job but we gained experience on how to deal with traders, compare prices and decide on the appropriate material to purchase.
>
> - Anusha, Community Facilitator.

As part of the organisational preparations, members of the CDC were equipped with financial management skills through trainings conducted by SSS. A tender board was also formed by the village for the procurement and purchasing of needed materials and equipment. Sub-committees were also formed for financial monitoring and to ensure transparency and improve efficiency in operations.

- *Effective partnerships*: The restoration of the irrigation canal allowed for the provision of water to 114 acres of paddy land, from which 22 acres were newly cultivated by 20 farmers. Through effective partnerships with the government and the NGOs, follow-up interventions were also observed: The canal construction process had built harmony among the villagers and CBOs as a result of working together for a common goal.

- *Social mobilisation*: Some sub-committees operated well; others, far from what were expected from them. In such cases, active CDC members took over the responsibility of all the other sub-committees to get the work done. There were also few leaders who committed to do the work and served as able guides to others. A learning that was drawn from these scenarios was people may be genuinely enthusiastic in joining community work at the start, but it is difficult to sustain the momentum. Hence, as what CDC did, active leaders and members should be ready to take on the task of other members who fail to function. On the other hand, some members may seem to be less interested and prefer to be mere observers. But when assigned and given duties by the leader, they may willingly do their share.

Effective partnerships

A Japanese NGO Jade Green volunteered to support the community's initiative by clearing the reeds growing in abandoned areas to prepare the barren land for cultivation. About 20 acres of land have been prepared and another 20 acres in progress. They also supported the farmers by supplying seed paddy and other necessities to start with. The whole process was a collaborative effort of the community, government and NGOs.

- Alahakoon, Thalalla Farmers' Organization.

3. Conclusion

With the restoration of the irrigation canal, the farmers cultivated their paddies again in less time. The farmers also observed an increase in their yield brought about by the improved supply of irrigation water. As a result, the farmers also enjoyed other benefits. Some were even able to purchase additional agricultural land from the income gained through improved intervention. This intervention was given special recognition during the District Secretariat's Coordinating Meeting. This goes to show that nothing beats collective effort particularly for development projects. Not only does an individual gain, but the impact had a ripple effect on the entire village as well.

The intervention through its participatory approach to canal restoration has not only addressed a major cultivation issue in the village. The exercise has strengthened the bond between SSS and the farmer organisation in the area, that has encouraged them to put collective efforts in addressing problems in their village. For farmers to continue their agricultural activities in the village, the farmer organisation is taking

Improved intervention, improved life

I still remember how we struggled bringing water down through the canal to cultivate 10-15 acres of land. Now, that time is saved. After the canal restoration, the yield from cultivated lands increased. In my case, all my fields are now cultivated. Of the 111 acres of land in our village which were not cultivated for many years, about 100 acres are now ready for cultivation in the coming seasons. There are farmers who were able to purchase land with the money they earned from the paddy. Thanks to those who came forward to help us solve a 20-year problem that we never thought of solving.

- Alahakoon, Thalalla Farmers' Organization.

the responsibility of maintaining the canal with the assistance from other community members also.

References

http://srtac.wordpress.com/about/ (accessed 3 November 2009).
http://www.sarvodaya.org/ (accessed 3 November 2009).
http://practicalaction.org/uk-about-us/whoweare (accessed 3 November 2009).

67

Reviving Traditional Paddy Farming in Andaragasayaya

Although yield of traditional paddy rice is low, market price is higher than that of the high breed rice.

Source:

Chintha J. Munasinghe Laymen's Den (PvT) Ltd., Sri Lanka.
Indika Dilhan Somarathne Sarvodaya, Sri Lanka, 2010.

1. Introduction

One of Sri Lanka's main suppliers of rice is Andaragasyaya which is located in Southeastern Sri Lanka in Tissamaharama. The farmers of Andaragasyaya harvest rain water for farming.

For many decades, the farmers of Andaragasayaya were cultivating hybrid varieties of paddy, which they believe to have better harvest than traditional varieties. Since the introduction of hybrid varieties, they are using pesticides which has increased their production costs, compared to the traditional varieties, which is proven to be more nutritional.

After the tsunami, Sarvodaya and Practical Action saw the need to introduce the traditional paddy varieties to Andaragasyaya Grama Niladhari Division (GND). Traditional rice varieties can withstand salinity and floods, have high nutrition value recommended by doctors for illnesses such as diabetes, arthritis, and hypertension. Therefore, it has a niche market that offers high prices to the produce.

2. Participatory intervention

Using Participatory Rural Appraisal, the post-tsunami project conducted community consultation meetings. The community identified traditional paddy and chena cultivation, canal restoration to prevent flooding and coir processing as key livelihood options. Based on these, the project team designed the intervention with the community.

3. Discovery-based awareness programme

An awareness programme was organised to educate the farmers on traditional paddy cultivation, its advantages, and disadvantages. An exposure visit to southern traditional paddy growing areas was organised by the project to help the farmers understand the advantages and disadvantages of traditional paddy cultivation and its suitability to their fields.

The 30 farmers who joined this visit learned that the cost of cultivating traditional rice paddy was low compared to that of hybrid paddy varieties; and that although harvest was low, market price was high. Aside from these, they also found out that the variety used almost no pesticides and insecticides and that water requirement was also low.

4. Co-farmer experience sharing

Four farmers interested in testing traditional paddy cultivation were given six traditional paddy varieties to cultivate in one acre of land. After the first season, they were able to identify four varieties (Rathdel, Kurulutuda, Pachcha perumal, Dahanala) most suitable to Andaragasyaya soil condition.

In the second season, encouraged by the economic gains of the four farmers, eight more farmers joined, increasing the cultivation area from 1 to 3.5 acres. In the third season, nine more farmers joined, thus increasing the land area allocated for traditional paddy to six acres.

5. Capacity building

A range of training programmes were conducted on different subject areas: land preparation; transplanting methods; weed and pest control; liquid fertiliser making, etc. In addition to the free seed paddy supplied by the project, equipment such as weeders, seeders, sprayers, plastic barrels

A range of trainings were conducted to equip the farmers in traditional paddy cultivation.

> *Recent history of traditional paddy farming*
>
> Only 38 per cent of farmers in Sri Lanka own more than two acres of paddy lands, who contribute to the market. The remaining 62 per cent are those small holders who own less than two acres of land, with paddy grown for consumption with more dependents in their families. Statistics shows that 49 per cent of paddy lands are dependent on large irrigation schemes; 26 per cent on rains and 25 per cent on small irrigation systems.
>
> Traditional paddy varieties were introduced to wet-zone small paddy holders as an alternative to existing practice of cultivating hybrid rice varieties, which was found to be not suitable for paddy lands that depend on rains and small irrigation systems.
>
>
>
> There are about 80 varieties of traditional paddy known to Sri Lanka. Few of the key characteristics of traditional paddy are its ability to be cultivated without weedicides, pesticides, and chemical fertilisers. Traditional paddy varieties are environment- friendly, less labour intensive with low production costs. Research has also found that traditional rice varieties are more nutritious compared to hybrid varieties.

were also provided. Part of the harvest was kept for producing seeds paddy for the next season.

6. Marketing

Practical Action and Rural Enterprise Network (REN) discovered that there is a niche market for traditional rice, which has been explored through Participatory Market Chain Analysis (PMCA) exercises conducted with the participation of stakeholders in this sector. REN came forward to buy traditional paddy from the Andaragasyaya village for 45- 50 LKR per kilogram. Farmers agreed to supply their surplus produce on the price proposed.

7. Progress monitoring

Farmer organisation usually hold meetings to discuss issues and arrive at decisions and actions, twice a season. The project team participated in the meeting as an opportunity to review the progress and learn about issues and analysis farmer observations.

The progress of traditional paddy cultivation was also shared at these meetings, which became more of an awareness session for other farmers. These generated interest among more farmers to join after the second season, thus increasing the land area allocated for traditional paddy.

8. The challenges

Farmers consider that traditional paddy cultivation is more time consuming and labour intensive than in the case of hybrid paddy, though the farmers participated in this intervention, revealed that it strengthened the family unity by working in the field together. Many believe that traditional paddy is suitable for smaller land plots and not for commercial based ventures. There is a market in the village itself, among other families, but to get higher price with a profit margin, they need to reach high and market as it produced for a niche market. Transportation is another challenge that farmers face, as they need to supply on time and in expected quantities. Only two organisations are promoting markets at present, which they also need to be expanded with increased awareness among the farmer communities. Intervention such as this need more time to address on going challenges as solutions unfold through the process.

68

Collective Action for Eco-Shrimp Farming in Sorlagondi

Collective harvesting and marketing allowed farmers to command a better price for their shrimp.

Source:

J.D. Sophia and R. Ramasubramaniam *M.S. Swaminathan Research Foundation, Tamil Nadu, India.*
P. Venkateswara Rao and E. Ram Babu *Praja Pragathi Seva Sangam, Andhra Pradesh, India,* 2010.

1. Introduction

Commercial shrimp farming gained popularity in the 1970s with the heightened demand for shrimps and prawns in the global market. The delta village of Sorlagondi *panchayat* of Nagayalanka mandal, Krishna district in Andhra Pradesh was, likewise, enticed to take part in the growing industry. Sorlagondi has a flat topography and is frequented by cyclones and flood. Primary livelihoods include fishing done through capture fisheries[1] and culture fisheries,[2] agriculture, and shrimp farming, done through capture fisheries and culture fisheries.

Aquaculture gained momentum in Sorlagondi in the early 1990s and as a result, about 230 acres of fertile agricultural land and 1,000 acres of saline areas (mud flats) were converted into shrimp farms. Shrimp farming proved to be a profitable endeavour as farm harvest brought in an annual profit of Rs.20,000 per acre. The following year, an outbreak of viral disease, coupled with poor management practices (i.e., overstocking, excessive use of artificial feeds, etc.) drastically affected the shirmp farms causing huge losses. As a result, farmers abandoned their farms. Since 1996, the farms were left fallow and it lasted for nearly a decade. Converting the farms to agriculture was not possible due to lack of freshwater sources. Continuing with shrimp farming was also not an option as the farmers were debt-ridden and lack financial and technical support to revive the ponds.

2. The evolution of farming practices in Sorlagondi

2.1 Extensive farming

At the start of their shrimp culture:

- Wild seeds were collected from the mangrove creeks, along the shore, and also purchased from the hatcheries.

---

1. Capture fishery refers to all kinds of harvesting of naturally occurring living resources in both marine and freshwater environments.
2. Culture fisheries are the cultivation of selected fishes in confined areas with utmost care to get maximum yield. The seed is stocked, nursed, reared in confined waters and the crop is harvested. Culture takes place in ponds, which are fertilised and supplementary feeds are provided to fish to get maximum yield.

Profit outweighs the input cost

A farmer requires 200 litres of diesel in a month to pump water for one pond of 2.5 acre. Therefore, about Rs.7,000 was spent on diesel alone which constituted 70 per cent of the total input cost of Rs.10,000. But the high cost of inputs was offset by the good harvest of two crops a year with 200 per cent profit i.e., with an investment of Rs.10,000/-, the income increased to Rs.30,000/- giving a net profit of Rs.20,000/- for each cropping season.

- Rice bran was used as feed.
- Brackish water was pumped into the ponds using diesel-powered engine. Farmers collectively bought the engines at equal contribution, with 2-3 farmers sharing for a five-hp engine. The farmers used the engine alternately based on an agreed schedule. Whenever the engine breaks down or requires general maintenance, the farmers also divided the expenses among themselves. The success of this system was attributed to the cooperative nature of the village people.

2.2 Extensive farming to semi-intensive farming

Eventually, people from other cities like Hyderabad, Vijayawada, Machilipatnam came to Sorlagondi village and started shrimp farming on a leased land while others merely encroached on the agricultural and forest lands. Both the local villagers and the transients had no technical know-how on shrimp farming; hence, they merely relied on their own limited understanding of it. The extensive farming practices were then altered with semi-intensive system.

- The farmers used commercial feed, chemicals, and aerator. This caused feed and chemical suppliers to flock the village.
- Seeds were obtained from hatcheries.
- Stocking density was increased.

The semi-intensive farming resulted in high input cost. But despite all these, the farmers still raked in very high profit reaching up to 200 per cent more than what they have invested. This attracted the local shrimp farmers to change their farming practices from extensive system to more or less similar to the semi-intensive system.

While the system provided high income to the farmers, the shift in farming practices led to major setbacks such as poor water and pond management practices and irregularity in the culture process due to inconsistency in the timing of seed stocking. Viral diseases started to set in later with the importation of viral-affected mother prawns. All these factors led to the decline in farm productivity translating to losses. Local villagers tried to save whatever is left while the transients started to leave the village.

The local farmers incurred huge losses, especially those in the saline areas, yet they remained unrelenting in their efforts to save the industry and keep their businesses afloat. This they did by reducing the input cost.

Since 70 per cent of the input cost was spent on diesel, the best solution was to shift from diesel-based to electrical engines. Applying the principle of collective action, the people organised themselves into groups so that they could acquire electric motors.

As part of the project, the MS Swaminathan Research Foundation (MSSRF) and Praja Pragathi Seva Sangam (PPSS) conducted a participatory rapid appraisal (PRA) and livelihood analysis in the village. During the exercises, it was discovered that most of the shrimp farms, though legally-owned, do not have the necessary licence from the aquaculture authority. Hence, the first step was to register the farms and have them licensed. Later on, the farmers were oriented on the pros and cons of the extensive and semi-intensive methods of shrimp farming and ascertained of their commitment towards the extensive farming practices. The community undertook the blend of extensive and semi-intensive farming methods.

Once the community came up with common consensus, the project extended its support by providing 22 electrical motor pump sets. The total cost was Rs.561,000 of which Rs.66,000/- (12 per cent) was from community contribution. Community people came forward to take up the cost of constructing the shed for the motor pump sets. Since 22 motors catered to 22 clusters comprising 180 farmers, each of them shared Rs.300/- and constructed 22 sheds. Eventually, all the 22 motors were connected and the process for first crop was started.

2.3 The eco-friendly shrimp farming practices

Drawing from the farmers' experiences and lessons learned, developments were observed and corrective measures were implemented in the village.

- With the establishment of the electric motor pump sets, water pumping became easy and cost effective.
- Seeds were purchased from certified hatcheries where the brood stocks were tested for viruses and diseases before they are sold. The seeds required by at least 180 farmers were purchased collectively by the representatives of the society.
- Instead of commercial feeds, all the farmers went back to using only rice bran as feed which is locally available. Optimum feeding

Cost sharing between societies and the government

A total of 22 groups were formed among owners of 365-acres shrimp farm. The farmers inquired on how they could avail of 22 electric connections to erect 22 electrical motor pumps and the needed transformers. The estimated cost for each connection was Rs.35,000/- and the total estimated cost was Rs.770,000/-. Of this 50 per cent came from community contributions while the remaining 50 per cent was covered by a Member of the Parliament under the local area development funds. They got the connection but could not utilise the facility as they have to spend an additional amount of Rs.25,500/- for purchasing an electrical motor pump set and erecting motor shed. For one and a half years they could not do anything. It was at this time the PPSS and MSSRF identified Sorlagondi as one community beneficiary under the International Development Research Center-Canadian International Development Agency (IDRC-CIDA) post-tsunami project.

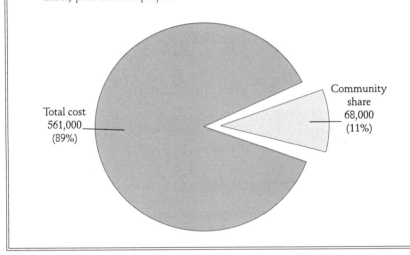

was done which not only reduced the cost but also helped in maintaining water and soil quality.

- Monitoring of water quality, disease incidences, and the feeding behaviours were done regularly to avoid risk. Two local persons, appointed and trained by the National Centre for Sustainable Aquaculture (NaCSA), did the monitoring. They test the water

Collective Action for Eco-Shrimp Farming in Sorlagondi

quality daily for its salinity, pH, and dissolved oxygen. When changes in water quality were observed or a disease occurrence was imminent, the concerned farmer was informed and provided with necessary advisories.

- Previously, harvesting and marketing were done individually and per demand. Hence, the farmers found it difficult to sell directly to exporters because of the high transportation cost for both parties. The harvests were then sold only in the local markets at a very low price. Under this eco-friendly farming system, collective harvesting and marketing became the practice. Exporters themselves come to the field to procure large quantities of produce and the farmers are able to command a better price. The normal harvest from one pond is 200 kilogram of shrimp and farmers get an additional amount of Rs.50/- per kilogram which is equal to Rs.10,000/- per crop per pond.

3. Other interventions

3.1 Institution building

To avail of benefits and entitlements from the government, the farmers were mandated by the PPSS to organise into local institutions or societies. So far, 10 societies with 20-50 members have been formed and registered under the Society's Act of 1860. All 420 families involved in shrimp farming are members of these societies. Membership and annual fees are being collected from the farmers. A contribution of 5 per cent from the farmer's total profit for each cropping season is also being collected and goes to a common fund to sustain the functions of these societies.

3.2 Networking and linkaging

The PPSS and MSSRF played a vital role in establishing linkages between the farmers and government departments that provide technical guidance, for licensing, and mobilisation of programmes and entitlements.

Networking and linkaging

Initial networking initiatives with the Fisheries Department enabled fund mobilisation for the acquisition of electrical supply. The Aquaculture Authority provided license to the farmers for shrimp farming. While NaCSA continues to provide supports like: (i) identifying certified hatcheries for quality seed, (ii) provide technical guidance on extensive farming as and when required, creating marketing tie up both nationally and internationally, and (iii) provide financial support to the societies. In addition, NaCSA is in the process of branding the product as organic shrimps and market them directly to the other countries where the shrimps will fetch higher price for the benefit of the farmers.

3.3 Eco-friendly farming and collective farming

The first cropping season begun in January 2008 immediately after the electric motor pump sets were installed. About 165 farmers implemented the eco-friendly shrimp farming in the 360-acres saline area. In the four-month (January-April 2008) farming period, the farmers adopted eco-friendly management practices such as use of disease-free seeds, regular water exchange process to maintain good water quality, and optimal feed consumption using rice bran.

The overall cost-benefit analysis of the first crop revealed that the total input cost per acre of shrimp farming was Rs.4,000/- and the average income was Rs.14,000/- per acre. The net profit from one acre of shrimp farm was Rs.10,000/-. The total profit from 360 acres was Rs.36,000,000/- as shown in Figure 68.1.

The second cropping season started in June 2008 after drying the pond for one month. All of the 360 acres were under farming from June to October. In addition to the eco-friendly farming practices that had been adopted in the first cropping season, collective farming and water quality monitoring were practiced by the community.

The whole process was institutionalised by organising farmers into societies. Though, collective farming system was insisted by the

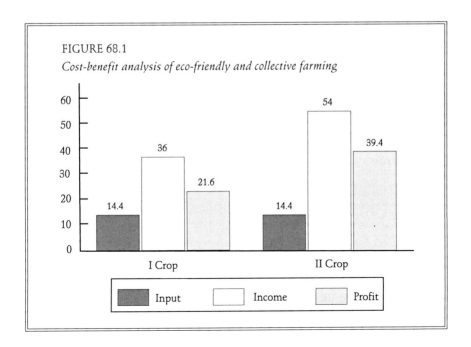

FIGURE 68.1
Cost-benefit analysis of eco-friendly and collective farming

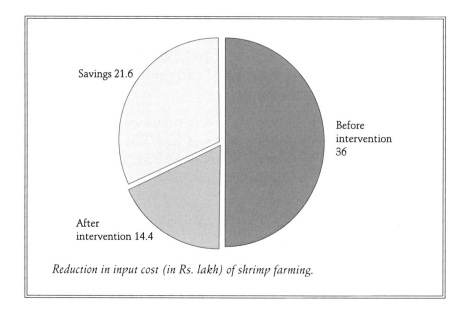

Reduction in input cost (in Rs. lakh) of shrimp farming.

concerned departments from the start of shrimp farming, it was not well-established anywhere. But this initiative proved that collective farming and water quality monitoring are effective management practices as they contributed to the reduction of production cost while increasing profit by 50 per cent more than the first crop with a total profit of Rs.54,000,000/-.

It unveiled the drastic debt reduction of up to 50 per cent among the villagers. After two years, the entire village became debt-free as a result of the intervention.

4. Outcome

Having gained good profit and realising the importance of networking with line departments, the farmers themselves have taken the initiative to replicate similar farming system in the saline lands other than the area covered by the project. As a result, one of the societies got electric supply for 18 electrical motor pump sets by availing 50 per cent subsidy i.e., Rs.250,000/- from Marine Products Export Development Authority (MPEDA). In addition, they mobilised Rs.540,000/- through PPSS from Diakonie Katastrophenhilfe (DKH), a Germany-based donor and erected

18 electrical motor pump sets and begun shrimp farming. Apart from that, the NaCSA has provided Rs.50,000 as Revolving Fund (RF) for three societies. The main purpose of the RF is for internal circulation to meet out the shrimp farming requirements and repay with 12 per cent interest. Process to provide RF to other seven societies is in the pipeline. The value of land before reviving these shrimp farms were Rs.5,000/- per acre but the current value is Rs.50,000/- per acre.

5. Conclusion

Across coastal India, there are thousands of hectares of abandoned shrimp farms. NGOs and the government can replicate the eco-friendly and collective shrimp farming model by providing forward and backward linkages. Reviving and bringing back such lands will not only increase land productivity but will also build assets that would enhance sustainable livelihoods and improve food productivity.

69

Facilitation makes a World of Difference in Danavaipeta

By using an ice box, farmers found a way to break the debt cycle.

Source:

S.V.V. Prasad, P.Suvarna Raju, N.Chitti Babu and R.Ramasubramanian *M.S. Swaminathan Research Foundation, Tamil Nadu, India, 2010.*

Danavaipeta is a village *panchayat* in Thondangi mandal of East Godavari district, Andhra Pradesh and Narsipeta is one of its hamlets. It is located 45 kilometres away from Kakinada, the district headquarters.

Danavaipeta is a homogenous village where except for eight Yerukula tribal households, the people are primarily fishers. In Narsipeta hamlet, around 70 per cent and 27 per cent of the households are Vadabalija and Settibalija community, respectively, while the remaining 3 per cent are Rajakka, a service caste of washers. The major occupation of Vadabalija is fishing, and for the Settibalija caste group, it is toddy tapping. During the lean season, the members of the latter caste work as agriculture labourers.

The demographics of Danavaipeta and Narsipeta show a greater number of people belonging to the lower rungs in terms of economic status. Together, the 'poor' and the 'poorest of the poor,' the women-headed households included, make up 80 per cent of the total number of families in Danavaipeta and 85 per cent in Narsipeta.

1. Fishing as a source of livelihood

Fishermen in these two villages either own non-motorised country boats, i.e., catamarans, or diesel-powered fiber glass reinforced plastic (FRP) boats. The fishermen either go on single day fishing from 3 am to 12 pm or deep sea fishing for 2 to 3 days. Small boat owners mostly go for single day fishing while the motorised boat owners stay for 2 to 3 days at sea. Around 6 to 7 fishermen usually go together for deep sea fishing.

Shore seine fishing, where nearly 25 to 30 families go fishing together, is also practiced in these villages. The fishing area starts right from the shore. Two persons go in a catamaran[1] and drop their nets in the sea in a semi-circle manner for about 5 to 8 kilometres stretch covering around 1.6 kilometres radius. The net is then dragged towards the shore to catch the fish. Around 80 families are dependent on this type of fishing during the

---

1. A catamaran (from Tamil *kattumaram*) is a type of multihulled boat or ship consisting of two hulls, or *vakas*, joined by some structure, the most basic being a frame formed of *akas*. Catamarans can be sail- or engine-powered.

Facilitation makes a World of Difference in Danavaipeta

Danavaipeta village	
Total households	: 4/4
Total families	: 789
Poorest of the poor	: 158
Poor	: 473
Middle	: 158
Women headed HH	: 76
Narsipeta hamlet	
Total household	: 104
Total families	: 252
Poorest of the poor	: 42
Poor	: 172
Middle	: 38
Women headed HH	: 26

Danavaipeta	
Country boat	: 32
Boat with motors	: 30
Narsipeta	
Country boat	: 21
Boat with motors	: 10

months of October to April. On other months, the fishers go as labourers on the big fishing boats.

According to the fishers, one of their primary problems is the lack of proper storage for their catch. Since fish prices are assessed partly according to the freshness of the catch, fishers are often forced to sell when they reach the shore, most often to middlemen who dictate lower than the prevailing buying price. The fishers are at the mercy of the middlemen because they would rather sell to the middlemen rather than end up with even less proceeds or possibly nothing at all if the catch loses its freshness. The lack

The fisher families of Danavaipeta and Narsipeta practise shore seine fishing especially in the months of April to October. They also fish in their non-motorised boats or diesel-powered reinforced plastic boats, or temporarily get employed as labourers on big fishing boats.

of storage facility and middlemen exploitation ultimately push fishermen into a debt trap, making life even more miserable than it already is.

Participatory rural appraisal carried out by the MS Swaminathan Research Foundation (MSSRF) unveiled information that the fishers did not have access to insulated ice boxes that the government of Andhra Pradesh provided at a subsidised rate in the year 2000. When MSSRF asked the State Fisheries Department (SFD) the reason for this, it was found that all the fisheries-related development interventions are implemented through the village level fishermen society promoted by the department. Unfortunately, fisher societies in these two villages are defunct because of their poor performance in repaying credit availed from the government. Hence, the government's reluctance to choose these villages for ice box distribution.

2. A problem, a solution

In order to ensure that the fishers will be able to acquire ice boxes, the MSSRF and the SFD agreed on a joint venture. Village level meetings were

Facilitation makes a World of Difference in Danavaipeta

Lacking preservation facilities, the fishers have no choice but to sell their catch to the middlemen at a very low price.

convened with the Fisheries Development Officer (FDO), a representative of the MSSRF, and the Village Development and Management Committee (VDMC) in attendance. The FDO explained the need for and the use of an ice box, and the government scheme that provides ice boxes at subsidised rate. The proposed scheme was for 50 per cent of the cost to be subsidised by the government while the remaining 50 per cent will be borne by the community. The criteria for identifying the beneficiaries are that: (i) they should be traditional full time fisherman, and (ii) should have registered their boats with the SFD.

Initially, around 150 fishermen who satisfied the defined criteria came forward with the need for ice boxes but when they learned about the 50:50 cost sharing, many backed out. Although the fishing community realised that the ice boxes were essential for quality fish and to ensure better income, their level of confidence towards the scheme was very low because of their bitter past experience. A few organisations had once collected money for the same purpose from these villages but failed to keep their promise.

Fishers store their catch in ice boxes for longer storage life.

To solve the problem, the MSSRF and VDMC members, in consensus with the community, decided on another cost sharing ratio of 1:1:2 or 25 per cent (Rs.48,624) by the community, 25 per cent (Rs.48,624) by the MSSRF, and 50 per cent (Rs.92,559) government subsidy through the SFD. In spite of the lower share, only 65 out of 150 members came forward and availed themselves of ice boxes within the stipulated date set by the implementing organisations. Because the MSSRF learned that the fishermen did not know how to store fish in the ice boxes, it facilitated a demonstration on the use of the storage equipment. Part of the training was the hygienic handling of the fish during storage.

3. Laggards catching up

The fishers have realised now how useful the ice boxes are in preserving the freshness of their catch. Their knowledge has even deepened with the realisation that the keeping quality of the fish is related to hygienic handling that was also taught to them during the training. Moreover, they have acquired some keen business sense because when they get a good catch of commercially valuable fishes like pomfrets, seer fish, sea bass and prawn, they preserve and take them straight to the harbour in order to sell at higher price. While comparing the prices between the shore and harbour, the fishers get from Rs.20 to 50 per kilogram price advantage for these fishes (Figure 69.1). Ice boxes are also used to store other types of fish species, namely, Rastrelliger (*Kanagurta*), Stoliferus anchois (*Netthallu*), and Trichiurus lepturus (*Chavudalau*).

The most interesting turning point is that after observing the benefits of the ice box, the remaining 85 fishermen who initially refused to avail themselves of the cost sharing scheme signified their desire to become project beneficiaries.

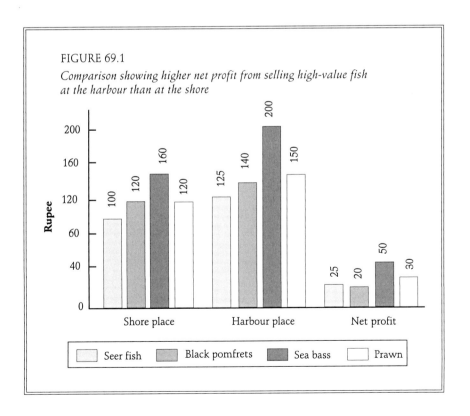

FIGURE 69.1
Comparison showing higher net profit from selling high-value fish at the harbour than at the shore

Focus group discussion (FGD) was held with 15 of the 65 fishers to discuss the benefits of using ice boxes. The participants said that on an average, they fish nine months in a year. They do not fish during the banned periods of April 15-May 31, on festivals such as sankranti, dussehra, and other local celebration, and during cyclones.

In these nine months, they used ice boxes for approximately 100 days to store fishes that have high commercial value. From using ice boxes for 100 days, the fishers received additional income of Rs.59,375, of which Rs.29,000 was used to cover the cost of transporting the fish to the harbour, labour, and the cost of ice. The net profit earned by a group of five people is Rs.30,375, and so individual fishers earned additional income of approximately Rs.6,075 in 100 days after using the ice boxes and following hygienic handling of fish. Not much, according to the fishers, but enough to help take them out of the debt trap.

4. Conclusion

This case has demonstrated how a simple intervention has made a world of difference in the lives of poor people like the fishers of Danavaipeta and Narsipeta in Andhra Pradesh. Using the ice boxes and knowing the hygienic way to preserve their catch has given them better control of their lives. A simple ice box has helped to break the debt cycle in which they have been entrapped by poverty and powerlessness.

Furthermore, the MSSRF and the SFD partnership have shown how to restore the confidence and trust of the people in institutions. At first hampered by the people's negative opinion and feelings towards the project, the implementers decided to probe the reason why the people did not want to avail themselves of the ice boxes. The implementers chose to be flexible and shouldered a big fraction of the cost in their genuine desire to help the fisher communities.

70

Weaving in Technology to the Coir Industry: The Thalalla Experience

The adoption of a new technology (motorised rope-making machines) improved efficiency and productivity of weavers in Thalalla.

Source:

Chinta J. Munasinghe *Laymen's Den (PvT) Ltd., Sri Lanka.*
Indika Dilhan *Somarathne Sarvodaya, Sri Lanka, 2010.*

1. Introduction

The coir industry has been one of the major sources of livelihood in Sri Lanka. In fact, Sri Lanka is the single largest supplier of brown coir fibre in the world market, with an aggregate production of almost 90 per cent of global coir exports together with India. Dozens of Sri Lankan companies export high-quality fiber and coir products such as geo-textiles, rubberised coir mattresses, and upholstery material.

But it is said that the machine-made choir is not as attractive as the beach-pit coir[1] being produced mostly in the southern and northwestern provinces of Sri Lanka. Unlike the machine-made brown coir, the traditional beach-pit method allows for waves to repeatedly wash away the brownish colour from the husks thus yielding clean and white coir.

White coir production remains to be a cottage industry employing mostly women. Being small-scale, the industry was limited to producing ropes, mats, brushes, and brooms merely for local consumption and could hardly compete against big producers.

2. The tale of Thalalla

Thalalla is a village located in Matara district in a province south of Sri Lanka. For generations, coir processing has been the main livelihood of about 62 per cent of 365 families in this village. They used to soak the coconut fibres in the beach-pits around the Thalalla lagoon; until the tsunami of 2004 struck most countries in Asia. Thalalla was not spared.

The tsunami destroyed the beach-pits, filling them with sand and debris. The rope-making machines were destroyed and experienced craftsmen died. The small coir producers struggled to rebuild their livelihood and the coir industry.

In 2003, the Sarvodaya Shramadana Society (SSS) was organised in the village. With the tsunami in 2004, it collapsed and started functioning

---

1. Beach-pit coir refers to the fibres extracted from coconut husks and soaked in sea water pits along coastal lagoons or water bodies. After a period of fermentation, the husks are fished out of the water, beaten to further soften the fibre and then sun dried. It could take 3-4 months to produce the clean and white coir.

We came to know about this programme through the village information centre (VIC). Officers from Practical Action and Sarvodaya came to our village and consulted us through small groups. Eighty people participated in the two workshops. During the workshop, we identified some problems, these are: low market price for our products, lengthy production process, and limited product lines.

- J.K. Mangalika, a farmer and an active member of Sarvodaya Shramadana Society at Thalalla.

again with the revival of activities when a group of women rejoined the coir sector. SSS was legally registered in 2006.

Recognising the need, the affected women unified to restore their livelihood. The first initiative of SSS Thalalla was a series of needs assessment workshops which aimed to: determine specific issues faced by the industry; identify alternative ways of improving coir production; and find the right market, local and international.

During this time, neighbouring villages already have two large fibre extraction machines owned by two traders in the area. Community women collect or purchase coconut husks and have the fibre extracted in the mill for a fee. As the mill is not in the same village, Thalalla women have to organise for the hauling of their coir. Sometimes, they also get their coir from the Kurunegala[2] district; but that system no longer applies at present.

The village women also rely on Nalini (a tsunami survivor and the main supplier of coir to the community), president of SSS since 2003. She continues to provide timely supply and easy access of coir to the working group. She purchases and transports about 2,000 kg of coir from Beliatte

---

2. Kurunegala district is located in the Northwestern Province of Sri Lanka, which is well-known for its vast coconut plantations. However, the expansion of the real estate business in Sri Lanka reduced the land area devoted to coconut, thus, bringing down coconut production as well.

Earlier, we solely depend on beach-pits for coir processing. I own 1,000 beach pits where I soak about 3,000 coconuts. Our income source is through direct selling of the husks for two rupees per coconut.

Majority of traders buy either the white coir, as raw material, or ropes. With the enhanced technology for coir production, I am now able to produce the brown coir. Now, I produce both white coir from my beach pits and brown coir that is machine processed.

- Synopsis of interview with Nalini Kurunanayake a coir processor, President of SSS since 2003.

(located 50 km away from Thalalla) every month, and distributes them to the coir processors at an affordable price. She covers about 75 per cent of the raw material demand of the village and she also collects the finished ropes for marketing.

Thalalla coir processors' sole product is coir ropes. Ropes used to be produced using manually-operated machines which were found to be time-consuming and labour intensive. The ropes are sold to traders as well as to tea manufacturers as packaging sacks for tea leaves. They also provide ropes to the fishing industry, especially in the production of fishing net. The ropes used for the production of fishing nets need to be thinner and stronger than the one they produce for common market. They have to do it using their hands. There are no machines available in the country for this purpose due to the high price. However, the industry requires for the production of thin and strong ropes which is time-consuming and undergoes a sophisticated process.

3. Technological innovation: How they did it?

- *Exposure trip*: As an initial activity, the project team organised an exposure visit bringing Thalalla coir producers to Kurunegala coir processing sites, where they met up with a group of about 10 women to share learning, issues, and experiences. Through the exposure trip, the group was able to discuss and propose ways to improve the coir industry in the south. One perceived solution was to improve their manually-operated rope making machines.

- *Technology inclusion*: It was the women's group who brought the idea of developing a motorised coir machine. Hence, they requested that a machine be developed in consultation with the coir processors. After designing and fabricating the machine, a society member, Mangalika, was asked to test run it at her home. Eventually, other members tested the same machine. As a result of the field test, the machine was proven to be viable and would most likely increase production. Fifteen motorised machines were then developed and distributed for sharing among 150 families. In addition, technical and business assistance were provided by the Rural Enterprise Network (REN) and National Design Centre (NDC) for three months among community members involved in the project.

- *Assessment and monitoring*: The group meets frequently to discuss their observations and issues in relation to the production and marketing of the product. The participating field coordinator reports to the project team the outcome of the meetings, group recommendations, and requests for any technical inputs.

4. Impacts of the technological inclusion

- The manually operated coir processing machine required three people to operate. In a day, these three produce about 150 coir ropes using a manually operated machine. In the case of new motorised machine, only two people are required to operate the machine and the output is more too i.e., in a day spending the same amount of hours, these two people could produce 200-250 ropes. In general, they sell to the trader a rope for LKR 5, and the income is equally divided among the women. In the case of manually operated machine, three women earn LKR 750/- (150 x LKR 5) per day and one person earns LKR 250/- (LKR 750/3). In the case of motorised machine, two women earn LKR 1,250/- (250 x LKR 5) per day where one woman earns about LKR 625/- (LKR 1,250/2). Thus, with the utilisation of motorised machine, the daily income of coir processors has increased from LKR 250/- to LKR 625/- (or by LKR 375/-).

- In terms of production cost *vis-à-vis* net income, the traditional or manual system has been found to be lower where the only expenditure is LKR 360/- for purchasing coir only. Motorised machine being efficient than the manual one, can process more coir worth LKR 580/-. But to run the motor they require electricity with a daily consumption cost of LKR 20/-. Therefore, machine processing cost is LKR 600/-. This translates to an increase of aggregate monthly net income of about LKR 4,000 even with the purchase of coir (Table 70.1). The people maintain their own beach-pits, which supply white coir at lesser production cost. Prior to tsunami, all these women had their own beach-pits which were destroyed. They used to collect coconut husks locally or purchased from traders at a lower price. They used to process fibre for rope making as well as selling to other coir processors who do not have access to beach-pits. Therefore, their production cost were even lower prior to the tsunami.

- Manpower utilisation has become more efficient as the new motorised machine requires only two women to operate compared to the old manually-operated machine which involves three people.

TABLE 70.1

Economic comparison between manual and motorised processing systems

Description	Manual system (SLRs.)	Motorised system (SLRs.)	Notes
Expenditure per day:			
Raw material: White coir from beach-pits	-	-	Beach-pits were damaged by tsunami
Raw material: Brown coir from traders	360.00	580.00	
Other cost (electricity)	-	20.00	
Labour cost	-	-	Generally, processors do not count their labour as it is not a full-time occupation (4-5 hours a day)
Total expenditure	360.00	600.00	
Income per day			
Sales of coir ropes (SLRs 5.00 per rope)	750.00	1,250.00	Manual machine produces 150 coir per day; Motorised machine produces 250 coir ropes.
Profit per month			
Net income per day	390.00	650.00	Unskilled labour charges per day in Sri Lanka is Rs. 400/-
Net income per month (14 days)	5,460.00	9,100.00	
Net monthly income per woman (14 days)	1,820.00	4.550.00	Manual machine requires three people to operate it; Motorised machine needs only two people for its operation
Net daily income per woman (4-5 hours of work in a day)	130.00	325.00	The labour charges per day in Sri Lanka is about Rs. 400/-
Net daily income per woman in US$	1.15	2.88	The unskilled labour charges per day in Sri Lanka is about Rs. 400/-

Note: The manually operated machine requires three people to operate it while the motorised can be operated with two people.

Machinery costs and depreciation have not been taken into account here.

Exchange rate as at 12.05.2010: US$ 1.00 = Sri Lankan Rs. 113.720.

- Since 10 families share the use of the machine, the production pace is inhibited. To address this issue, the processors manage their work effectively by employing both the manual and machine process alternately. The proven motorised machine was developed

in a local workshop and not readily available in the market. The price of the machine has not been worked out due to the delay in distribution. They requested Sarvodaya's assistance as a grant or under a subsidy scheme.

5. Conclusion

The adoption of new technology has improved the efficiency and productivity in the rope-making process. In analysing the gains from the project interventions, the coir processors found that it was worth investing on a motorised machine for each family, which can be supported by microfinance initiatives operating in the province. For example in Sri Lanka, Sarvodaya's economic development wing, SEEDS (Guarantee) Ltd. is one of the largest microfinance and business development service provider. A credit product with a business support package is one option to ensure increasing the project benefits to the total community which is traditionally a coir processing village. Popularising/commercialisation of the proven machine can serve more numbers of people under the leasing or loan scheme. With the improvement in income comes the creation of a village environment that fosters group work and cooperation to achieve common goals. This is clearly illustrated with the increase in membership of SSS from 55 to 100 as of 2008. The project evidently demonstrates ways to increase the income of coir processors, while effectively utilising and empowering human resources.

71

Applying Market-Systems Approaches in Wanduruppa

Source:

Chopadithya Edirisinghe, Ramona Miranda, Jayantha Gunasekera and Chaminda Jayalath *Practical Action, Sri Lanka.*
Nilantha Atapattu *Rural Enterprise Network, Sri Lanka, 2010.*

1. Introduction

The participatory market systems development (PMSD), methodology (detailed in Transforming the value chains, business overview of concepts and principles in Part 3 of this sourcebook) was engaged in Wanduruppa in the post-disaster livelihood development programme. The project used the PMSD in two different ways: first approach was direct facilitation for market system development of selected sub-sectors and the second one was through training facilitators of other development agencies to carry out the PMSD process effectively.

Wandurruppa, is a village close to the southern tip of Sri Lanka (280 kilometres from the capital city).

In Wanduruppa, Practical Action began with the sub-sector selection which was based on the request of the village development committee (VDC) and engaging the view of various segments of the community. Growing a green belt was a project priority and the potential was identified for developing a market for products from green belt plants. In order to initiate the project work with the producers, it was important to have discussions with different stakeholder groups.

During the discussions with producer groups, the key market actors and issues related to chain actors at service provider level and enabling environmental level was identified from the producers' point of view. It was used as a platform to get information for cost benefit analysis and to create an interest for collaboration. Focus group discussions (FGDs) with other stakeholders like National Crafts Council (NCC), Industrial Development Board (IDB), Agromart, Consortium of Humanitarian Agencies (CHA), Sarvodaya, Chamber of Commerce, Pradeshiya Sabha, DS office and Coastal Conservation Department (CCD) helped identify issues and roles and responsibilities of key service providers/facilitators. There was a high involvement of women throughout the process.

Based on the findings and information collected from these discussions, a preliminary market map was developed to illustrate relationships within the sub-sector of handicrafts based on green belt plants and to identify the possible hooks such as 'reducing the cost of production' of small producers, 'producing quality products and reaching the upper level of

Applying Market-Systems Approaches in Wanduruppa

the market chain' while also highlighting the problems and issues (Figure 71.1).

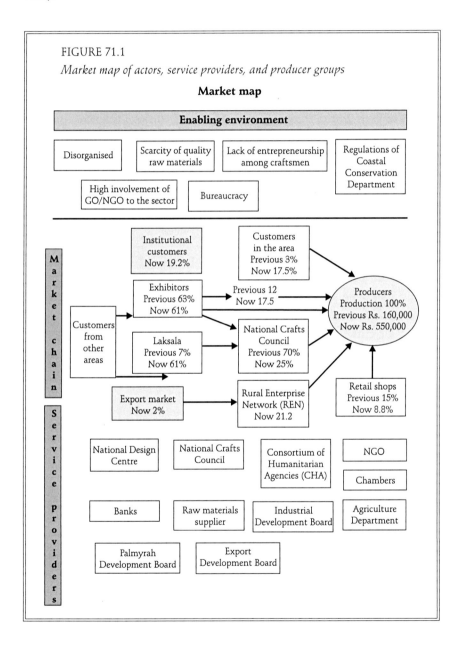

FIGURE 71.1
Market map of actors, service providers, and producer groups

2. Market map

These small-scale producers had faced many issues and did not do well in the mainstream market system as a result of poor quality products. This was due to lack of quality materials, lack of storage facilities, high cost of transport for supplying raw materials from distant villages, individual purchasing, lack of working capital, no collaboration among members, lack of business thinking, and low entrepreneurial culture.

The business services they had to depend on had not provided a reliable service. They had to deal with the low prices offered and payment delays by NCC, and Laksala (Sri Lanka Handicrafts Board). On the other hand, NCC which was a main actor in this market system, was not a business entity, and for these small-scale producers the only source for the information on marketing opportunities was the NCC. Thus, they had no choice than to depend on the inadequate service.

The business environment was also not conducive for them to develop. Sometimes their lack of awareness made them harvest their raw material in unsustainable ways. For example, the CCD had to ban cutting the pandanas grown alongside the beach as they began to cut the entire tree instead of cutting only the leaves. Furthermore, the producers had to

Participatory market mapping

The market actors (the handicraft producers, intermediary sellers i.e., shop owners from the area, NCC, IDB, the local authorities—*grama niladhari*) were then gathered together to a central place to carry out the participatory market mapping (or PMCA). This mapping exercise brought out not just the linkages, but also the issues and the bottlenecks in the different market sectors. Some of the bottlenecks and the potentials of the enabling environment and the service sector emerged as a result. This then led to the training of the entrepreneurs/MSPs and who were also linked with new raw material (reed) collectors who joined the market opportunity group, and a group of master trainers on weaving skills was created. The market system gained momentum due to this development. Interests groups developed in the series of discussions that continued.

depend on microcredit schemes with high interest rates. To purchase a small quantity of good quality dye, they had to travel a long distance or be content with low quality dye from local markets.

During the next phase of the multistakeholder discussion, an action plan was developed. This created the opportunity to elaborate on issues identified, fine tune the market map with the support of all stakeholders, and to learn and discuss potential solutions to identified issues.

Implementation of the action plan included following actions:

- a) Exposure visit to Kurunegala (sales centres for handicrafts) and Galgamuwa (to study on functionality of producer groups).
- b) Market study with producer groups.
- c) Introduction of new market links to the producers through Rural Enterprise Network (REN).
- d) Technology training to the small producers on new designs and colourings, use of alternative materials (banana fibres, *Eichhornia crassipes*).
- e) Training on business planning, develop individual and group business plans.
- f) Facilitation of the producers to regroup.
- g) Developed leadership and entrepreneurship skills of the group members.
- h) Facilitation of a women's group to register as a cooperative of handicraft producers.
- i) Facilitation to set up marketing centre for handicraft producers.
- j) Facilitation to purchase raw materials and to collectively sell the end products.
- k) Provision of market information through the REN.
- l) A district level market mapping exercise to introduce new market actors and service providers.
- m) Linking the producer group with district level coordination body of small producers (Hambantota Product Promotion Committee or HPPC).

n) Facilitation of the HPPC to develop strategic plan which included handicraft sector as a prominent sector.

Significant changes of this sub-sector was observed after the PMSD approach was carried out. New market channels were introduced through the REN which facilitated links to institutional buyers (18 per cent) as well as export markets (3 per cent). A producers' society was formed which took many positive initiatives such as registering themselves as a cooperative society, managing a sales centre which also acts as a coordination centre. Producers were motivated to establish a fund for purchasing raw materials using membership fees that were levied to be part of the society. The fund is grown with the 10 per cent received from sales and training income (current balance is LKR 20,000). Sales centre functions as a contact point for market actors and facilitators. This has increased access to information for the producers e.g.: an order of LKR 75,000 received recently through these links. Consequently, sales have increased by 243 per cent (previously 160,000 pa and now it is LKR 550,000 pa).

The high cost of transportation of raw materials has been reduced by collective purchasing of raw materials, using alternative raw materials (banana fibre—five per cent, water hyacinth—*Eichhornia crassipes*—still in the process of testing); linking with new suppliers; adapting proper harvesting/post-harvesting technologies. For example, a simple machine was used to produce quality materials and to reduce waste (waste reduction was by 60-80 per cent); organising common storage for bulk purchased materials tide over scarcity periods for materials. There are also other benefits such as the producers' society getting an additional income through this bulk purchasing.

Members of the society now are working collaboratively. They meet monthly in order to conduct reviews against the individual and group business plans. Producers have become trainers and train new members. As one of their marketing strategies, they form small groups to sell the products outside the village, especially in exhibitions and trade fairs.

With this intervention, the producers have improved the relationship and understand with the CCD while learning to adopt proper and non-destructive harvesting methods, as well as cultivating raw material plants in their home gardens.

A favourable business environment is also being created as alternative commercial buyers have entered the market system. As a result, producers have stopped supplying their products to Laksala through NCC. The need for microcredit is lesser now as they have the option of using the cooperative fund established for raw material purchasing.

3. Conclusion

The PMSD has enabled the poor small producers to link up with other market players in a sustainable manner. Practical Action played only a facilitation role to make this process successful, which is one of the keys to success in this approach, as the market transformation process that should be led by market actors themselves in order for it to be sustainable.

Some of the key benefits of the PMSD approach, as seen in its application in the project, are that it increases:

- Competitiveness: The community member is better exposed to the sectoral view and has a more holistic perspective of the livelihood option of their choice (business). She/he also knows her/his competing edge/s as well as that of the rivals and understands the need to be competitive to be sustainable.

- Linkages and bridging gaps: The participatory process helps to build linkages amongst the actors of value chain. It not only helps to understand the role of other players (other than those who deal with community directly) in the value chain, but also of the need to meet some requirements of the others' needs and priorities to ensure/secure their livelihood.

- Collaboration and participatory nature of the process may also:
  - lead to collaboration across the value chain among different actors.
  - help the community understand other community members who are engaged at their own level. There could be many instances where they may see the value of collaborating during some functions rather than competing that will lead to increased bargaining power and mutual advantages (e.g. purchasing or marketing).

- Access to services: Linkages established with government regulatory and advisory bodies, e.g., CCD and IDB.
- Creating an enabling environment.
- Value addition to technical content: When the problems and gaps in the production and market chain are identified market chain actors are driven to find solutions. Most of the time, this helps to improve the productivity by developing or acquiring new technologies.

We also need in this process to be aware of the constraints of PMSD as well as some of the principles that guide our activities.

Among the constraints is the limitation of the sub-sectoral approach where the sub-sector defines who belongs to the system and what issues or opportunities are important. This has both advantages (e.g. better communication among actors) and disadvantages (e.g. risk of neglecting external factors to the system but critical to its functioning). Therefore, few sectors have to be selected by the livelihood development programme at the project site. Poor groups who are not engaged in any of the selected sectors were included to skill development, home gardening, and small business development.

Another limitation is that some of the market actors are far from the producer and are not interested or willing to participate in the PMSD. For example, promotions are held in the capital city, for which these suppliers participate, but organisers of such promotions may not be interested in being part of the process even though they may be a key player as the handicraft industry is not so much a demand led sector.

Examples of activities that development practitioners must avoid in the use of PMSD for poverty reduction through livelihoods improvement of small-scale rural producers is to remember that they are only facilitators of the market transformation processes that should be led by market actors themselves in order to be sustainable. However, the practitioner facilitating the process also needs to have adequate knowledge and expertise about the markets that are to be transformed in order to have credibility to promote successful dialogue and collaboration.

72

Integrated Mangrove and Fishery Farming: A Model for Saline Transformed Lands

Source:

V. Selvam, A. Sivakumar and A. Sivanesan *M.S. Swaminathan Research Foundation, Tamil Nadu, India, 2010.*

The increase in sea level rise due to climate change will further endanger the livelihood security of 100 million people and the ecological security of coastal zones. In India, sea level rise will inundate 5,700 square kilometre of coastal land and is predicted to directly affect nearly 7.0 million coastal families. Many fishery-dependent communities already live a precarious existence because of poverty and lack of social services and essential infrastructure, overexploited fishery resources, and degraded ecosystems. With the increase in sea level rise, farming families, fishers and coastal inhabitants will suffer even more due to less stable livelihoods, changes in the availability and quality of fish, and rising risks to their health, safety and homes.

However, sea level rise and the consequent salinisation of land provide an opportunity to increase fish production through aquaculture. The Coastal Zone Management Subgroup of the Intergovernmental Panel on Climate Change (IPCC) predicts that in many coastal areas, people would modify landuse pattern and subsystems to adapt to salinisation and flooding. One of the major landuse change predicted is the conversion of saline-affected agriculture lands into aquaculture farms. However, the current aquaculture practice in India warrants more responsible and sustainable systems and practices.

1. Aquaculture *vs.* Integrated Mangrove Fishery Farming System

More than 150,000 farmers grow prawns in about 160,000 hectares of brackish water on the east and west coasts of India. Semi-intensive culture system had increased prawn production at the rate of 8.4 per cent per year until the mid-1990s, after which coastal aquaculture suffered setbacks in production, value and acreage. Prawn production, at around 1.06 million MT in 2007-08, came down to 0.76 million MT in 2008-09. Monoculture, disease, poor seed quality, excessive use of artificial feed, increased input costs and decreased market value contributed to the decline. Poor environmental management and lack of options to diversify livelihoods within the aquaculture farming system were also considered responsible for the decline of coastal aquaculture in India.

Integrated Mangrove and Fishery Farming: A Model for Saline Transformed Lands

The social impact of the decline in prawn farming is enormous. Prawn farmers now do not earn income from either agriculture or from aquaculture, and many have either migrated temporarily or permanently in search of livelihood. The Integrated Mangrove Fishery Farming System (IMFFS) project, wherein cultivation of mangroves, halophytes (salt-loving plants) and culture of fish, crab and prawn are integrated into mangrove farming, provides a tangible solution to make coastal aquaculture sustainable and to strengthen the resilience of coastal communities.

2. Design

Most of the prawn farms in India are clusters of 5-20 rectangular shaped earthen ponds in a hectare of land. Brackish water is periodically pumped in and out of the ponds to maintain water. The prawns are put in ponds, fed with artificial feed until they reach marketable size for about 4 to 6 months. As a result of this practice, input and artificial feed costs are

The livelihood of families in the coastal zones is endangered by disastrous floods, cyclones and tsunamis.

always high. It also increases the organic load, consequently resulting in pollution.

In the IMFFS, earthen ponds are designed in such a way to provide space for growing saline-tolerant vegetation including mangroves and halophytes. Space for planting is created by constructing bunds inside the pond in a zigzag manner or as small mounds (Figure 72.1 and 72.2). These bunds and mounds are created by digging the soil from the bottom of the pond. Tidal water will fill the pond by gravitation during high tide and drain out during low tide. The tidal water inlet and outlet may be established at opposite ends or a single structure can be used both as inlet and outlet. About 30 per cent of the space is for planting mangroves and halophytes whereas the remaining space is left for holding seawater for fish culture.

3. Advantages of IMFFS

Mangrove trees can be grown along the inner and outer bunds and mounds while halophytes such as Sesuvium or Salicornia can be cultivated as commercial crops on top of the bunds. Fish can be grown in the water spread area. The mangrove plants will provide nutrients and feed to the fish, crab and prawn, and serve as bioshield when they grow big. Leaves, twigs and other plant matter, which fall into the water, will degrade and release nutrients and result in the formation of detritus (finely decomposed particle). The detritus become substrate for bacteria and fungi that convert nitrogen in the detritus into protein. In addition, a variety of enzymes are also produced during decomposition. Nitrogen, carbon, protein, enzymes, fungi, and bacteria increase the nutritive value of the detritus and since both crabs and prawns get nourishment from detritus, they get naturally balanced feed.

Another important advantage of IMFFS is the limited energy required for the operation since water is exchanged daily through gravitation and not by pumping. Also there is no need to use aerator to increase oxygen content of the water. The daily exchange of water brings in lots of fresh food in the form of planktons, eliminating the use of artificial feed. Zero use of energy and artificial feed greatly reduce input cost and prevent pollution.

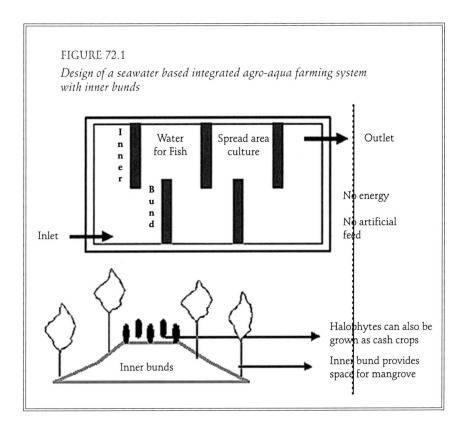

FIGURE 72.1
Design of a seawater based integrated agro-aqua farming system with inner bunds

4. Adaptive research on fish culture

About 2,500 fingerlings of the commercially important fish, Lates calcarifer (seabass), each approximately 2.5 centimetres long, were cultured in September 2007. They were initially kept in fine nylon net bags (*happa*) in the farm water for acclimatisation. Survival after three months was only about nine per cent because the fingerlings were still too small to adapt to their new environment. In December 2007, 500 fingerlings of 8 centimetres length were again acclimatised in the farm. At the end of January 2008, nearly 85 per cent of the fingerlings survived and reached a length of 13 centimetres. They were released into the farm in February 2008. In October 2008, 125 kilogram of seabass was harvested along with about 161 kilogram of milk fish, mullets, tilapia and prawns.

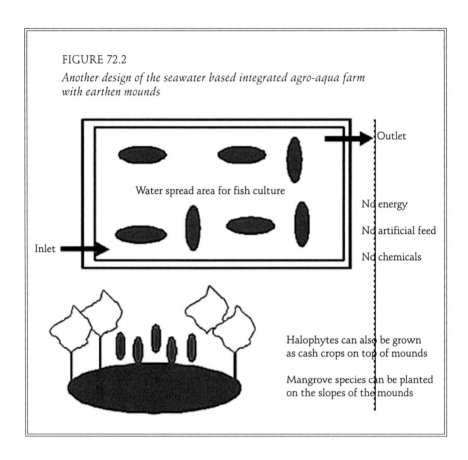

FIGURE 72.2
Another design of the seawater based integrated agro-aqua farm with earthen mounds

The local fishing community was satisfied with the mangrove and halophyte cultivation on the bunds. However, the people said that income from seabass culture was limited and suggested growing of fish mullet and tiger prawn. For instance, in a one-hectare pond, 10,000 juvenile prawn and 2,000 fingerlings of mullet can be cultured with a stocking density of 1 prawn and 0.2 fish per square metre. A 60 per cent survival of prawn and 50 per cent for fish at an average body weight of 50 gram and 1 kilogram respectively, would fetch a profit of about LKR 40,000 in eight months.

In another proposed experiment, about 4,000 fingerlings of seabass will be cultured in a one-hectare pond with a stocking density of 0.4 fish

Integrated Mangrove and Fishery Farming System

- Ensures livelihood security of poor coastal families.
- Ensures ecological security of coastal areas.
- Enhances adaptive capacity of coastal communities to sea level rise and climate change.

per square metre. The size of the stocking fingerlings is 10 centimetres. All the fingerlings will first be acclimatised for the IMFFS in a *happa* in which brooders of tilapia fish will be introduced to be used as biofeed by supplying its young to the seabass. Approximately 50 per cent of the fish is expected to survive at the end of eight months and body weight increase to about 750 grams. The computed net profit at the end of 8 to10 months is LKR 70,000 to 80,000.

Integrated mangrove, fishery and farming system takes care of environment and livelihood concerns of the coastal zone.

5. Observations of the community and other stakeholders

More than 1,500 fishermen including fish farmers, officials from the Fisheries Department and specialists in coastal livelihood gave the following observations of the IMFFS model:

- it has huge replication potential both at the national level and in the global arena;
- it would take care of both livelihood and ecological need of the coastal areas;
- it is suitable for high saline areas and not suitable for sandy coastal areas;
- it is most suitable for crab culture. The root system of mangrove trees would provide shelter and food to the commercially valuable mud crab or mangrove crab. Raising mud crabs is difficult in open aquaculture ponds because the crabs can easily climb out of the bunds. Moreover, unmoulted mud crabs kill and eat the weak moulted crabs. The roots of grown mangrove trees provide a natural environment for the crabs, prevent their migration as well as protect the moulted crabs. Most importantly, crabs feed on detritus and mangrove trees ensure the availability of enough detritus for crab culture;
- main weakness is its initial cost (about LKR 200,000 per acre);
- it should be tried in large scale and lessons learned should be used for replication;
- availability of land is the biggest issue though large tracts of saline-affected lands are available in the coastal area of Tamil Nadu, most of them belonging to government agencies;
- there should be policy support for allotting land to IMFFS;
- carnivorous species should be avoided;
- rope culture of mussel, which is of great demand in Kerala, can be tried in the pond along with fish culture; and
- a thorough study on economics of the farming system should be worked out as an important requirement to start the replication process.

6. Conclusion

The lack of direct economic benefit is one of the major issues against mangrove bioshield management and the IMFFS model effectively addresses this concern. Secondly, though initial investment in IMFFS is more or less equal to shrimp farming, the latter can be considered as an improved extensive type of shrimp farming, because it is environmental-friendly, is low cost, has zero energy demand, and does not require use of artificial feed and chemicals. IMFFS can also be used to restore many of the abandoned shrimp farms and help make them sustainable. Adopting IMFFS can enhance resilience of coastal communities to increase in sea level rise and salinisation. Increased sea level rise may have adverse impacts in the coastal areas but it offers an opportunity for the people who live in the coasts or make a living from fisheries to benefit from the situation.

73

Raising Mangrove Nurseries in Muthuregunathapuram Village

Sea water inundates farmlands in Tamil Nadu during extreme weather.

Source:

J.D. Sophia, A. Sivakumar and V. Sivenesan *M.S. Swaminathan Research Foundation, Tamil Nadu, India, 2010.*

Extreme weather events such as cyclones, storm surges, and floods are common phenomena in the coastal villages of Tamil Nadu. When they occur, sea water inundates the farmlands, damages the standing crop, and increases soil salinity, adversely affecting the people's livelihood and their food security. Because of this, MSSRF helped develop mangrove bioshield plantations in the area as a mitigating measure against floods.

Depending on the biophysical conditions of a prospective mangrove bioshield, establishing the mangrove plants may be done by direct seeding or by transplanting nursery-raised saplings. After conducting a field survey, the MSSRF saw that the mangrove plantation area in Muthuregunathapuram, a village along the coast in Tamil Nadu, is an open mud flat in the mid intertidal region that gets submerged in two-metre high water for 3 to 4 months during the monsoon season. Therefore, planting seeds directly in the plantation is not feasible as they will be submerged by water and get carried away by the strong monsoon current. The MSSRF decided that the mangrove must first be grown in nurseries before they are planted in the plantation area. It successfully enlisted the participation of four self-help groups that saw the opportunity to earn income from mangrove nursery establishment.

How the MSSRF accomplished this is a story of community participation, women empowerment and involvement in social mobilisation, consideration of the local leadership structure, the power of concept ownership by the stakeholders, consensus setting and the importance of responding to felt needs, and of using technical expertise.

1. Identifying Muthuregunathapuram

The MSSRF deployed a team of mangrove scientists and NGO leaders to 10 villages in the 35 km stretch between Devipattinam and Thondi of Palk Bay, Ramanathapuram district, to see where it is best to establish mangrove plantations. A suitable village should have experienced the effects of natural hazards like cyclone and flood and must have land suitable for a plantation with good intertidal water exchange, tidal amplitude and silty clay soil. The group also assessed the community's willingness to implement mangrove restoration in their village, and the availability of labour both within and outside the villages.

The team found the village of Muthuregunathapuram, geographically located in a low lying area and highly vulnerable to natural phenomenon, to have nearly 100 hectares land suitable for mangrove plantation. Moreover, the team found it to be suitable for bioshield plantation given the above criteria.

The MSSRF team discussed with the community how a mangrove bioshield can protect the village from natural hazards and how it provides livelihood benefits through a consequent increase in fishery resources. Really getting Muthuregunathapuram into the picture not only meant listing the village into the MSSRF programme but also enlisting the people's participation so that they will make the concept of mangrove bioshield development their own.

Concept sharing was aimed at motivating the villagers to pass a resolution asking the *gram sabha's* permission to allocate land where the mangrove bioshield could be established. It was done at two levels: (i) at the village, and (ii) at the *gram sabha* level. Concept sharing was facilitated because the community had been frequent victims of cyclone hazards and had a negative experience of a top-down approach to mangrove development that the Forest Department attempted to implement in their village. The implementing team further conducted concept sharing activities at the level of the Vennathur *panchayat gram sabha* to where Muthuregunathapuram belongs and also passed a resolution supporting the initiative.

2. Laying the groundwork

An exposure visit to Pichavaram, one of the biggest, most well conserved and protected mangrove forests was organised to enlighten villagers on community-based mangrove development and management. The implementing team decided to get equal representation from both genders in the fieldtrip. To ensure women participation, the leaders decided to take both spouses of a family to the trip. In a culture that discourages women participation, this was a positive development that encouraged other women to join. In the end, 22 women, 26 men, and two NGO staff participated in the fieldtrip.

Women's self-help groups operated the mangrove nursery while a village leader acted as monitor.

MSSRF scientists oriented exposure trip participants by focussing on the successful models of participatory mangrove forest restoration, bioshield development and management. Interactive presentations, field visits and interaction with the community, were the tools used for the exposure programme. Leaders and youth from villages where the mangrove management programme had already started acted as resource persons. Technical aspects on bioshield development were also discussed in relation to disaster risk reduction and livelihood development.

An awareness programme later brought out a concern among the villagers: they wanted to fight widespread illiteracy through a tuition centre for their children. MSSRF saw this as an entry point activity and in December 2007, the tuition centre was established. Being an entry point activity based on community's needs, fees were not collected from the children.

To enhance the local people's capacity to analyse and act in a participatory manner with other implementing agencies towards restoring, conserving and sustaining mangrove wetlands, village level institutions (VLI) were formed in the community. One of these VLIs was the Village Disaster Management Committee (VDMC) which was tasked to plan, implement, and monitor project interventions. As a testament to the increasingly

No. of children enrolled in the Tuition Centre - 62
13 boys and 14 girls were in primary classes and 16 boys and 19 girls were studying in 6th to 12th standard.

important role of women, VDMCs are now comprised by 8 men and 8 women where before women only numbered half the men.

The VDMC has three tiers: the General Body (GB), Executive Committee (EC), and Office Bearers (OB). Both the spouses, i.e., husband and wife from 91 households are GB members, 8 men and 8 women representatives identified from GB constitute the EC, and three representatives serve as office bearers. EC meetings are held monthly while GB meetings, once a year. Following the VDMC formation, two joint accounts in the local bank were created. The first account is for transacting project funds and the second is to save the repayments and reutilise at later stage for the community development activities, with the approval of VDMC.

Another VLI are the women self-help groups (SHG) that played a very significant role in the mangrove bioshield project as they were tapped to establish and run the mangrove nursery. At first, the women were hesitant to form SHGs as it would require frequent travel outside the village. However, the project team convinced the women that the SHG would help provide need-based support for enhancing livelihood activities or to start new ones. Four SHGs were subsequently formed who fixed their monthly savings according to their capacity. They opened separate bank accounts and agreed to meet fortnightly to agree on granting credit to members in need at the rate of two per cent monthly.

3. Setting up the mangrove nursery and transplanting the saplings

Mangrove nursery establishment and management is process-oriented and comprises the following: distribution of raw materials including seeds, preparation of seedbed for the nursery, preparation of soil mixture, pocketing the mixture, planting of seeds in poly bags, fencing of nursery grounds, watering, disease management, and transplanting of the mangrove saplings. Most of the activities were women-led except for fencing and transplanting.

Before receiving the raw materials, the SHG members prepared the nursery bed. The 20 members of each SHG mixed the raw materials, packed them in the poly bags and planted the seeds. Two women on rotation basis, took care of weeding and watering the seedlings each day. Overall monitoring was a function volunteered by the VDMC president.

To protect the saplings from damage caused by human and animal intruders, the VDMC members suggested that the nursery be fenced off using locally available main stems of palm leaves. The VDMC also suggested engaging skilled persons to construct the fence so that it can help provide employment opportunities to the local community. In all, 16 men were able to get labour opportunity for three days with each of them earning about LKR 600.

After three months, saplings from the nursery yard were transplanted in the mangrove plantation. Saplings raised by all the groups were transferred to the shore using a tractor, and from the shore, these were taken in a boat

Transplanting the mangroves is usually done by skilled men.

and placed in different locations. Four skilled persons from Muthupet and the villagers transplanted the saplings at the interval of two metres. The skilled persons from Muthupet marked the planting points using a nylon rope for marking each point, followed by a local person who dug the pits, and by the women who put the sapling in the pit after removing the poly bag. All the poly bags were collected and safely disposed of. One sapling costs LKR 3.60 to raise. The SHGs received LKR 92,060 as income from operating the mangrove nursery.

4. Conclusion

The MSSRF-Muthuregunathapuram experience is full of lessons that intervention planners should take note of:

- It is always important to seek the expertise of qualified people. The presence of the mangrove scientist to get first hand information on the biophysical suitability of the area for mangrove plantation is crucial to the success of the project.
- Any agency involved in community development should not use top down methodology and should be cautious in imposing their own decisions on other people.
- Addressing the basic problems of the community as an entry point activity increases the community's trust on the implementing agencies and extends their full cooperation and participation.
- Exposure visits provide an eye opener for similar initiatives that are being pursued by the community.
- A major initiative that worked was engaging the spouses of the same family. It enabled the implementers to ensure gender participation. Whenever women are hesitant to come out of the village without their husbands, the above strategy could be adopted to ensure their participation.

Nursery establishment was implemented not only to meet the project requirements but to make it a viable income generating activity through strengthened social and human livelihood assets of the village. The social asset was built by promoting women SHGs while the human asset, by

Steps in establishing a community-managed mangrove bioshield
The Muthuregunathapuram village, Tamil Nadu experience

Steps	Activity	People involved		Process	Purpose	Criteria
		MSSRF	Community			
1	Feasibility survey	Mangrove scientists implementing team	Community leaders and contact persons	One-on-one interaction	Explore availability of suitable areas for mangrove plantation	Community willingness towards mangrove restoration, land and labour availability, and biophysical suitability
2	Selection of suitable village	Mangrove scientists implementing team	Community leaders and contact persons	One-on-one interaction	Determine technical suitability of a village	Prone to hazards, good intertidal mud flats, tidal amplitude inundating mud flats, silly clay substratum, community interest
3	Concept sharing	Implementing team	Community *gram sabha*	One-on-one interaction, village and *gram sabha* meetings	To make the community aware of how the mangrove bioshield protects the village from the effects of natural hazards and provide livelihood benefits; To stimulate interest and develop concept ownership among the community people.	Passage of resolution at the hamlet and *gram sabha* levels; Community participation; Linking with *panchayat*
4	Exposure visit	Mangrove scientists implementing team	Key village persons, village members including women, *panchayat* leaders, *panchayat* clerk, traditional leaders, and youth	Field visits, interactive presentations, interaction with community, orientation on successful models of participatory mangrove restoration	To enlighten the visit participants on community based mangrove development and management; Orient them on successful models of participatory mangrove forest management, community participation, and technical aspects and benefits of a mangrove bioshield	Participation of stakeholders men and women, traditional leaders, *panchayat* members and clerk, youth, NGO

contd...

...contd...

Steps	Activity	People involved		Process	Purpose	Criteria
		MSSRF	Community			
5	Enhancing participation	Implementing team	Community children	Issue emerged in talks among community members and implementing team	To determine appropriate point of entry by which the needs of the community will be addressed in order to enhance their participation in mangrove bioshield management.	Tuition centre for school children
6	Formation of village level institutions	Implementing team	Leaders and other key members of the village, women	One-on-one interaction, meetings	To enhance capacity of local community to analyse and act in a participatory manner in the mangrove bioshield project.	Village development and management comm.ttee Self-help groups composed of women
7	Demonstration on establishing a mangrove nursery	Implementing team	Women, experts in mangrove nursery establishment	Women themselves signified interest in mangrove seedling production/nursery management to provide needed saplings and as a source of income; Onsite demonstration on mangrove nursery establishment	To teach women how to establish and manage a mangrove nursery	Technology on mangrove sapling production and management
8	Process of establishing and maintaining a mangrove nursery	Implementing team	80 women members of SHGs	Women-led initiative	To teach women how to establish and manage a mangrove nursery	Technology on mangrove sapling production and management
9	Mangrove transplanting	Implementing team	Skilled mangrove transplanters, villagers	Field work	To transplant mangrove saplings in the bioshield area	Transplanting process

A restored mangrove forest protects the people from the strong waves and the flooding that often take place during the monsoon months.

capacitating the women SHG members as skilled mangrove nursery raiser. Special attempt was made to strengthen their financial asset by cultivating the habit of savings among them through the SHGs. These livelihood assets, built either directly or indirectly, enhanced the resilience of villagers, especially the women, to face extreme weather events. The overall success can be attributed to the small and homogeneous nature of the villages/hamlets by caste and occupation coupled with good democratic and traditional control mechanism.

74

Viability of Bioshield Development in Sri Lanka

Bioshields provide multiple benefits to the community.

Source:

Bhathiya Kekulandala *Practical Action, Sri Lanka.*
Chinta J. Munasinghe *Laymen's Den (PvT) Ltd., Sri Lanka, 2010.*

The South Asian Tsunami of 2004 severely affected many coastal communities in Sri Lanka. Early observations at tsunami-affected sites indicated that the places that had less damage during the 2004 tsunami were 'protected' by a coastal front covered by dense mangrove and sand dune vegetation. Initiative was taken by Practical Action and Sarvodaya through the post-tsunami project funded by the International Development Research Centre (IDRC), to pilot bioshield development as an intervention aimed at providing multiple benefits to the people.

The approach taken by the project to establish bioshield plantations include initial community awareness creation programmes, consultation and mobilisation, participatory decision-making, establishment of community nurseries, participatory monitoring and maintenance. It also aimed at offering links to alternative source of income by the people through handicraft making, home gardening, increase land value, enhance local environment to ecotourism and diversify access to natural resources. Experiences from these locations in Sri Lanka is shared back.

1. In Andaragasayaya

Sarvodaya Shramadana Society (SSS) took the leadership in establishing the bioshield plantation. Members of the society and other community members selected the land to be planted through a consultative process. Narrow strip of land between the sea and private lands and sand dune area was selected as the suitable site. Project team got the assistance of experts to train the community members in bioshield establishment and nursery management. Community members were helped to set up nurseries and plants for the bioshields were bought from them. Several community members were appointed to monitor the progress of activities.

2. In Wanduruppa

An environmental committee composed of members from the biodiversity society and the Village Development Society of Wanduruppa was organised to oversee this aspect of the project. Technical experts and community members participated in selecting suitable species for the bioshield. They proposed using plants that are naturally occuring in the village; have strong buffering effects against wave and wind; capable

Viability of Bioshield Development in Sri Lanka

> Sunethra, the Secretary of SSS, recounted how the bioshield project benefited Andaragasyaya and its people:
>
> We gained knowledge on nursery management and home gardening. We have never thought that run-off water during rainy season could be used for the dry season. During project implementation, the labor requirement also provided additional income to the people. SSS used to provide savings and credit as its main service. This was the first time that the SSS in this village collaborated for a development intervention. There had also been an observed perspective change among society members through years of learning. We were taught the great lesson of being prepared ahead to overcome the impacts of disasters.

of soil stabilisation; adaptable to saline conditions; and could provide multiple products and services to people. Sand dune, river bank, and beach area was selected for plantation.

3. Challenges

Bioshield plantations established in open access areas such as narrow beach strips and sand dunes were very difficult to maintain over a long period of time. Harsh dry and arid environmental conditions in the area (especially the dry salty winds) were a major threat to the plants at the sapling stage. The mortality rate was high in the sand dune areas. Fifty thousand palmyra seeds were planted in high sand dune and open beach stretches in Andaragasyaya area but the germination rates were very low (less than 10 per cent). Germinated saplings of palmyra and Casuarina also perished due to the dry salty winds. The lack of soil organic matter and difficulties faced by the community members in watering young sapling during the extensive dry season would have contributed to the high mortality rate. However, saplings planted in areas that were not directly exposed to the dry winds blowing across the dune area survived but showed retarded growth. This was clearly seen in the coastal sand dune plantations in Wanduruppa.

Community members were trained on bioshield establishment.

It has been observed that plantations have to be managed for at least three years to ensure its viability through minimal inputs. However, this was impeded by the restricted human and financial resources available to the project. Although some community members volunteered to support the maintenance work, it was not possible to carry out regular maintenance. The high survival rate observed in the plantation in Wanduruppa and some areas in Andaragasyaya is due to incentive based mechanism for local communities, where they were paid nominal labour cost for being involved in the maintenance work.

Project evaluation found that a manmade bioshield is not cost-effective. It also brought to light the lack of a justifiable relationship between bioshield development and livelihood enhancement in the short term. This means that the villagers were not able to generate livelihood from the plants in the bioshield, leading to minimal community participation. Moreover, the project proved that most of the plants unique to the area will grow over time, and therefore, using resources to establish bioshield

The project helped community members set up nurseries and bought the plants for bioshields from them.

will only serve to fast track the establishment of the protective vegetation. These findings and observations showed that the bioshield model used in this project is not replicable in other similar areas especially in Sri Lanka. It must not be forgotten, though, that the use of bioshield for the specific purpose of enhancing the status of coastal environment or to provide long-term livelihood benefits to adjacent communities is still viable as bioshields bring benefit in the long term.

75

Participatory GIS for Village Development

Source:

Gnanappazham Lakshmanan and V. Selvam *M.S. Swaminathan Research Foundation, Tamil Nadu, India, 2010.*

Pictorial representation of the village with most accurate details in terms of location, distance and elevation can be obtained, instead of viewing spread sheet data and a rough village map, when spatial technologies are used. This will help in decision-making process of the community for the development of both natural and human resource by considering spatial entities.

The resource mapping done using manual method can be improved in terms of accurate location, distance and areas of land features when Global Positioning System (GPS) and Geographic Information System (GIS) are integrated along with Participatory Rural Appraisal (PRA) output. Overlaying different status of the village community with natural resource map will help to identify the hidden issues related to land, water and vegetation and also to derive suitable solutions to that.

Similarly, when the household data collected through PRA is added in GIS with location of houses on the village map, it will help analyse the socioeconomic status of the community, along with the spatial indicators such as markets, hospitals, business, government institutions, banks, etc. Thus, it enables the problem identification, identifying causes of problems, and developing possible solutions.

Updating the resource database and household information either once in a year or once every six months will become easy by not repeating the whole exercise. Similarly, updating cadastral map with new divisions within the farms along with the modifications in details related to owners of the farms and other cultivation and cropping pattern. This enables the comparative analysis of the village and community over a period of time.

As coastal environment not only encompasses the coastal habitats but also the habitations surrounding that, the management has to take care of both habitats and habitations.

1. Thachampathu and Pannimunda, tribal villages: A case study

Thachampathu and Pannimunda are two villages which are under the umbrella of village knowledge centre (under VRC programme of MSSRF) located in Thachampathu. The PRA was conducted in the villages and

Participatory GIS (PGIS)

In some developing countries including India, aerial photography, high resolution satellite imagery and official large scale topographic maps were not accessible to a common man for his land planning and development because of national security concerns. The situation changed in the late 90s because of the development in Spatial Information Technologies including GIS, GPS and remote sensing, open access to various kinds of data through Internet and decreased cost of computers. Spatial data, previously controlled by government institutions became progressively more accessible to and used by non-governmental, community-based organisations, and minority groups. Thus, Geospatial Technologies made an entry in community-centered initiatives.

Variety of technologies is adopted to support two-way communication and widen public participation across socioeconomic contexts. This resulted in community-based management of spatial information through Participatory GIS (PGIS). PGIS is public rural appraisal done with the help of RS, GIS and GPS. This composes peoples' spatial knowledge in the forms of virtual or physical, two or three-dimensional maps used as interactive vehicles for discussion, information exchange, analysis and as support in advocacy and decision-making. Users employ the outputs in the form of different maps and reports mainly as media to support their arguments (Rambaldi et al., 2004).

PGIS harness the local knowledge with the help of modern Geographic Information Technologies to integrate that knowledge in the planning process for the village development. It promotes interactive participation of stakeholders in generating and managing spatial information and it uses information about specific landscapes to facilitate broadly-based decision-making processes that support effective communication and community advocacy.

simultaneously spatial information about the villages was also collected. The spatial information includes the village map with cadastral boundary, location of infrastructure facilities like health centre, open well, hand pump, temples, thrashing floor, shops, location of village boundary, places of ecological, religious importance, irrigation channels, and location of houses using GPS (Table 75.1). The GPS readings are entered in GIS environment and overlaid on the digital form of the village map. The household data tabulated from PRA exercise was integrated with location of houses on the village map. The series of household data like education, literacy, community, wealth status are mapped individually and also overlaid with one another by suitably selecting the parameters which are relevant to assess the status of community, identification of prevailing problem in society and in natural resources.

Overlay of community map on the wealth map (Figure 75.1) shows the poor status of a particular group of community (Paniyas) in Pannimunda

TABLE 75.1

List of PRA tools which are done by involving geospatial technologies

Participatory tools	GIS involvement	Community involvement
1. Transect walk	GPS survey of important features in the village GIS mapping of above	Survey done and relevant data collected with community
2. Timeline/trend line analysis	Location of (existing and demolished) the hostoric events of importance, hazards, developments in village map	Map of transect walk is given to community to locate the details
3. Social mapping	Household survey using GPS GIS mapping of HH location and integration of HH data	Survey done with the help of community first and then followed by representatives of the community
4. Resource mapping	Remote sensing data, secondary spatial data, village and cadastral maps (spatially and quantitatively near accurate)	Mapping is jointly done and verified with community
5. Ven diagram	Location of infrastructure are spatially depicted in order to analyse in terms of distance and direction (shortest route, nearby markets, nearby industries)	Mapping is jointly done and verified with community

Outputs
- Village maps, village resource map, map of each houses with details, possible superimposition of different maps.
- Community feels that they are also the authority of their village maps and can come front with right perception of the problems and solutions.

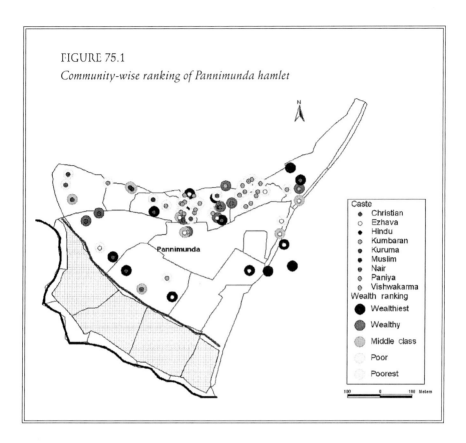

FIGURE 75.1
Community-wise ranking of Pannimunda hamlet

village and also shows that the community is isolated. Similarly, the location of infrastructure facilities and community map are overlaid to identify the access of the facility to the community.

Geographic Information Technologies (GITs) provide helping hand in assessing the coastal environment and in supporting the planning, monitoring and evaluation process. However, challenges exist in applying the technologies in coastal environment.

2. Challenges and issues in application of RS and GIS in coastal management

Though remote sensing (RS), GIS, and GPS are in use by the coastal zone management authorities, they are either available at higher level and or not used to their full extent. There are issues related to the coastal environment management in applying the technologies. If they are overcome, coastal environment can be managed in a successful manner and the loss of coastal habitats can be minimised. Some of the issues to be tackled are:

1. Getting cloud-free satellite data in a year is difficult when we intend to monitor the seasonal coastal dynamics such as condition of river mouths and estuaries, monitor the seasonal changes in healthy and stressed mangroves, etc.

2. Non-availability of large scale maps of village map, land use/land cover map, soil map, geology/geomorphology, temperature and rainfall hold back the deployment of RS and GIS to their full extent.

3. East coast of India is topographically very flat. Available maps show up to an elevation difference of 20 metres. They do not provide local undulations which is very crucial for the coastal zone management. Though sophisticated GPS provides accurate estimation of elevation, it is not affordable to the local organisations both technically and economically.

4. Periodic update of the large scale spatial database of the coastal environment is necessary for their effective and timely application in the process of planning and monitoring.

5. Making GIS/RS easily accessible to multiple stakeholders improving transparency of primary and secondary data.

6. Administrative bodies at ground level should take initiative in updating the data base of regional coastal environment so that they can be immediately used in decision-making process in subsequent administrative authorities. Hence, it is not just required to create awareness of technical knowhow but also to regularise the usage of techniques in daily activities such as taking GPS during the field survey, collecting periodic satellite data of the region and making use of the data in field.

7. Lack of awareness in effectively utilising RS and GIS outputs in planning, management, and monitoring of coastal resources at different levels.

76

Village Resource Centres: The Spokes of the Information Village

Source:

Nancy J. Anabel and S. Senthilkumaran *M.S. Swaminathan Research Foundation, Tamil Nadu, India, 2010.*

1. Introduction

Rural families in remote communities whether they are in the rainfed areas or dryland areas should be able to access up-to-date and relevant information if they are to be empowered and participate actively in the development processes. With this in mind, the principles of social inclusion, gender equity, reaching to remote areas (last mile and last person connectivity) and remedying regional imbalances were built into the concept of the village knowledge centres (VKCs) of the M.S. Swaminathan Research Foundation (MSSRF). To do this, the promotion of functional literacy among adults without formal education to learn and enjoy the benefits of interactive pedagogic methodologies was another major consideration in setting up the VKCs.

The 'Hub and Spokes' model for information/knowledge transfer was adopted by the MSSRF in Tamil Nadu, India initially through its Information Village Project in Puducherry. In this project, the village resource centre (VRC) is the hub that processes and responds to the community's information requirements and connects to several VKCs which are the spokes.

This paper presents the processes and requirements in establishing, operating and managing the VKC based on the MSSRF experience.

2. Pre-VKC operations

The VRC staff leads the conduct of the pre-VKC operations. These involve three major steps or requirements leading to the establishment of the VKC: (i) community orientation at district/block level; (ii) community needs assessment; and (iii) feasibility studies and planning.

3. District/block level orientation

To popularise the concept and create the demand for VKC in the communities, the local government council (*panchayat*) members and representatives of the other key local stakeholder groups (traditional *panchayat* leaders, farmers, fishers association, and women self-help groups) are invited for an orientation meeting through the Block Development Officer (BDO) and Assistant Director—*panchayat*. Project staff may also promote the concept of VKC by requesting to put VKC into the agenda of the BDO when she/he organises calls for or organises their own meetings. These orientation meetings focus on sharing the concept of VRC and VKC, their establishment procedures and requirements, need for community participation, benefits to the different target groups, and expectations from the community in establishing, managing and sustaining the operation of the centre. In this process, 'boundary partners' are identified.

The persons or groups who express willingness to respond and comply with the requirements of VKC establishment, operations and management, would be the potential boundary partners.

Boundary partners are those individuals, groups, or organisations with whom the programme (VRC) interacts directly concerning the VKC and with whom the programme can anticipate some opportunities for influence.

4. Community needs assessment

Methodologies adopted by the MSSRF in doing community needs assessment leading to planning the VKC are the village profiling, participatory rural appraisal (PRA) and feasibility studies—using both primary and secondary data sources and involving stakeholder groups at different levels. It is very crucial also that in using these methodologies the different segments of the community based on caste, gender, age and occupation are fairly represented; otherwise, the needs assessment results could be misleading. The findings are to shed light on the situation of the unreached groups in particular—their conditions, problems, issues and information needs. These would thus facilitate the planning of subsequent VKC activities and interventions that are locale-specific and demand-driven towards empowering the vulnerable people to improve their lives and livelihoods.

The village profile would normally generate information from secondary sources and would include: location; demography (population, literacy

rate and related data); social stratification (age, education, caste, gender); economic stratification (occupation, landholding details, fishing assets); land use and land cover; soil type and rainfall details; market and credit facilities; and infrastructure (communication networks, health, education, water and sanitation, and institutions involved in these infrastructure facilities). Primary information are generated through direct observation and engagement of the people through the use of participant observation techniques, household surveys and PRA tools such as transect walk, time and trend analysis, social mapping, wealth ranking, venn diagram, resource mapping, seasonality calendars, mobility mapping and problem tree analysis. Focus group discussions (FGDs) are conducted to ensure changing and emerging needs, issues and priorities of the community are considered.

5. Feasibility studies and planning

To ascertain the suitability of the village in terms of social, institutional, technical and physical appropriateness for setting up the VKC, several feasibility studies have to be conducted—social, technical, physical and institutional. The social feasibility looks into the existing power structures in the community to establish whether the VKC will cater to the needs of the entire community without any bias on gender, caste and class. The technical feasibility is on the electrification of the VKC room/building including provision of plug points to transfer the data. The physical feasibility recommends the venue for establishing VKC considering its access and reach to the entire community, room ventilation and floor plan for the computer systems and users. The institutional feasibility explores potential cooperation and collaboration with different organisations in the community including the community-based organisations (CBOs).

After going through the community needs assessment and feasibility studies, planning with the community follows. Planning is facilitated by the VRC staff to decide the suitability of the village to set up the VKC, its function and management; and for the community to decide on their participation throughout the process of VKC establishment.

6. Establishment of the VKC

Actual establishment of the centre includes: (i) signing of a Memorandum of Understanding (MoU), (ii) recruiting and training staff (knowledge workers) to run the centre, and (iii) setting up the centre's physical structures as per the technical and physical feasibility studies.

Signing the MoU is between VRC and boundary partner who will be the direct contact of the VRC in the VKC. The MoU incorporates the validity

Planning with the community commences after results of assessment and feasibility studies are obtained.

of the agreement, community contribution such as infrastructure like rent-free building, required furniture, electricity bill payment, safeguarding the equipments, identifying the knowledge workers, and commitment of support to the VRC.

The knowledge workers are oriented and trained on the concept of VKC, knowledge workers' responsibilities, general management of VKC, basic computer literacy, local data base, etc. They also get informal training such as asking them to make presentations in the monthly meetings and take quizzes on various government schemes and local database management.

The VKC is equipped with the required infrastructures such as 2 to 3 computers, a printer, UPS, notice board, name board, GSM-based public address (PA) system and software by MSSRF, furniture and telephone unit.

Signing of MoUs between the VRC and the boundary partners is the first step in the establishment of a VKC.

7. VKC services and management

7.1 Services of the VKC

The list of needs, issues and problems derived from the community needs assessment are used to define the services to be provided by the VKC according to: (i) content—initial categories are agriculture, health, education, government schemes/entitlements, fisheries, microenterprises; (ii) form of delivery e.g., awareness/campaign, training and video conferencing activities; and (iii) creating institutional linkages. Existing local database at VKC and VRC is first referred to in accessing information, and then through the strategic partners. Facilitation of two-way communication between experts and farmers of different VKCs and VRCs may also be part of the linkage services provided. Disseminating weather forecast/cyclone warning/announcing government schemes is done through PA system; while season-based information like pest and disease management of paddy crop, foot and mouth disease of cattle and malaria control is through the community newspaper.

7.2 The VKC management committee

Members of the VKC management committee are selected by representatives from the different segments of the community. The members of the committee should come from the local government unit, CBOs, representatives from the different government agencies (school teacher, extension worker, public health worker, etc.), and community representatives based on caste, gender, age and occupation.

7.3 Monitoring and evaluation

Based on the annual and quarterly plans of the VRC (on which the monthly plans of the VKC are based), the monitoring and evaluation (M&E) system of the VRC and VKC are evolved in a participatory manner through brainstorming sessions and workshops. From the results chain (output, outcome and impact) established for the Information Village, key indicators are accordingly identified to measure and track these results. As such, the VKC monitoring system at the village level synchronises with the M&E of the VRC at the block level which in turn synchronises with the M&E of the district/state or national Level VRC.

77

Village Resource and Knowledge Centre as Hubs for Disaster Preparedness

Technological innovations and interventions help coastal villages prepare for disasters and reduce risks.

Source:

Nancy J. Anabel and S. Senthilkumaran *M.S. Swaminathan Research Foundation, Tamil Nadu, India, 2010.*

The fishing communities along Tamil Nadu and the rice farmers of Andhra Pradesh once lacked early warning systems and access to knowledge that are very crucial in making appropriate decisions that would help them in their livelihoods and keep them and their livelihood assets safe during disasters.

Under these circumstances, the M.S. Swaminathan Research Foundation, (MSSRF) through its National Virtual Academy, established village resource and knowledge centres (VRCs and VKCs) to help bring information on disaster preparedness and disaster risk reduction, among many others, to the rural communities in the area using information communication technology (ICT). This was implemented under the post-tsunami project in Tamil Nadu and Andhra Pradesh.

The Public Address system saves the day

Remembering the tsunami on December 26, 2004 never fails to put terror into the hearts of everyone in the coastal communities. However, there were stories of whole villages being saved too. What did they do?

The people in Veerampattinam village were able to evacuate in time. Out of 6,200, only one died in the tsunami. In another village named Nallavadu, all 3,600 people survived due to timely information disseminated through PA system. The story goes that it being a Sunday, a few fishermen were knitting their nets on the shore when they sensed that the sea was surging. They rushed to the VKC and using the PA system announced that their villages had to be evacuated fast due to an impending tsunami. The people paid attention and evacuated the villages at once.

In Singapur, a resident heard about the tsunami warning and phoned the village. The village elders quickly used the PA system to ask the villagers to evacuate. Many were also saved.

This experience highlighted the importance of ICT tools in the rural context. Many ICT tools are now being piloted by various agencies who are trying to empower and build capacities of the rural communities.

The VRC and VKC became an information hub that benefited the coastal communities by connecting them to information and knowledge that safeguarded their lives and livelihood especially in times of disasters. At the start, people expressed doubt that 'a computer and technology' could bring development to the rural populace. But today, the people rely on the VRC and VKC because they have brought remarkable changes in their lives and livelihoods in different realms.

The MSSRF experience has shown how a blend of technological innovations and interventions along with content development and delivery mechanism help in disaster preparedness and risk reduction in the coastal villages. For the fishers, it has shown the use of ICT in ensuring better, if not a really good fish catch.

1. The role of the VKC and the current scenario in the fishing community

In the past, the VKCs used a public address system to broadcast information to fishermen on whether it is safe to go out to the sea and fish. Those early days, very high frequency[1] (VHF) and spread spectrum[2] were the technologies used as tools to provide invaluable data on weather forecast. Of late, growing research brought in multiple technologies that are being piloted among communities to demonstrate effective utilisation of satellite-based digital boards, mobile, WiFi, GSM-based public address system, wireless telephone and many others.

After the tsunami, most fishermen became fearful of navigating into the water due to unpredictable weather. Moreover, they were fearful of mistakenly crossing into international waters and being imprisoned for it. Many of them did not also have the resources to go on exploratory trips in search of potential fishing zones using traditional knowledge because they consider it to be wasteful.

---

1. VHF is a technology used for two way communications to transfer data (information, content) and voice. It is working with lower bandwidth and maximum speed of 244 kbps.
2. Spread Spectrum (802.11b) is a WiFi technology used for two way communication to transfer data (information, content), audio and video along with video conferencing facility. It works with the higher bandwidth and maximum speed of 11 mbps.

Apart from this, they used motorised fibre boats that cannot withstand disasters at mid-sea. When they fish, they carried all the types of fishing nets as they were uncertain on which species are in season. Also, the tsunami created vast disturbances along the shore and the sea, hence even fish availability became very unpredictable for the first couple of months.

2. Lab to lab and land to lab connectivity

MSSRF's long-term association with the Indian National Centre for Ocean Information Services (INCOIS) enabled it to leverage its services to the villagers for it to receive wave height and weather information since 2003. The INCOIS used an early warning information board, which displayed crucial information such as potential fishing zone valid for three days in a week, ocean state forecast (wave height and its direction, wind speed and its direction), and tsunami alert through siren about two and a half days before its effect. It also has an LCD screen and LED scroll to show informative video shots on fisheries-related aspects or development information in general.

ICT has not only served as early warning devices but have also directed fishers to exact spots where fish colonies could be found. This meant more harvest for less diesel used looking for fishing areas.

The General Pocket Radio Service (GPRS) and world space satellite radio service showed their usefulness in giving early warning alert and receiving satellite signals from satellites, even during cloudy days. In the past few years there was a significant shift in both technological options and level of advancement in obtaining information and knowledge through regular communication between and among land to lab, lab to land, and lab to lab. Among these were innovative technologies like digital board, GSM (Global System for Mobile Communications)-based public address system, internet radio, world space SMS server, Wifi technology and mobile technologies for establishing first and last mile connectivity.

MSSRF, with the support of QUALCOMM, TATA and Astute Technologies piloted the use of mobile technology for disseminating information to the fishing community. The novelty of this Fisher Friend Mobile Application[3] (FFMA) is that enables fishers to access crucial information needed to take immediate and appropriate decision at sea. For instance, it can help them save lives during disasters, indicate fish shoals where fish is bountiful, conserve diesel and time, and alert them in advance to enable them to evacuate in the event of a tsunami.

Other information communication technologies such as WiFi 802.11g,[4] internet radio, audio conferencing using fixed wireless phone, satellite based video conferencing, web based SMS server are also used to keep the villagers informed and make right decisions. Internet radio is used to broadcast audio content on animal husbandry, fisheries, education, health, microenterprise, and others.

Audio conference using fixed wireless phone (FWP) is also a popular technology used with the help of an amplifier, mike speaker for interfacing the community with the experts and other community members. Video conferencing using satellite connectivity provided by Indian Space

---

3. Evidences prove that the reach of mobile application in test phase was done in the coastal line of entire Tamil Nadu. Also they are planning to scale up. Similarly the IFFCO Kisan Sanchar Pvt. Ltd. reached more than 1,30,000 farmers with agriculture and its related activities.
4. WiFi 802.11g is a technology used for two way communication to transfer data (information, content), audio and video along with video conferencing facility. It works with the higher bandwidth and maximum speed of 54 mbps.

Research Organisation (ISRO) and internet-based video conferencing enables face to face interaction, discuss with experts on various subjects. Web portal is another sphere wherein the community can access value added information and content that are developed by the VRCs and VKCs.

3. Voices from community stakeholders

Nagalingam, 40 years old, hails from Panithittu, a small fishing village in the Union territory of Puducherry. He owns a small fibre fishing boat.

> *Before ICT: The plight of fisher women in Tamil Nadu*
>
> In the fishing communities of Tamil Nadu, it is the women who are burdened with selling the fish on top of housekeeping and child rearing. Vimala Peiyandi of Veerampattinam said that the daily catch of marginal fishers is often meagre because the sea is unpredictable. As the fishers stay longer at sea to search for fish, the boats consume more diesel and the fish catch deteriorates and looses market value. Vimala is forced to sell the fish at a low price to be competitive. The loss of income affects the education, nutrition, and health of her children. Vimala suffers further as she decides to eat less just so her children and husband could eat enough. Vimala is trapped into a health hazard. To avoid this, the fishermen try to work harder and longer on the sea. But every fishing trip is an agonising time for Vimala as she looks out to the sea hoping that her husband, father or brother will return safely ashore.
>
>

One day, he came across news through the public address system about the FFMA. Nagalingam registered as a pilot test user. On October 23, 2008, the application predicted high waves accompanied by severe winds. Based on this, he decided not to go to the sea. Later he found out that conditions were really rough as predicted. Nagalingam marvelled at how technology prevents loss of lives and further economic loss.

4. Conclusion

The ICT-enabled VRC and VKC plays an important role as a hub that provides timely information to the community, information that spells the difference between life and death and gain and loss in their livelihoods. The VRC and VKC is not only important in the area of disaster preparedness and risk reduction, but it also continues to play a vital role in addressing issues on health, education, microenterprise,

---

ICT can save lives

Chandran, aged 27 is a fisherman from Chinangudi village in Nagapattinam district of Tamil Nadu. For about a decade, Chandran with other fishermen have been using traditional knowledge to forecast the state of sea using techniques like: (a) storm is imminent, if the body of a live whale fish appeared in deep sea fishing, (b) there will be very high wind speed if they come across Lora fish. Meanwhile, Machandran, his brother, was selected and trained to access and interpret the data on ocean forecast. One day, on a fine morning on June 2009, Chandran, Machandran, and 50 other fishermen ventured out into the sea with 11 to 20 boats. The purpose was to lay purse seine type net, which requires team effort of 11-20 boats to haul the catch. After accessing information from the VRC and VKC, they came across critical information concerning very high wind speed and high waves. This piece of information made them turn back with their boats to the shore. They immediately disseminated the same information using mobile to the fishermen on sea. As predicted, the wind started blowing at very high speeds, creating huge waves. Chandran says none would have survived in their fibre boats had they not received the timely message in their mobile phones.

Village resource and knowledge centres have brought the people in disaster-prone areas together to implement a community-based disaster alert system.

Preventing accidents and uneccessary costs through ICTs

Jegan is a fisherman from Olaikuda, a small fishing village in the island of Rameshwaram, Tamil Nadu. He has been a fisher at a very young age and now provides for his parents and family—wife and two young children through fishing using a small fiber boat. A mobile application user in its test phase, it is his normal routine to check for updates on ocean forecast and the sea status before venturing into the sea. One day, Jegan confirmed the weather status and headed off to fish where he cast his nets near a rocky formation beneath the sea. Jegan noticed after awhile that the wind speed had increased up to 40 km/hour. He checked his mobile application immediately for wind speed predictions and it indicated that winds were expected to blow at a very high speed towards the southwest. Jegan realised that the undersea rocks were situated towards the southwest and there was a very good possibility that the wind could make his net drift towards the rocks. Jegan immediately hauled back his net and returned ashore. He now says that timely information enabled him to protect his nets which are quite expensive to buy and repair. He is ready to pay in order to receive invaluable information through his mobile after realising its benefits.

government entitlements and schemes, and others, targeting users from children to the elderly. The power of partnership is also demonstrated during the process of connecting people with knowledge by linking with various public and private, and research and academic institutions.

The initiative also shows that content development is as important as technology and it is crucial to source the right kind of information, translate and interpret the same into a non-jargon format and deliver it in appropriate time, when it is most useful to the local communities. While the process of finding the most suitable, low cost, and effective communication mechanisms for the poor and vulnerable is ongoing, there are still more areas to improve the current services provide by the VRC and VKC and other technology-oriented interventions based on the emerging needs and issues.

78

Livelihood Avenues: Reducing Economic Vulnerabilities

Alternative livelihood help reduce the people's socioeconomic vulnerability.

Source:

Nancy J. Anabel *M.S. Swaminathan Research Foundation, Tamil Nadu, India, 2010.*

1. Introduction

Economic conditions can have a direct bearing on community resilience to disaster. Livelihood opportunities provide economic security that can help communities become more resilient to disasters.

By analysing the community situation and identifying needs, people can discover sustainable livelihood opportunities that can provide them with additional income. It is also important to be able to recuperate the livelihood assets from disaster impacts through programmes like land reclamation and agriculture rehabilitation. Samples of these types of programmes are: (a) strengthening microenterprises like *sari* business, goat rearing, fish vending and lime making, (b) community-based grocery shops, (c) reclamation of agriculture lands from prawn farms, (d) Integrated Mangrove Fishery Farming System (IFFS) Eco-friendly Aquaculture, (e) Revolving fund assistance to SHGs, (f) Diary farming etc. Each one of these are documented in detail in this chapter.

2. Building livelihood avenues on local demand as economic resource building: The Muthuranganathapuram experience

A good example of analysing and identifying community needs for a livelihood opportunity is the situation in Muthuranganathapuram (one of the project villages in Tamil Nadu). They tried to build the livelihood based on local demand, which is a key for embracing success. Upon recognition of the need for mangrove nursery raising and it was decided in the village level institution (VLI) that the four self-help groups (SHGs) could take up the nursery task. This opened up a window of opportunity to persuade SHG members to take up new challenges beyond cultural barricades by gaining a new skill through exposure and training on mangrove nursery raising and management. The required raw materials were provided to the SHGs and after three months, they sold the mangrove saplings. Though the target fixed was 40,000, about 50,000 saplings were raised altogether by all the SHGs, with each group gaining about Rs.20,000 to 30,000 as an additional income in the venture.

Even if the livelihood entailed a new skill and knowledge from the community, they were able to produce quality sapling with 99 per cent survival rate of mangrove sapling with an average height of 60 cm. Not only were they able to earn, they were able to upgrade their capabilities as well with new skill and knowledge.

3. Impressive development out of experience in Kattumavadi, Tamil Nadu

After the SHG members benefited economically by using revolving fund of Rs.3 lakh, they paid back the Village Development Management Committee (VDMC) who decided to use the money to benefit the villagers. They collectively decided to give each person a maximum loan of Rs.10,000 with a payback period of 10 months. Each month they had to pay Rs.1,100. Within four months Rs. 80,000 was disbursed as loan to 38 persons. They also set guidelines for the transactions and fixed a date for disbursements and presented balances in the VDMC meetings.

The experience of project has impact on other places too such as Muthuranganathapuram. After the success of the Revolving Fund concept, the village administration changed their interest rate from Re.1 for Rs.100 to Rs.2. The village administration also provided loan facilities to the villagers.

4. Land reclamation and agriculture rehabilitation

The intervention of land reclamation and agriculture rehabilitation stopped people migrating from the village Nali in Andhra Pradesh for agriculture works and generated employment opportunity by enabling people to gain economic benefit within. This gave out a ripple effect unveiling livelihood avenues such as cattle rearing due to the availability of fodder.

Realising the importance of investing in children's education, women stepped up and headed their families in dairy farming income generating activities. As a result, 22 milking animals were identified and bought by the women and VDMC. The partner organisation Praja Pragathi Seva Sangham (PPSS) has arranged veterinary care and insurance to all the animals that produce an average of five litres of milk per day each. Even if the market potential for selling most of the milk was within, they sold some to milk co-operative societies too and earned Rs.1,000 as profit from each animal per month.

79

Customised Process-Oriented Facilitation of Project Monitoring

Source:

Khilesh Chaturvedi *Association for Stimulating Know-How (ASK).*

1. Introduction

This is a process document of how a monitoring system was developed and used in the project, Strengthening Resilience among Tsunami Affected Community (SRTAC), implemented by Sarvodaya and Practical Action in Sri Lanka, M.S. Swaminathan Research Foundation (MSSRF), and three NGOs (SPRIT, PAD and PPSS) in India. CIDA and IDRC of Canada funded the project. IDRC was also the legal project holder *vis-à-vis* CIDA and played a major coordination role.

The purpose of this process document is to highlight two main aspects, the 'process' and the 'learning' from the process. This process document may be useful for project managers, monitoring facilitators, and general development practitioners in the implementation of their own projects.

What follows is a description of the steps that were taken in the monitoring process and the learning that emerged. It must be noted that steps number 7, 8, and 9 were not envisaged earlier but evolved and were considered influential factors in the entire process, thus their inclusion.

Step 1. Orientation of the monitoring facilitator to the project and interface with project holders

The IDRC was the one that selected the monitoring facilitator. To be acquainted with the implementing partners, the monitoring facilitator acted as an observer in one of the workshops of the project. He also played the role of a co-facilitator in some sessions related to monitoring

> *Learning:*
>
> It is important for the monitoring facilitator to understand the project, its context, objectives, status, and challenges, and know the partner organisations to help facilitate a customised process. The monitoring facilitator should enter the process as a learner and not an expert. It is also important from the point of view of the implementing organisation that it must not see the person as imposed by the donor. The monitoring facilitator and implementing organisation must also strive to build bridges of understanding between themselves.

and also held discussions at the sidelines of the workshop with a relevant staff. It was a favourable coincidence that the facilitator had a positive work experience on another project with one of the organisations. The facilitator was also familiar with the community contexts of both countries due to previous work experience.

Step 2. Delineation of the role of the monitoring facilitator and agreement among the stakeholders

The IDRC and the implementing organisations agreed on the following roles, scope of work, and involvement of the monitoring facilitator:

1. Guide and support the project partners and the IDRC representative(s) with the monitoring and evaluation (M&E) system for progress markers (PMs); support project partners on quantitative and qualitative data collection approach and process, data compilation and analysis, semi-annual technical progress reporting to the IDRC, participating in follow-up project meetings and field visits, and regularly communicating with project leaders and M&E staff;
2. Provide support on aggregation of results across sites and providing inputs in cross-site data analysis for reporting to CIDA;
3. Support project evaluation processes in consultation with the IDRC representative(s) by identifying evaluation topics, introducing evaluation concepts; facilitating and accompanying in participatory evaluation with the teams; and collaborating with external evaluators;
4. Communicate with the IDRC representative(s) for project meetings and updates;
5. Identify the gaps and strengthen the overall M&E system;
6. Identify next steps for effective M&E implementation;
7. Assist and guide the IDRC representative(s) with reporting of the logical (LFA) and PMs to CIDA;
8. Provide assistance to the project, as advised by the IDRC representative(s); and

9. Regularly submit detailed reports of field visits to the IDRC representative(s).

Step 3. Introductory workshop on concept and purpose of monitoring

Project monitoring begun one year into project implementation. The workshop in which the monitoring facilitator was introduced to the organisation was followed by another where the concept and purpose of monitoring was discussed.

This workshop provided an opportunity to the implementing organisation's staff to develop a deeper understanding of monitoring. Some of the perceptions on monitoring being a 'policing exercise', 'fault finding exercise' came to surface and the participants gained insights into the roots of such a perception. The monitoring facilitator could bring the 'learning' (improving) dimension of monitoring as an equally important feature as the 'accountability' (proving) dimension. It also became clear that learning could as much be from good practice or success as from failure.

The first user of monitoring was the staff and then their supervisors at the higher levels. The workshop downplayed the usefulness of monitoring

to the donor. The project staff considered as desirable the value of monitoring in being able to involve communities and in improving transparency and accountability.

> *Learning:*
>
> The extent to which monitoring would be accepted by implementing organisations, especially in cases where the discussion has been begun by the donor organisation, depends upon how much clarity of purpose is established before entering into the development of the system for monitoring. The implementing organisations and the staff members are keen to find out what is in it for them. The learning dimension of monitoring is more appealing to implementing organisations, especially the staff.
>
> It is very important for the facilitators to not position monitoring as a mechanism for ensuring accurate reporting to the donor. While this may be a byproduct, it cannot and should not be the starting point. Equally important is for the facilitator to clarify that while the donor may have been instrumental in introducing him/her to the implementing partners, the facilitator does not represent the donor. The facilitator must be careful to adopt an independent position and show the implementing organisation that it is the client first and foremost.

Step 4. Development of data collection plan and monitoring plan

This was the next logical step to take once the concept and purpose of monitoring was agreed upon. Even though the past experience of the team was limited, the facilitator took time to move at the pace of the participants and engaged them fully in identifying the steps in monitoring. First of all, they developed a data collection plan and answered questions on what data to collect, how (methods and tools), how often, from where (source/respondents), and by whom.

In order to provide a complete picture of the monitoring process, discussion was also held and tools and formats were developed for data compilation, data analysis and decision-making. This gave a clearer idea to the implementing organisation staff that monitoring was much more than data collection in the field and that analysis and decision-making were key issues.

The monitoring system was developed in a way that analysis and decision-making could take place firstly at the village level and then move upwards.

An important aspect that came forth was also that the project did not have clear targets attached to the various results. The implementing agencies utilised this opportunity to set their targets and later reported that this step together with the monitoring system helped provide much focus to the project.

Step 5. Pre-testing of tools and formats and adapting them

The staff of the implementing organisation and the monitoring facilitator pre-tested and then fine-tuned the tools and formats based on the results of the pre-test.

The modification process carried on well into the monitoring facilitation process specifically on the compilation and analysis formats. One of the modifications was made when the Sarvodaya monitoring officer began the compilation of data at the project level. She not only provided the added up figures but also presented figures from each village and district. This sparked off a very healthy competition among grassroots staff as they could compare their performance with that of other village. At a later stage, the project manager even applied the monitoring learning to develop a rating system for the staff and linking their incentives to the performance.

Learning:

It is important to differentiate between data collection plan and the overall monitoring plan, especially highlighting the importance of data analysis in order to assess the level of achievement of results and making decisions to improve the effectiveness of the project.

There is obvious value in making the monitoring process participatory for reasons of ownership. Once this is done, the project also benefits from local insights to make it realistic. Participatory monitoring, where commu-

contd...

...contd...

nity participants get to provide their feedback on various aspects of the project, and later also sit with the staff to analyse the compiled findings and finally make decisions that they can make at their level or influence decisions of the project leadership is truly empowering.

The participatory process also enables the project manager and monitoring facilitator to take stock of the level of understanding of different levels of staff on monitoring related concepts and tools, and provide facilitation accordingly.

Developing a monitoring system is an opportunity to look back into the plans and identify missing links or any unrealistic plans. If the donors are open to relook into planning and negotiation with the implementing organisation, the monitoring process helps in bringing greater realism into projects.

Step 6. Implementation of monitoring plan

The implementation of the monitoring system was moved from simple to complex. Looking into relative newness of the experience of gathering qualitative data on aspects such as level of community participation, gender, etc., it was decided to first gather quantitative data. Qualitative data was gathered at a later stage. The facilitator himself demonstrated the collection of qualitative data in the field and then provided support to the staff in data gathering, providing on-the-job training, and facilitation. Once data collection and compilation was properly organised, the facilitator focussed on observing and guiding in the process of data analysis and decision-making.

Learning:

Provision of accompaniment support monitoring system development, especially with organisations and projects with earlier informal ways of monitoring, must have a provision of accompaniment support by the monitoring facilitator. This in effect means that the monitoring facilitator

contd...

...contd...

> will not always conduct trainings and workshops. Instead the monitoring facilitator must accompany/visit the project staff in the field or sit with staff engaged in processing the data. The monitoring facilitator should (be allowed to) keep open/broad agenda for the follow-up visits.
>
> During the accompaniment visits, the monitoring facilitator should be able to demonstrate the use of a method or tool; critically observe the processes of monitoring, identify strengths and limitations, and provide forthright feedback with clear illustrations; analyse situations and identify root causes for certain limitations; challenge the staff if levels of achievement are low and bring in senior project staff into the picture to ensure that decisions regarding monitoring system are implemented; listen to the staff engaged in the process of monitoring, identify reasons for problems, and be able to address them as necessary; look for new initiatives taken by the team in the monitoring domain and react to them appropriately; and answer queries of related to monitoring.

Step 7. Handling of the feedback to the donor

One of the tricky parts of a tripartite arrangement is deciding on whether the donor will be informed and involved. The implementing agency, for obvious reasons, is not always keen on having the donor too much involved or informed of all details including the failings. On one occasion the facilitator erred in forwarding, without prior sharing with the implementing organisation, a mission report to the donor. The mission report was critical about the staff of the implementing organisation for

> *Learning:*
>
> The monitoring facilitator gets an inside view into the functioning of the implementing organisation. It is a conscious decision on the part of the implementing organisations and they expect sensitive handling of information. It is a good idea for the facilitator to take the implementing organisation into confidence before informing the donor on the perceptions held by the facilitator.

not having done enough towards implementing the monitoring system. The implementing partner was not too happy with this and it took a face to face meeting to clear the air.

Step 8. Capacity building processes emerging from or feeding into monitoring

This was another step which had not been planned but emerged as a logical fallout of the accompaniment and facilitation process. While collecting data on aspects of gender balance, it was realised that the staff was drawing quick conclusions without going sufficiently deep into the discussion and analysis. This highlighted the need for implementing organisation's staff to undergo a process of gender sensitisation and developing understanding on practices in the community that reflect gender balance and imbalance.

> *Learning:*
>
> There is a need to keep the options open and the practice of a monitoring system may highlight the need for capacity building on concepts and subjects other than monitoring.

Another accompaniment situation highlighted that the staff members went through the administration of tools in a pre-decided manner, unable to understand the actual situation and not fully well adapting to it and not being very skillful in their facilitation (i.e., not changing the sequence of questions as needed and inability to moderate the people who dominate the discussions). This brought forth the need for facilitation skills training for the staff.

Step 9. Experience sharing workshop

Users of the monitoring system from the different organisations gathered to share experiences with each other. This step provided helpful feedback. Some of the major reflections shared by the implementing organisations on monitoring are mentioned below.

- Helped the village institutions to do a self-evaluation on their progress.
- Provided decision-making tool for management and the field teams.
- Served as a learning experience and capacity building for staff members.
- Helped bring transparency in the implementation process.
- Helped to streamline the focus to the underprivileged and poor women as target groups.
- Brought more focus into the work, and
- Inspired organisations to improve the monitoring system of their other projects as well.

2. Challenges and constraints

The community was not interested to forego the income of the day to participate in meetings. Their involvement in the district level meetings was less. Less time consuming and more fun and informal tools to elicit the perception of the community would have been better.

The monitoring process should have begun at the very beginning of the project. This would have saved anxiety, contributed to keeping the project on track, and made the project staff perceive monitoring as an integral part of the project instead of an extra burden.

Information Resources

Key References

Abarquez, Imelda and Zubair Murshed (2004). *Community-based disaster risk management field practitioners handbook.* ADPC. Available at: http://www.proventionconsortium.org/themes/default/pdfs/CRA/CBDRM2004_meth.pdf

Action Aid International (2006). *The evolving UN cluster approach in the aftermath of the Pakistan earthquake: An NGO perspective.* Available at: http://www.actionaid.org/docs/evolving%20un%20cluster.pdf

Actionaid International (2005). *Participatory vulnerability analysis: A step-by-step guide for field staff.* Available at: http://www.proventionconsortium.org/themes/default/pdfs/CRA/PVA_ActionAid2005_meth.pdf. *Adaptation Strategies.* Available at: http://www.aiaccproject.org/resources/ele_lib_docs/adaptation_decision_tools.pdf

ADRC. *Application of DRR tools: Sharing Asian experiences.* Available at: http://www.adrc.asia/publications/Tool_Book/

Allison, Edward H., Neil L. Andrew, Jamie Oliver (2007). *Enhancing the resilience of inland fisheries and aquaculture systems to climate change.* WorldFish Center, Malaysia. Available at: http://ejournal.icrisat.org/SpecialProject/sp15.pdf

ALNAP (2006) *Tsunami evaluation coalition: Links between relief, rehabilitation and development in the Tsunami response.* Available at: http://www.alnap.org/initiatives/tec.aspx

Ariyabandu, Madhavi Malalgoda and Maithree Wickremasinghe (2005). *Gender dimensions in disaster management: A guide for South Asia.* Available at: http://practicalaction.org/?id=disaster_gender_book

Asian Disaster Preparedness Centre (2006). *ADPC Coastal Community Resilience (CCR) Initiative.* Available at: http://www.adpc.net/v2007/Programs/EWS/CCR/CCR.asp

Benson, Charlotte and John Twigg (2007). *Tools for mainstreaming disaster risk reduction: Guidance notes for development organisations.* Available at: http://www.preventionweb.net/files/1066_toolsformainstreamingDRR.pdf

Bettencourt *et al.* (2006). *Not if but when – Adapting to natural hazards in the Pacific islands region: A policy note.* Washington, DC: The World Bank. Available at: http://go.worldbank.org/TAH2IM2X90

Bradshaw, S. (2004). *Socio-economic impacts of natural disasters: A gender analysis.* Available at: http://www.eclac.cl/publicaciones/xml/3/15433/lcl2128i.pdf

Burton, I. and M. van Aalst (2004). *Look before you leap: A risk management approach for incorporating climate change adaptation into World Bank operations.* Washington, DC: The World Bank. Available at: http://go.worldbank.org/M02PN5Y730

———. (1999). *Come hell or high water - Integrating climate change vulnerability and adaption into bank work.* Washington, DC: The World Bank. Available at: http://go.worldbank.org/ZIS4OTK7J0

Coastal Community Resilience (2009). *Coastal Community Resilience (CCR): Case Studies.* Available at: *http://community.csc.noaa.gov/ccr/index.php/case-studies*

Coastal Resources Centre (2004). *Coastal Community Resilience.* Available at: *http://www.crc.uri.edu/index.php?actid=328*

Commission on Climate Change and Development (2009). *Closing the gaps: Disaster risk reduction and adaptation to climate change in developing countries.* Available at: *http://www.ccdcommission.org/Filer/report/CCD_REPORT.pdf*

Courtney, Catherine A., Atiq K. Ahmed, Russell Jackson, David McKinnie, Pam Rubinoff, Adam Stein, Stacey Tighe and Alan White (2008). *Coastal Community Resilience in the Indian ocean region: A unifying framework, assessment, and lessons learned.* Available at: *http://cedb.asce.org/cgi/WWWdisplay.cgi?162876*

DFID (2006). *Disaster reduction policy paper.* Available at: *http://webarchive.nationalarchives.gov.uk/+/http://www.dfid.gov.uk/pubs/files/disaster-risk-reduction-policy.pdf*

———. (2005). *Natural disasters and disaster risk reduction measures: Desk review of costs and benefits.* Available at: *http://webarchive.nationalarchives.gov.uk/+/http://www.dfid.gov.uk/pubs/files/disaster-risk-reduction-study.pdf*

———. (2004) *Disaster risk reduction: A development concern.* Available at: *http://webarchive.nationalarchives.gov.uk/+/http://www.dfid.gov.uk/pubs/files/disaster-risk-reduction-scoping.pdf*

Dietrich, Kathleen (2010). *Participatory tools to aid adaptation.* Available at: *http://wikiadapt.org/index.php?title=Participatory_tools_to_aid_adaptation*

Dilley, Maxx (2006). *Setting priorities: Global patterns of disaster risk.* Available at: *http://rsta.royalsocietypublishing.org/content/364/1845/2217.full*

Enarson, Elaine (2000). *Gender and natural disaster.* Available at: *http://www.unisdr.org/eng/library/Literature/7566.pdf*

FAO (2006). *The role of local institutions in reducing vulnerability to recurrent natural disasters and in sustainable livelihoods development.* Rome, FAO. Available at: *http://www.fao.org/docrep/009/a0879e/a0879e00.HTM*

———. *Coastal community resilience assessment process.* Available at: *http://www.fao.org/Participation/ft_more.jsp?ID=8305*

Feenstra, Jan F., Ian Burton, Joel B. Smith and Richard S.J. Tol (1998). *Handbook on methods for climate change impact assessment and adaptation strategies.* Available at: *http://research.fit.edu/sealevelriselibrary/documents/doc_mgr/465/Global_Methods_for_CC_Assessment_Adaptation_-_UNEP_1998.pdf*

Few, R. *et al.* (2006). "Linking climate change adaptation and disaster risk management for sustainable poverty reduction", Synthesis report. Vulnerability and Adaptation Resource Group. Available at: *http://ec.europa.eu/development/icenter/repository/env_cc_varg_adaptation_en.pdf*

GDN (2008). *Gender and disasters sourcebook.* Available at: *http://www.gdnonline.org/sourcebook.htm*

Giuliani, Alessandra, Ruth Wenger and Susanne Wymann von Dach (2009). *Disaster risk reduction: A gender and livelihood perspective.* Available at: *http://www.inforesources.ch/pdf/focus09_2_e.pdf*

Hellmuth, M.E., A. Moorhead and J. Williams (eds.) (2007). *Climate risk management in Africa: Learning from practice.* New York, IRI, Columbia University. Available at: *http://portal.iri.columbia.edu/portal/server.pt/gateway/PTARGS_0_2_1171_0_0_18/Climate%20and%20Society%20No1_en.pdf*

http://www.proventionconsortium.org/themes/default/pdfs/CRA/Tearfund2006_meth.pdf

Huq, S. and J. Ayers (2009). "Linking adaptation and disaster risk reduction", in Brainard, L. et al. (eds.), *Climate change and global poverty: A billion lives in the balance?*. Washington, DC: Brookings Institution Press. Available at: http://www.brookings.edu/reports/2007/1010_disaster_risk_reduction.aspx

Id 21. (2008). *Disaster risk reduction and climate change adaptation closing the gap.* Available at: http://www.id21.org/publications/climate_change_1.pdf

IISD, IUCN, SEI and Inter-cooperation (2003). *Livelihoods and climate change: Combining disaster risk reduction, natural resource management and climate change adaptation in a new approach to the reduction of vulnerability and poverty.* Available at: http://data.iucn.org/dbtw-wpd/edocs/2003-034.pdf

International Save the Children Alliance (2008). "In the face of disaster: Children and climate change", Save the Children. Available at: http://www.savethechildren.org/newsroom/2008/climate-change.html

International Strategy for Disaster Reduction (ISDR) (2009). *Disaster risk reduction in United Nations.* Available at: http://www.unisdr.org/preventionweb/files/9866_DisasterRiskReductionintheUnitedNat.pdf

———. (2005). *Hyogo framework for action 2005-2015: Building the resilience of nations and communities to disasters.* Available at: http://www.unisdr.org/eng/hfa/docs/Hyogo-framework-for-action-english.pdf

IUCN- The World Conservation Union (2006). *Ecosystems, livelihoods and disasters: An integrated approach to risk management.* Available at: http://app.iucn.org/dbtw-wpd/edocs/CEM-004.pdf

Katon, Brenda M., Robert S. Pomeroy, Marshall W. Ring and Len R. Garces (2000). "Rehabilitating the mangrove resource of Cogtong bay, Philippines: A comanagement perspective", *Coastal Management* 28(1): 29-37. Available at: http://www.co-management.org/download/reprint9.pdf ð

Kumpa, Ladawan (1998). *Coastal resource management with local participation: Case study of Surat Thani, Thailand.* Available at: http://srdis.ciesin.columbia.edu/cases/thailand-001.html

Lavell, Allan (2008). "Relationships between local and community disaster risk management & poverty reduction: A preliminary exploration", A Contribution to the 2009 ISDR Global Assessment Report on Disaster Risk Reduction. Available at: http://www.preventionweb.net/english/hyogo/gar/background-papers/documents/Chap6/Lavell-C-DRM-L-DRM-&-Poverty-Reduction.doc

Maguire, Brigit and Sophie Cartwright (2008). *Assessing a community's capacity to manage change: A resilience approach to social assessment.* Available at: http://adl.brs.gov.au/brsShop/data/dewha_resilience_sa_report_final_4.pdf

Mitchell, Tom and Maarten van Aalst (2008). *Convergence of disaster risk reduction and climate change adaptation.* Available at: http://www.preventionweb.net/files/7853_ConvergenceofDRRandCCA1.pdf

Mulekom, Leo Van (2008). *Reflections on community based coastal resources management (CB-CRM) in the Philippines and South-East Asia.* Available at: http://www.oxfam.org.uk/resources/downloads/FP2P/FP2P_Philippines_Fish_Coast_Resc_CS_ENGLISH.pdf

O'Brien, K. et al. (2008). "Disaster Risk Reduction, Climate Change Adaptation and Human Security", *A Commissioned Report for the Norwegian Ministry of Foreign*

Affairs. GECHS. Oslo, Norway: GECHS. Available at: *http://www.gechs.org/ downloads/GECHS_Report_3-08.pdf*

OCHA. "Capacity assessment and capacity building/development", Available at: *http:// ocha.unog.ch/drptoolkit/PCapacityAssessment&Building.html*

OECD (2006). *OECD studies in risk management - Denmark, assessing societal risks and vulnerabilities*. Available at: *http://www.oecd.org/dataoecd/36/18/36099961.pdf*

——————. (2009). *Policy guidance on integrating climate change adaptation into development co-operation*. Paris, OECD. Available at: *http://www.oecd.org/document/40/0,3343,en_ 2649_34421_42580264_1_1_1_1,00.html*

Olmos, Santiago (2001). *Vulnerability and adaptation to climate change: concepts, issues, assessment methods*. Available at: *http://www.cckn.net/pdf/va_foundation_final.pdf*

Overseas Development Institute (2006). *Disasters* 30 (1): 1 – 150. Available at: *http:// www3.interscience.wiley.com/journal/118596602/issue*

Patwardhan, Anand (2006). *Assessing vulnerability to climate change: The link between objectives and assessment*. Available at: *http://www.ias.ac.in/currsci/feb102006/376. pdf*

Provention Consortium. "Community risk assessment toolkit". Available at: *http://www. proventionconsortium.org/?pageid=39*

Pomeroya, Robert S., Blake D. Ratnera, Stephen J. Halla, Jate Pimoljindab and V. Vivekanandan (2006). *Coping with disaster: Rehabilitating coastal livelihoods and communities*. Available at: *http://ideas.repec.org/a/eee/marpol/v30y2006i6p786-793. html*

Publications from the Institute of Development Studies (IDS), The University of Sussex. Available at: *http://www.i-s-e-t.org/index.php?option=com_ content&view=section&la yout=blog&id=5&Itemid=9*

Quiamco, Madeline Baguio (2006). *Implementing isang bagsak: Community-based coastal resource management in central Viet Nam*. Available at: *http://www.idrc.ca/en/ev-105014-201-1-DO_TOPIC.html*

Reid. H. *et al*. (eds.) (2009). *Community-based Adaptation to climate change*. Available at: *http://www.iied.org/pubs/pdfs/14573IIED.pdf*

Schneider, S.H. *et al*. (2007). "Assessing key vulnerabilities and the risk from climate change", in M.L. Parry *et al*. (eds.) *Climate change 2007: Impacts, adaptation and vulnerability, contribution of working group II to the fourth assessment report of the intergovernmental panel on climate change*. Cambridge, UK: Cambridge University Press, 779-810. Available at: *http://www.ipcc.ch/pdf/assessment-report/ar4/wg2/ar4-wg2-chapter19.pdf*

Sperling, F. and F. Szekely (2005). "Disaster risk management in a changing climate", Discussion Paper Prepared for the World Conference on *Disaster Reduction* on behalf of the Vulnerability and Adaptation Resource Group (VARG). Available at: *http://www.preventionweb.net/files/7788_ DRMinachangingclimate1.pdf*

Sphere Project (2004). *Sphere humanitarian charter and minimum standards in disaster response handbook*. Available at: *http://www.sphereproject.org/index.php?option=content&task =view&id=27&Itemid=84*

Srinivasa, Hari and Yuko Nakagawa (2008). *Environmental implications for disaster preparedness: Lessons Learnt from the Indian Ocean Tsunami*. Available at: *http://www.sciencedirect. com/science?_ob=ArticleURL&_udi=B6WJ7-4PS63X6-1&_user=4164700&_cov*

erDate=10%2F31%2F2008&_rdoc=1&_fmt=high&_orig=search&_sort=d&_ docanchor=&view=c&_rerunOrigin=google&_acct=C000062460&_version=1&_ urlVersion= 0&_userid=4164700&md5=892c3aef83a566fef6f7432e5baa6078

The Inter-Agency Task Force of the ISDR (2005). *Disaster risk reduction tools and methods for climate change adaptation*. Available at: http://www.preventionweb.net/files/5654_ DRRtoolsCCAUNFCC.pdf

The WorldFish Centre (2008). *Waves of change: Lessons learned in rehabilitating coastal livelihoods and communities after disasters*. Available at: http://www.worldfishcenter. org/v2/files/LessonsLearned1866.pdf

Tompkins, E.L. and W.N. Adger (2004). "Does adaptive management of natural resources enhance resilience to climate change?", *Ecology and Society*. Available at: http:// www.ecologyandsociety.org/vol9/iss2/art10/

Trobe, Sarah La and John Twigg (2005). *Mainstreaming disaster risk reduction a tool for development organizations*. Available at: http://www.unisdr.org/HFdialogue/download/ tp2-Tearfund-Mainstreaming-drr.pdf

Tsunami Evaluation Coalition (2006). *Impact of the Tsunami response on local and national capacities*. Available at: http://www.alnap.org/resource/3532.aspx

Turner, B.L., Roger E. Kasperson, Pamela A. Matson, James J. McCarthy, Robert W.Corell, Lindsy Christensen, Noelle Eckley, Jeanne X. Kasperson, Amy Luers, Marybeth L. Martellor, Colin Polsky, Alexander Pulsipher and Andrew Schiller (2003). *A framework for vulnerability analysis in sustainability science*. Available at: http://www. pnas.org/content/100/14/8074.full

Twigg, J. (2001). "Disasters and poverty reduction/sustainable livelihoods", *Working Paper 2*, Sustainable livelihoods and vulnerability to disasters Benfield Hazard Research Centre. Available at: http://www.eird.org/cd/on-better-terms/docs/Twigg-Sustainable-livelihoods-and-vulnerability-to-disasters.pdf

UNDP (2004). *A global report on reducing disaster risk: A challenge for development*. Available at: http://www.undp.org/cpr/whats_new/rdr_english.pdf

———. (2004). *Global report on local level risk management*. Available at: http://www.undp. org/cpr/disred/rdr.htm

———. (2002). "A climate risk management approach to disaster reduction and adaptation to climate change", Collection of Papers from Expert Group Meeting in Havana 2002. Available at: http://www.undp.org/cpr/disred/documents/wedo/icrm/ riskadaptationintegrated.pdf.

UNFCCC (2008). "Integrating practices, tools and systems for climate risk assessment and management and strategies for disaster risk reduction into national policies and programmes", *Technical Paper*. Available at: http://unfccc.int/resource/docs/2008/ tp/04.pdf

———. *Climate change: Impacts, vulnerabilities, and adaptation in developing countries*. Available at: http://unfccc.int/resource/docs/publications/impacts.pdf

UN-ISDR (2004). *Living with risk: A global review of disaster reduction initiative*. Available at: http://www.unisdr.org/eng/about_isdr/bd-lwr-2004-eng.htm

UNISDR (2009). *Applying disaster risk reduction for climate change adaptation: Country practices and lessons*. Available at: http://gfdrr.org/docs/ISDR_Applying_DRR_For_ CCA.pdf.

United Nations (2008). *Gender perspectives: Integrating disaster risk reduction into climate change adaptation, good practices and lessons learnt*. Available at: http://www.unisdr.

org/eng/about_isdr/isdr-publications/17-Gender_Perspectives_Integrating_DRR_CC/Gender_Perspectives_Integrating_DRR_CC_Good%20Practices.pdf

USAID/ASIA (2007). *How resilient is your coastal community? A guide for evaluating coastal community resilience to tsunamis and other hazards.* Available at: http://apps.develebridge.net/usiotws/13/CoastalCommunityResilience%20Guide.pdf

Vakis, Renos (2006). "Complementing natural disaster management: The role of social protection", *Social Protection Dicussion Paper* 0543, 2006. February. Available at: http://siteresources.worldbank.org/SOCIALPROTECTION/Resources/SP-Discussion-papers/Social-Risk-Management-DP/0543.pdf

Venton, P., and S. La Trobe (2008). "Linking climate change adaptation and disaster risk reduction", *tearfund*. Available at: http://www.tearfund.org/webdocs/Website/Campaigning/CCA_and_DRR_web.pdf

Venton, Paul and Bob Hansford (2006). *Reducing risk of disaster in our communities: Guidance Notes.* Available at: http://www.proventionconsortium.org/themes/default/pdfs/CRA/Tearfund2006.pdf

Wattegama, Chanuka (2007). *ICT for disaster management.* Available at: http://www.apdip.net/publications/iespprimers/eprimer-dm.pdf

World Bank (2006). "Managing climate risk: Integrating adaptation into World Bank group operations." Washington, DC: The World Bank. Available at: http://go.worldbank.org/9NFE3IAWZ0

Yohe, G.W. *et al.* (2007). "Perspectives on climate change and sustainability", in M.L. Parry et al. (eds.) *Climate change 2007: Impacts, adaptation and vulnerability. contribution of working group II to the fourth assessment report of the intergovernmental panel on climate change.* Cambridge, UK: Cambridge University Press. p. 811-841. Available at: http://www.ipcc.ch/pdf/assessment-report/ar4/wg2/ar4-wg2-chapter20.pdf

CCA and DRR Sites

Action Aid: http://www.actionaid.org.uk/100036/the_right_to_security_in_conflict__emergencies.html

Asia Pacific Network for Global Change Research: http://www.apn-gcr.org/newAPN/indexe.htm

Asian Disaster Preparedness Center (ADPC): http://www.adpc.net/v2007/Default.asp

Asian Disaster Reduction Centre. http://www.adrc.asia/project/index.html

Climate Change Knowledge Network: http://www.cckn.net/about.asp

Commission on Climate Change and Development (CCCD): http://www.ccdcommission.org/

Community Based Adaptation Exchange (CBA-X): http://community.eldis.org/cbax/

Community-based Risk Screening Tool (CRISTAL): http://www.cristaltool.org/

Eldis: http://www.eldis.org/

Emergency Capacity Building Project: http://www.ecbproject.org/PCVA

Eupopean Commission : http://ec.europa.eu/index_en.htm

Food and Agriculture Organization (FAO): http://www.fao.org/

Global Environmental Change and Human Security (GECHS): http://www.gechs.org/

Information Resources

Global Platform for Disaster Risk Reduction. http://www.preventionweb.net/english/hyogo/GP/
Huairou Commission: www.huairou.org
IAPAD: http://www.iapad.org/
Institute for Social and Environmental Transition (ISET): http://www.i-s-e-t.org/
Institute of Development Studies (IDS), The University of Sussex: http://www.ids.ac.uk/go/home
Inter-Agency Network for Education in Emergencies: http://www.ineesite.org/index.php/post/disaster_risk_reduction_tools
Intergovernmental Panel on Climate Change (IPCC) - http://www.ipcc.ch/
International Institute for Environment and Development (IIED): http://www.iied.org/
International Institute for Sustainable Development (IISD): http://www.iisd.org/
Knowledge Networking for Rural Development in Asia Pacific (ENRAP). www.enrap.org
Learning for Sustainability.net : http://learningforsustainability.net/
National Alliance for Disaster Risk Reduction (NADRR), India. http://www.nadrrindia.org/
National Institute of Disaster Management, India. http://www.ndmindia.nic.in/
NOAA Coastal Service Centre: http://www.csc.noaa.gov/index.html
Open Forum on Participatory Geographic Informtion Systems and Technolgies. http://www.ppgis.net/pgis.htm
Organisation for Economic Co-Operation and Development (OECD) http://www.oecd.org/home/0,2987,en_2649_201185_1_1_1_1_1,00.html
Overseas Development Institute (ODI): http://www.odi.org.uk/work/themes/
Participatory GIS: http://www.proventionconsortium.org/themes/default/pdfs/CRA/PGIS_Sept08.pdf
Prevention web: www.preventionweb.net
ProAct Network: http://www.proactnetwork.org/proactwebsite/
SpringerLink: http://www.springerlink.com/
Telecentre.org. http://www.telecentre.org/
The Community-Based Natural Resources Management (CBNRM) Learning Center: http://www.cbcrmlearning.org/
The World Bank Climate Change - http://beta.worldbank.org/climatechange/
United Nations Development Group: http://www.undg.org/index.cfm?P=1093
United Nations Development Programme (UNDP): http://www.undp.org/
United Nations Framework Convention on Climate Change (UNFCCC): http://unfccc.int/2860.php
United Nations International Strategy for Disaster Reduction (UNISDR): http://www.unisdr.org/index.php
Uplift Mutuals - micro insurance programmes: www.upliftmutuals.org
World Watch Institute. Natural Disasters & Peacemaking. www.worldwatch.org/taxonomy/term/435